INDUSTRIAL ACCIDENT PREVENTION

FIFTH EDITION

INDUSTRIAL ACCIDENT PREVENTION

A SAFETY MANAGEMENT APPROACH

H. W. Heinrich
Late Superintendent
Engineering and Loss Control
The Travelers Insurance Company

Dan Petersen, P.E., CSP
Safety Management Consultant
Fort Collins, Colorado

Nestor Roos, D.B.A.
Professor of Insurance
Director, Safety Management
Arizona Center for Occupational Safety and Health
University of Arizona

McGRAW-HILL BOOK COMPANY
New York   St. Louis   San Francisco   Auckland   Bogotá   Hamburg
Johannesburg   London   Madrid   Mexico   Montreal   New Delhi
Panama   Paris   São Paulo   Singapore   Sydney   Tokyo   Toronto

This book was set in Melior by Black Dot, Inc.
The editors were Julienne V. Brown and Susan Hazlett;
the designer was Joan E. O'Connor;
the production supervisor was Donna Piligra.
The drawings were done by Burmar.
R. R. Donnelley & Sons Company was printer and binder.

INDUSTRIAL ACCIDENT PREVENTION

2 3 4 5 6 7 8 9 0 D O D O 8 9 8 7 6 5 4 3 2 1

Library of Congress Cataloging in Publication Data

Heinrich, Herbert William, date
  Industrial accident prevention.

  Bibliography: p.
  Includes index.
  1. Industrial safety. 2. Factories—Safety
measures. I. Petersen, Dan, joint author. II. Roos,
Nestor R., joint author. III. Title.
HD7262.H4   1980   614.8'52   79-16878
ISBN 0-07-028061-4

Contents

Preface

H. W. Heinrich in his preface to the fourth edition of this book stated that accident prevention is both a science and an art. Then he stated:

> As such, one expects to find and will find, clearly covered by the text, what accident prevention is, what it accomplishes, and how it is accomplished.

> As in all art and science, one should expect to find governing laws, rules, theorems, or basic principles. These exist. They are stated, described, and illustrated. After understanding comes application. What industry and its accident preventionists want most to know is how best to apply their knowledge. Examples showing practical application are given to supply this need.

> Largely because of the recognition of basic principles, such as originally given in the first edition of this book and as herein reiterated, accident prevention has progressed from uncoordinated and arbitrarily selected activities, often ineffective and wasteful, to interrelated steps based on knowledge of cause, effect, and remedy, i.e., to an effective, practical, scientific approach.

Soundly established basic principles must withstand tampering, notwithstanding the natural temptation to discard the old, adopt the new, and to change merely for the sake of change. The amazing technological advances in the industries—in automation, metallurgy, chemistry, electronics, and nucleonics—provide an exacting test.

It is precisely because of the principles that he laid down in the early editions of this text that his book became so influential to industrial safety practitioners. Few of us perhaps realize just how influential this book was. It was and still is the basis for almost everything that has been done in industrial safety programming from the date it was written until today. *Industrial Accident Prevention* was the only text in industrial safety that laid down principles; and those principles still guide practitioners today; rightly or wrongly, they still exist.

It is true as Heinrich stated that the amazing technological advances provide exacting tests for principles. And that is what this book, this revision, attempts to do—to look at how the principles have stood up to these exacting tests.

Heinrich went on to state that:

Purposely, therefore, emphasis herein is placed on principles, facts, and method rather than on incidental detail. In view of the continued progress of industry it is clear that there can be no end to the development of new materials, tools, and equipment and of new processes and procedures. These call continually for the dissemination of additional information with regard to the hazardous properties and characteristics of materials and with regard to ways and means of detection and safeguarding.

This revision adheres to this approach. Our emphasis remains on principles and method rather than on incidental detail. Heinrich wrote:

Individual texts are available which treat exhaustively of one specific subject, such as prevention of eye injury, safe handling of acids, nuclear fission, the hazards in foundries, machine shops, construction, or any one of hundreds of other subjects. Any attempt to cover the subject of accident prevention completely from the viewpoint of such detail as referred to would be more bewildering than constructive. It also has been found confusing to the student (until he has selected an area in which he wishes to specialize) to include exhaustive detail of *any one area* in a general course or text. Furthermore, no text featuring *all* the details of mechanical and physical safeguarding or *all* the pertinent details relating to the control of personal attitudes and actions could possibly provide complete coverage for more than a brief period. But the person who has a sound knowledge of the *principles* of accident occurrence and prevention is well equipped to deal constructively with the safeguarding of *all exposures*. If, for example, he encounters unfamiliar hazards created by the use of radioisotopes or by the dangers inherent in nuclear power reactors, he has only to fill the gap of his detailed knowledge and proceed exactly as in other situations by applying established principles and using proved and tested methods.

If principles and methods are basically correct, satisfactory results must follow even though the hazards vary and the details of corrective procedure differ.

Here in this revision we are directing the readers' attention even more to the principles and methods than the earlier editions. Probably the greatest change in approach in this revision is to treat the section on special subjects differently from the earlier editions. Early editions focused on such specifics as machine guarding, personal protective devices, radiation, illumination, etc. This revision eliminates these sections in the belief that they are well covered in many other places—in the National Safety Council's manual, in OSHA standards, in a myriad of places that attempt to explain OSHA standards, etc.

Rather than looking at these specific areas, we have chosen to put our emphasis in the special subjects area on such things as safety law, insurance and risk management, professional liability, etc. Many of these are not often treated in safety texts. Similarly we have retained only two of the eight appendices from the last edition. We felt the appendices on chronology and background (combined here as Appendix 1) fit our purposes in the revision, but the other six dealing with specific items no longer fit the changed directions of the text.

Other changes will be apparent throughout to the reader.

Finally, we wish to express special thanks to E. R. Granniss, Former Manager, Loss Prevention and Engineering, Royal-Globe Insurance Group, for his assistance in the preparation of previous editions.

Dan Petersen
Nestor Roos

Acknowledgments

The authors wish to thank the following people for their contributions to the present work.

Ed Adams, P.E., CSP, Director of Loss Control, Pet, Inc., St. Louis, Missouri.

Leslie Ball, D.Sci., Consultant, Former Director of Safety, NASA, Huntsville, Alabama.

Frank Bird, Jr., P.E., CSP, Executive Director, International Loss Control Institute, Loganville, Georgia. President, International Loss Control College. Former President, National Safety Management Society.

Hugh M. Douglas, Senior Loss Control Coordinator, Imperial Oil Ltd., Toronto, Ontario, Canada. President, Industrial Accident Prevention Association of Ontario.

Russell Ferrell, Ph.D., Professor of Human Factors, Systems Engineering Department, University of Arizona, Tucson, Arizona.

Robert Firenze, P.E., CSP, President, RJF. Inc., Consultant, Bloomington, Indiana.

Acknowledgments

John P. Gausch, P.E., CSP, President, Asset Control, Inc., Consultant, Greenwich, Connecticut.

Joseph Gerber, J.D., Associate Professor, Insurance, University of Arizona, Tucson, Arizona. Former Insurance Commissioner of Illinois.

John V. Grimaldi, Ph.D., P.E., CSP, Director of Degree Programs, The Safety Center, University of Southern California, Los Angeles, California.

Jerry Goodale, Ed.S., Manager of Safety Services, Adolf Coors Co., Golden, Colorado.

W. G. Johnson, Consultant, Willoughby, Ohio. Former General Manager, National Safety Council.

David V. Maccollum, P.E., CSP, Consultant, Sierra Vista, Arizona. Former President, American Society of Safety Engineers.

Harold O'Shell, P.E., CSP, Vice President, ESIS, Inc., Los Angeles, California.

William Pope, P.E., CSP, President, Safety Management Information Systems, Inc. Consultant, Alexandria, Virginia. Founder, National Safety Management Society.

D. A. Weaver, CPCU, CSP, Manager of Safety Dynalectron, Inc., Pueblo, Colorado.

Michael Zabetakis, Ph.D., Director, Training Academy, MESA, Beckley, West Virginia.

PART ONE

The Basis and Philosophy of Accident Prevention

ONE

Basic Philosophy of Accident Prevention

In Part 1, our purpose is to examine the basic philosophy of safety management and the techniques of accident prevention and to identify principles upon which we can build techniques that will give us results: an improved accident record individually, corporately, and nationally.

Industrial accident prevention has come of age. The long struggle for its recognition, for supporting interest and effective action, waged over a period of years by individuals and organizations, bears fruit. Need and value being well-established, it follows that emphasis must be placed on practical methods of accomplishment. The text is in harmony with this thought. Method is stressed at the sacrifice of detail concerning the many reasons why unnecessary chance taking and unnecessary mechanical hazards should be corrected. Historical background is relegated to App. 1 so as to provide space for the vital elements of successful practice. Specific detailed examples are given, sufficient for illustration, but no attempt is made to provide the complete details of any one of the myriad areas of application.

Our purpose is to discuss and illustrate concepts, principles, and philosophies. In this chapter our intent is to look at some initial thoughts about what accident prevention is, what it accomplishes, and how it accomplishes it. We'll also examine the term "safety management," for the text now includes not only the accident-prevention concepts of previous editions, but also the newer, more encompassing concepts of safety management.

WHAT ACCIDENT PREVENTION IS

Accident prevention is both science and art. It represents, above all other things, *control*—control of worker performance, machine performance, and physical environment. The word "control" is used advisedly because it connotes *prevention* as well as *correction* of unsafe conditions and circumstances.

As thus defined accident prevention is a vital factor in every industrial enterprise, one which if ignored or practiced unskillfully, leads to needless human suffering and business bankruptcy.

In the mid-1920s a series of theorems were developed which are defined and explained in the following chapter and illustrated by the "domino sequence." These theorems show that (1) industrial injuries result only from accidents, (2) accidents are caused directly only by (a) the unsafe acts of persons or (b) exposure to unsafe mechanical conditions, (3) unsafe actions and conditions are caused only by *faults* of persons, and (4) faults of persons are created by environment or acquired by inheritance.

This domino theory of accident causation has, since the 1920s, been quoted at length. It has been almost, until recently at least, universally accepted as a real description of the accident process. There are, now, other equally good, perhaps better, models of accident causation as is discussed in the next chapter. Nonetheless, the domino theory has held up from its inception, almost till the present, before being challenged by newer theory. The domino theory even today is the basis for most accident investigation techniques in industry, for most inspection techniques, etc.

Most newer theories of causation focus more intensely on the worker as a primary causative factor, but the reader should keep in mind that even in the early editions of this book, the major emphasis has been on the worker as a prime accident cause.

Thus from this sequence of steps in the occurrence of accidental injury it is apparent that worker failure is the heart of the problem and that methods of *control* must be directed toward worker failure.

Worker failure is thus justified as the basic cause and the one deserving of major consideration even though the application of a *mechanical* remedy remains the obvious *immediate* remedial action.

IMMEDIATE AND LONG-RANGE APPROACHES

The accident-prevention task in industry requires *both* the immediate approach (direct control of personal performance and environment) and the longer-range approach of instruction, training, and education. Whenever there is opportunity to make a dangerous condition foolproof mechanically, this should be done whether or not personal unsafe action exists. Whenever an unsafe personal action causes or may cause an accident and there is no practical remedy of a mechanical nature, action should at once be taken to prevent its repetition. These are *immediate* and practical remedies. However, it is imperative that sight not be lost of the all-important basic cause of worker failure else the corrective action will treat only that which is superficial and ignore the deep-seated.

One further thought deserves special mention in this section—it is, above all, the work of eliminating the mechanical hazards of environment and the unsafe actions of persons *before* the accident and the injury occur.

In subsequent chapters we will amplify the statement that the causes of accidents are identical with the causes of inefficient production, and that the remedies are similar. Thus in the work of identifying and eliminating the causes of accidents, there is simultaneous improvement of industrial productivity. Even more important is the fact that the objectives of accident prevention are humane as well as economic and that the methods emphasized herein influence the attitudes of persons. This cannot but mean that improvement in the accident-injury record through the elimination of worker-failure causation has a beneficial effect on human behavior in general. Indeed it must be recognized that there is more to safety than meets the eye. Both foundational and specific education are indicated. Selection and placement of personnel are essential. Engineering must continue to play a vital part. Tying all together in application, there must be practical and effective method.

Both philosophy and psychology are underlying sciences, supporting and guiding the selection and application of accident-prevention methods—philosophy, because it includes recognizing facts, finding reasons or causes, and drawing realistic conclusions; psychology, because each step of the control program is affected by and affects human behavior.

Basically, the program is the same for large or small establishments, and all kinds of industrial activities. Modification and adaptation are required in individual cases but not structural change. There are three basic principles: *Interest*, sufficient to initiate prompt action, is a first and a continuing principle. Pertinent *facts* must be found and analyzed. *Corrective action* must then be taken on the basis of well-selected remedies.

Thus accident prevention may be defined as an integrated program, a series of coordinated activities, directed to the control of unsafe personal performance and unsafe mechanical conditions, and based on certain knowledges, attitudes, and abilities. In this text these several steps or factors in the program are given identifying names, defined, and illustrated by examples of application.

SAFETY MANAGEMENT VERSUS ACCIDENT PREVENTION

In recent years, several terms have emerged and become popular that are similar to, perhaps are even synonymous with, the term "accident prevention." They are terms like "loss prevention," "loss control," "total loss control," "safety engineering," "safety management," etc. Of these the term "safety management" seems to be the most widely accepted by safety professionals. The above definition of accident prevention as an integrated program or a series of coordinated activities directed to the control of personal performance and of unsafe conditions is not too dissimilar to some commonly accepted definitions of safety management: getting things done through others with a minimum of error, an integrated approach in the organization to achieve error-free performance, etc.

In short our definition of accident prevention is almost synonymous with current-day definitions of safety management. Thus any lengthy discussion that we might embark on here about the differences between terms would be primarily a semantic discussion rather than a discussion of tasks, missions, or even duties to be performed.

WHAT ACCIDENT PREVENTION ACCOMPLISHES

What accident prevention *can* do is best described by stating what it *has* done. Hundreds of examples can be given showing spectacular achievements in the saving of life and limb, insurance premium reduction, continuity of service, increased production, decreased labor turnover, and improved labor-management relations. The illustration in Fig. 1-1 should be sufficient to justify the assertion that results can be achieved with good accident-prevention effort.

an ANSI rule
American national standards Inst.

BEST NO-INJURY RECORDS KNOWN IN INDUSTRY

The following list shows, for major industry groups, the largest number of continuous hours worked without a disabling Z16.1 injury, as reported to NSC. For additional records, see Work Injury Rates pamphlet.

Industry	Company and Plant or Location	Hours Worked
Chemical	E. I. Du Pont de Nemours & Co., Kinston Plant, Kinston, N.C.	66,645,399
Aerospace	Hughes Aircraft Company, Space & Communications, El Segundo, Calif.	53,163,698
Electrical equipment	Western Electric Company, North Carolina Works, Greensboro, N.C.	38,004,393
Textile	Monsanto Company, Nylon Plant, Pensacola, Florida	27,261,284
Automobile	General Motors Corp., Assembly Div., Wilmington Plant, Wilmington, Del.	20,719,687
Sheet metal products	Remington Arms Co., Lake City Arsenal, Independence, Mo.	20,023,455
Communications	Southwestern Bell Telephone Co., St. Louis, Mo.	18,624,242
Machinery	International Business Machines Corp., Plant No. 2, Poughkeepsie, N.Y.	17,604,263
Rubber & plastics	Uniroyal, Naugatuck Footwear Plant, Naugatuck, Conn.	17,422,846
Clay & mineral products	Monsanto Company, St. Peters Plant, St. Peters, Mo.	17,157,282*
Steel	United States Steel Corp., Gary Steel Works, Gary, Indiana	17,133,243
Tobacco	Bayuk Cigars, Inc., Philadelphia, Pa.	14,314,436
Petroleum	American Oil Company, Whiting Research Labs., Whiting, Indiana	14,300,000
Iron & steel products	Remington Arms Company, Ilion, New York	13,824,446
Fertilizer	Monsanto Company, El Dorado Plant, El Dorado, Arkansas	11,709,490
Mining	Reserve Mining Company, Silver Bay Div., Silver Bay, Minn.	10,659,100
Foundry	General Motors Corporation, Central Foundry Div., Saginaw, Mich. Plt.	10,112,268
Printing & publishing	R. R. Donnelley & Sons Company, Crawfordsville Div., Crawfordsville, Ind.	9,611,563
Pulp, paper & related products	Buckeye Cellulose Corp., Cellulose & Spec. Div., Memphis, Tenn.	9,428,814

*Record continuing.

Figure 1-1 From *Accident Facts*, National Safety Council, 1977 ed.

Construction	E. I. Du Pont de Nemours & Co.,	
	Old Hickory Const., Old Hickory, Tenn.	9,165,858
Electric utilities	Gulf States Utilities Company,	
	Baton Rouge Division, Baton Rouge, La.	8,986,638
Shipbuilding	New York Naval Shipyard,	
	Naval Base, Brooklyn, New York	8,509,572
Glass	PPG Industries, Inc.,	
	Works Seven, Cumberland, Maryland	8,447,386
Nonferrous metals	Aluminum Company of America,	
	Warrick Operations, Newburgh, Indiana	7,566,115
Leather	Genesco, Cowan Plant, Cowan, Tennessee	7,310,972
Gas	Washington Gas Light Company, Washington, D.C.	7,248,332
Food	National Distillers Products Co.,	
	K. D. & W. Plant, Louisville, Kentucky	7,229,953
Transit	Virginia Transit Co.,	
	Trans. Dept., Richmond Div., Richmond, Va.	5,683,281
Cement	Lehigh Portland Cement Co., Oglesby, Illinois	5,487,376
Railroad equipment	General Motors Corp.,	
	Electro-Motive Div., Cleveland, Ohio	5,294,960
Meat Packing	Wilson & Company, Inc., Los Angeles, Calif.	5,051,451
Air transport	General Electric Co.,	
	Aviation Ser. Oper., Ontario, Calif.	4,819,740
Marine	Interlake Steamship Co.,	
	A Division of Pickands Mather & Co.	4,176,709
Wood products	Kroehler Mfg. Co., Plant #3, Kankakee, Illinois	4,100,615
Quarry	U.S. Steel Corp., Michigan Limestone Div.,	
	Calcite Plant, Rogers City, Mich.	3,634,588
Lumber	Pope & Talbot, Inc.,	
	Oakridge Sawmill, Oakridge, Oregon	2,394,809

Figure 1-1 (*Continued*)

Countrywide possibilities are indicated by the following, which shows the magnitude of the problem and its impact on society: Working time lost during 1974 as a result of disabling injuries was about 245 million work-days, according to the National Safety Council, equivalent to the working time of over 1 million people for a full year. Occupational injuries cost the nation approximately $5.064 billion.

The cost saving to society in general (of accident prevention) can be estimated but vaguely. Savings in physical and mental anguish are immeasurable. Accounting methods cannot be established to portray accurately the tremendous beneficial effect on production, morale, and the national economy.

HOW ACCIDENT PREVENTION IS ACCOMPLISHED

Accident prevention is accomplished through five separate steps as shown in Fig. 1-2; as shown in the figure, the steps are built on a foundation.

Figure 1-2 is very similar to several decision models in common use in industry in the past and even today. For instance, the Kepner-Tregoe model shown in Fig. 1-3 is a common approach used and adapted in many industries for problem solving of all types. In its analysis, decision (selection), and execute (application) steps it is very close to the steps of Fig. 1-2.

Two other models of decision making are also mentioned here for comparison purposes to the five steps of accident prevention, and to

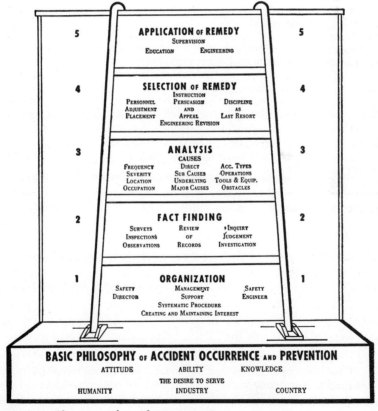

Figure 1-2   The steps of accident prevention.

KEPNER — TREGOE

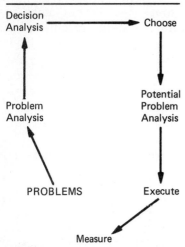

Figure 1-3 The Kepner-Tregoe model (Kepner, C., and B. Tregoe, *The Rational Manager*, McGraw-Hill, New York, 1965.)

help in evolving a model updating these original five steps. The first one is the model for improving human performance discussed by Johnson in MORT (management oversight risk tree). This is shown in Fig. 1-4. Perhaps the primary difference in this, and in the Kepner-Tregoe model, is that there is a measurement and feedback which tend to provide the needed feedback loop for an ongoing safety system.

Johnson, in describing the model, also lists six rather interesting criteria:

1 Errors are an inevitable (rate-measurable) concomitant of doing work or anything.
2 Situations may be error-provocative—changing the situation will likely do more than elocution or discipline.
3 Many error definitions are "forensic" (which is debatable, imprecise, and ineffective) rather than precise.
4 Errors at one level mirror service deficiencies at a higher level.
5 People mirror their bosses—if management problems are solved intuitively, or if chance is relied on for non-accident records, long-term success is unlikely.
6 Conventional methods of documenting organizational procedures seem to be somewhat error provocative.

IMPROVING HUMAN PERFORMANCE
(ERROR REDUCTION)

Philosophy, Practice

Assistance

Study

Situation Change

Participation

ERRORS

Error Rate Reductions

Measure

Figure 1-4   Improving human performance model. [Johnson, MORT (management oversight risk tree), U. S. Government Printing Office, Washington.]

Johnson, in summing up the area of decision models in his MORT, also provides one other model that closely approximates our original accident-prevention steps. It is a performance cycle model, and is shown in Fig. 1-5.

Out of the original five-step process of accident prevention, and utilizing the insights from other decision models, the updated safety management approach has evolved as shown in Fig. 1-6. The remainder of this part of the book looks at the steps of this model.

PERFORMANCE CYCLE
(A REPERTORY OF MANAGEMENT METHODS USEFUL FOR SAFETY)

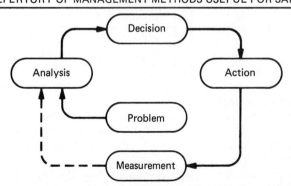

Decision

Analysis

Action

Problem

Measurement

Figure 1-5   The performance cycle model. (Johnson, MORT.)

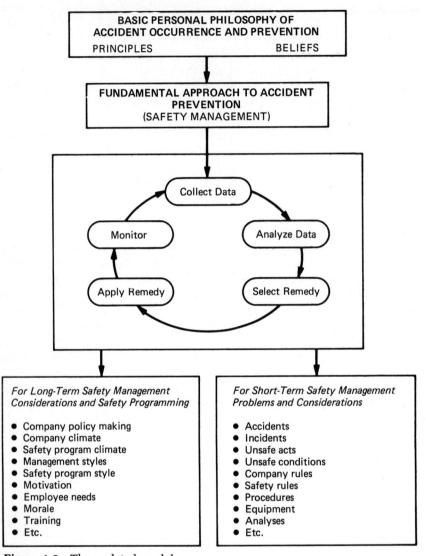

Figure 1-6 The updated model.

BASIC PHILOSOPHY

The model begins with the basic philosophy or the theorems of accident occurrence and prevention. Familiarity with this philosophy is a "must" for every safety engineer, supervisor, and executive. Success in safety work depends upon a sound knowledge of *what* an accident is and *how* and *why* it occurs, upon the knowledge of *reasons* and *incentives* for prevention, and upon *opportunities* and *practical methods* of achievement.

Following the initial development of the personal philosophy, principles, and beliefs, the model then focuses on a rather simple model of managerial decision making similar to those discussed earlier (and not too dissimilar from the original five-step process). The steps are to collect data, analyze those data, select a remedy, apply that remedy, and then monitor or measure results. The model is circular, representing an ongoing process. Following the monitoring of previous remedies there usually will be a need to recollect data to see if fine tuning or different approaches might be in order.

Collect Data

The need and value of this step require little if any defense or explanation. It should suffice to say that workers should know where they intend to go, why they need to go, the means of transportation, and what they expect to do when they get there before starting on a trip. Facts and fact finding are covered in Chap. 4. In this work the investigation of accidents for cause and remedy by the supervisor is the keystone. When tied in with the findings resulting from surveys, inspections, and observations, and when supplemented by conclusions based on experience, judgment, and inquiry, it completes the outline of methods whereby pertinent information on accident causation and remedy is obtained.

Analyze Data

In logical sequence, the determination of the facts of accident causation is followed by *analysis* of the facts disclosed. This is defined as the work of drawing conclusions from assembled data. In small establishments it may be done without recourse to records. In larger operations, manual or machine coding methods are used. The conclusions must invariably identify the principal *direct* causes and types of accidents; the kinds of injuries; the locations, operations, and equipment chiefly involved; and the persons who are responsible or who are affected. Analysis also includes the study and solution of problems as when obstacles to progress are encountered. Specifically this means that *subcauses*, or reasons *why* persons persist in unsafe action or *why* unsafe mechanical conditions repeatedly exist, should be determined when, for example, the goggles which are provided for persons who are endangered by flying particles or by acid splash, are not worn.

Select Remedy

When the facts of accident causation have been so analyzed as to indicate *what* it is that needs to be corrected, it then becomes necessary to give thought to the selection of an effective remedy.

In many instances accident-prevention methods are adopted without conscious selection, sound reasoning, or knowledge that they fit the case to which they are applied—and sometimes they work satisfactorily.

In one case of record the general superintendent of a large construction company issued and followed up a single executive order and thereby reduced his accident frequency and cost over 30 percent. Becoming alarmed because of the rising trend of accident experience, he called a meeting of all superintendents, their assistants, and the supervisors and stated in no uncertain terms that he wanted them to stop accidents. The order was repeated several times, records were kept, and responsibility was placed. Despite the fact that there was no knowledge of accident causes, no safety organization, and no specific educational work, satisfactory results were achieved.

A sales manager succeeded in increasing sales by the simple method of injecting her personality into the matter in a most forceful way. She held a series of meetings at which she demanded of her salespeople that they "go out and bring in the orders or else."

In countless other cases, however, these methods do not succeed. The sales manager who simply demands that her salespeople produce results, at the same time being totally unaware that deliveries are delayed, competitors' products are better and less expensive, commissions are lower than average, and salespeople are inexperienced or incompetent, would be more likely to increase sales volume if she knew these conditions and adopted methods based on her knowledge. The industrial executive, likewise, who in the example cited was fortunate enough to get results by the issuance of a single forceful order would be more certain of success if he were better informed about accident causes and remedies.

Apply Remedy

The next step is *application* of the selected remedy. Here, progressive simplification again clarifies the task. If the three prior steps have been taken from a full appreciation of *basic philosophy*, there is nothing to *application* except to go ahead and do the work indicated by these steps and to "follow through" for continuity and permanency. If tools, machines, structures, procedures, or equipment are unsafe, they must be guarded, replaced, revised, or otherwise made mechanically safe. Safety engineers may suggest, recommend, or order these changes according to the degree of their authorization. Management has the responsibility, and the supervisors see to it that the necessary work is done.

If personal performance is unsafe, employees must be selected, instructed or trained, cautioned, persuaded, convinced, appealed to, or enthused. Certain cases require placement; others, medical treatment or advice. In rare cases and as a last resort, some form of disciplinary action is applied. As in the case of unsafe mechanical conditions, the safety engineer suggests, recommends, etc., according to the limits of his or her authority, while management through its line and staff supervisors does the actual work of application. Medical treatment or advice is given by authorized company-employed or private practitioners.

Application of selected remedial measures is both immediate and long-term. In other words, existing unsafe conditions and circumstances are attacked at once while at the same time the program includes procedures devised to anticipate and prevent situations of a similar nature.

Monitoring

The final step in the process is to monitor results, to see whether or not improvement is achieved as a result of the remedy applied. This can be done in any number of ways, depending on the situation. The most common is to see whether or not there are fewer accidents occurring. This may well not be the best way of monitoring, however, as will be discussed later.

Monitoring is a feedback mechanism; it tells us how we are doing, what progress we are making. It not only is the final step in the process, but is often also an intermediate step. As a result of our measurements, we can spot additional difficulties to deal with, and by some additional measures or remedies we might affect better results. This fine tuning is often an integral part of the total process. Our monitoring dictates additional needs; we then collect additional data, analyze it, select additional remedies, apply them, monitor, etc.

SCOPE AND FUNCTIONS OF THE SAFETY PROFESSIONAL

Figure 1-7 outlines the functions of the safety professional as determined by the American Society of Safety Engineers (ASSE). Their four-step process is very similar to the model discussed in this chapter. The only real difference is that they combine the data-gathering and analysis functions into one function. Figure 1-8 details each of the four areas in the ASSE's Scope and Function definition.

Each will be discussed in some detail in later chapters.

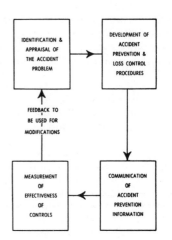

Functions of the Professional Safety Position

The Scope of the Professional Safety Position

The safety professional brings together those elements of the various disciplines necessary to identify and evaluate the magnitude of the safety problem. He collects and analyzes the information essential to the solution of the problem. He is concerned with all facets of the problem, personal and environmental, transient and permanent, to determine the causes of accidents or the existence of loss producing conditions, practices or materials.

Based upon the information he has collected and analyzed, he proposes alternate solutions, together with recommendations based upon his specialized knowledge and experience, to those who have ultimate decision-making responsibilities.

The functions of the position are described as they may be applied in principle to the safety professional in any activity.

The safety professional in performing these functions will draw upon specialized knowledge in both the physical and social sciences. He will apply the principles of measurement and analysis to evaluate safety performance. He will be required to have fundamental knowledge of statistics, mathematics, physics, chemistry, as well as the fundamentals of the engineering disciplines.

He will utilize knowledge in the fields of behavior, motivation, and communications. Knowledge of management principles as well as the theory of business and government organization will also be required. His specialized knowledge must include a thorough understanding of the causative factors contributing to accident occurrence as well as methods and procedures designed to control such events.

The safety professional of the future will need a unique and diversified type of education and training if he is to meet the challenges of the future. The population explosion, the problems of urban areas, future transportation systems, as well as the increasing complexities of man's every day life will create many problems and extend the safety professional's creativity to its maximum if he is to successfully provide the knowledge and leadership to conserve life, health and property.

The major functions of the safety professional are contained within four basic areas. However, application of all or some of the functions listed below will depend upon the nature and scope of the existing accident problems, and the type of activity with which he is concerned.

The major areas are:

A. Identification and appraisal of accident and loss producing conditions and practices · and evaluation of the severity of the accident problem.

B. Development of accident prevention and loss control methods, procedures, and programs.

C. Communication of accident and loss control information to those directly involved.

D. Measurement and evaluation of the effectiveness of the accident and loss control system and the modifications needed to achieve optimum results.

Figure 1-7 ASSE Scope and Functions.

A   *IDENTIFICATION AND APPRAISAL OF ACCIDENT AND LOSS PRODUCING CONDITIONS AND PRACTICES AND EVALUATION OF THE SEVERITY OF THE ACCIDENT PROBLEM*

These functions involve:

1   The development of methods of identifying hazards and evaluating the loss producing potential of a given system, operation or process by:

   a   Advanced detailed studies of hazards of planned and proposed facilities, operations and products.

   b   Hazard analysis of existing facilities, operations and products.

2   The preparation and interpretation of analyses of the total economic loss resulting from the accident and losses under consideration.

3   The review of the entire system in detail to define likely modes of failure, including human error and their effects on the safety of the system.

   a   The identification of errors involving incomplete decision making, faulty judgment, administrative miscalculation and poor practices.

   b   The designation of potential weaknesses found in existing policies, directives, objectives, or practices.

4   The review of reports of injuries, property damage, occupational diseases or public liability accidents and the compilation, analysis and interpretation of relevant causative factor information.

   a   The establishment of a classification system that will make it possible to identify significant causative factors and determine trends.

   b   The establishment of a system to insure the completeness and validity of the reported information.

   c   The conduct of thorough investigation of those accidents where specialized knowledge and skill are required.

5   The provision of advice and counsel concerning compliance with applicable laws, codes, regulations and standards.

6   The conduct of research studies of technical safety problems.

7   The determination of the need of surveys and appraisals by related specialists such as medical, health physicists, industrial hygienists, fire protection engineers and psychologists to identify conditions affecting the health and safety of individuals.

8   The systematic study of the various elements of the environment to assure that tasks and exposures of the individual are within his psychological and physiological limitations and capacities.

B   *DEVELOPMENT OF ACCIDENT PREVENTION AND LOSS CONTROL METHODS, PROCEDURES AND PROGRAMS*

In carrying out this function, the safety professional:

1   Uses his specialized knowledge of accident causation and control to prescribe an integrated accident and loss control system designed to:

   a   Eliminate causative factors associated with the accident problem, preferably before an accident occurs.

   b   Where it is not possible to eliminate the hazard, devise mechanisms to reduce the degree of hazard.

Figure 1-8   Detailed scope and functions from ASSE.

  c Reduce the severity of the results of an accident by prescribing specialized equipment designed to reduce the severity of an injury should an accident occur.

2 Establishes methods to demonstrate the relationship of safety performance to the primary function of the entire operation or any of its components.

3 Develops policies, codes, safety standards and procedures that become part of the operational policies of the organization.

4 Incorporates essential safety and health requirements in all purchasing and contracting specifications.

5 As a professional safety consultant for personnel engaged in planning, design, development, installation of various parts of the system, advises and consults on the necessary modification to insure consideration of all potential hazards.

6 Coordinates the results of job analysis to assist in proper selection and placement of personnel, whose capabilities and/or limitations are suited to the operation involved.

7 Consults concerning product safety, including the intended and potential uses of the product as well as its material and construction, through the establishment of general requirements for the application of safety principles throughout planning, design, development, fabrication and test of various products, to achieve maximum product safety.

8 Systematically reviews technological developments and equipment to keep up to date on the devices and techniques designed to eliminate or minimize hazards, and determine whether these developments and techniques have any applications to the activities with which he is concerned.

C *COMMUNICATION OF ACCIDENT AND LOSS CONTROL INFORMATION TO THOSE DIRECTLY INVOLVED*

In carrying out this function the safety professional:

1 Compiles, analyzes and interprets accident statistical data and prepares reports designed to communicate this information to those personnel concerned.

2 Communicates recommended controls, procedures, or programs, designed to eliminate or minimize hazard potential, to the appropriate person or persons.

3 Through appropriate communication media, persuades those who have ultimate decision making responsibilities to adopt and utilize those controls which the preponderance of evidence indicates are best suited to achieve the desired results.

4 Directs or assists in the development of specialized education and training materials and in the conduct of specialized training programs for those who have operational responsibility.

5 Provides advice and counsel on the type and channels of communications to insure the timely and efficient transmission of useable accident prevention information to those concerned.

Figure 1-8 *(Continued)*

D MEASUREMENT AND EVALUATION OF THE
EFFECTIVENESS OF THE ACCIDENT AND LOSS
CONTROL SYSTEM AND THE NEEDED MODIFICATIONS
TO ACHIEVE OPTIMUM RESULTS

1 Establishes measurement techniques such as cost statistics, work sampling or other appropriate means, for obtaining periodic and systematic evaluation of the effectiveness of the control system.

2 Develops methods that will evaluate the costs of the control system in terms of the effectiveness of each part of the system and its contribution to accident and loss reduction.

3 Provides feed back information concerning the effectiveness of the control measures to those with ultimate responsibility, with the recommended adjustments or changes as indicated by the analyses.

Figure 1-8 *(Continued)*

REFERENCES

Accident Facts, National Safety Council, Chicago, 1975.

Johnson, W., *The Management Oversight and Risk Tree—MORT*, U.S. Atomic Energy Commission, Washington, 1973.

Kepner, C., and B. Tregoe, *The Rational Manager*, McGraw-Hill, New York, 1965.

Scope and Functions of the Safety Professional, American Society of Safety Engineers, Park Ridge, Ill., 1966.

TWO

Principles
of Accident Prevention

SECTION 1   The Accident Sequence

In the first edition of this book, as in all succeeding editions, Heinrich presented a set of theorems called the "axioms of industrial safety." These axioms, or so-called self-evident truths, are shown as originally written in Fig. 2-1. These axioms were the first set of principles or guidelines ever set down in industrial safety, and as such did in fact guide all safety activity and effort from the day of the first edition until the present. Naturally today, some 50 years later, we question some of the axioms, and even doubt if they are in fact "self-evident truths." Many are no longer self-evident; some are no longer believed to be truths.

    The axioms dealt with the most important areas in industrial safety: with the theory of accident causation, with the interface of worker and machine, with the relationship of frequency to severity, with the underlying reasons for unsafe acts, with the relationship of accident control to other management functions, with fundamental

---

AXIOMS OF INDUSTRIAL SAFETY

1 The occurrence of an injury invariably results from a completed sequence of factors—the last one of these being the accident itself. The accident in turn is invariably caused or permitted directly by the unsafe act of a person and/or a mechanical or physical hazard.

2 The unsafe acts of persons are responsible for a majority of accidents.

3 The person who suffers a disabling injury caused by an unsafe act, in the average case has had over 300 narrow escapes from serious injury as a result of committing the very same unsafe act. Likewise, persons are exposed to *mechanical* hazards hundreds of times before they suffer injury.

4 The *severity* of an injury is largely fortuitous—the *occurrence* of the *accident* that results in injury is largely preventable.

5 The four basic *motives* or *reasons* for the occurrence of unsafe acts provide a guide to the selection of appropriate corrective measures.

6 Four basic methods are available for preventing accidents—*engineering revision, persuasion and appeal, personnel adjustment,* and *discipline.*

7 Methods of most value in *accident prevention* are analogous with the methods required for the control of the quality, cost, and quantity of *production.*

8 Management has the best opportunity and ability to initiate the work of prevention; therefore it should assume the responsibility.

9 The supervisor or foreman is the key man in industrial accident prevention. His application of the art of supervision to the control of worker performance is the factor of greatest influence in successful accident prevention. It can be expressed and taught as a simple four-step formula.

10 The humanitarian incentive for preventing accidental injury is supplemented by two powerful economic factors: (1) the safe establishment is efficient productively and the unsafe establishment is inefficient; (2) the direct employer cost of industrial injuries for compensation claims and for medical treatment is but *one-fifth* of the total cost which the employer must pay.

Figure 2-1 The original axioms.

responsibilities for getting things done in an organization, with the costs of accidents, and with the relationship between safety and efficiency. These areas are still the most important areas in safety. As we look at these axioms in the light of today's knowledge and beliefs, we find ourselves looking at exactly the same areas, and in fact using the axioms as the basis for current thought over and over again.

The intent of this chapter is to look at each of the above areas, to explain Heinrich's original thinking, and to then update or change the axioms in the light of the theories of today. Since most safety texts in recent years have started from the axioms to develop their own newer theory, we also shall start from the axioms and describe newer theory.

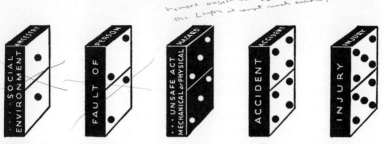

Figure 2-2   The five factors in the accident sequence.

The first axiom deals with a theory of accident causation now labeled the "domino theory." This section deals with this and newer theories of causation.

A preventable accident is one of five factors in a sequence that results in an injury. See Figs. 2-1 to 2-4.

The injury is invariably caused by an accident, and the accident in turn is always the result of the factor that immediately precedes it.

In accident prevention the bull's-eye of the target is in the middle of the sequence—an unsafe act of a person or a mechanical or physical hazard.

The several factors in the accident-occurrence series are given in chronological order in the following list:

Accident Factors	*Explanation of Factors*
1 Ancestry and social environment	Recklessness, stubborness, avariciousness, and other undesirable traits of character may be passed along through inheritance. Environment may develop undesirable traits of character or may interfere with education.

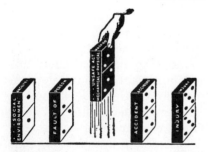

Figure 2-3 The unsafe act and mechanical hazard constitute the central factor in the accident sequence.

Figure 2-4 The removal of the central factor makes the action of preceding factors ineffective.

		Both inheritance and environment cause faults of person.
2	Fault of person	Inherited or acquired faults of person, such as recklessness, violent temper, nervousness, excitability, inconsiderateness, ignorance of safe practice, etc., constitute proximate reasons for committing unsafe acts or for the existence of mechanical or physical hazards.
3	Unsafe act and/or mechanical or physical hazard	Unsafe performance of persons, such as standing under suspended loads, starting machinery without warning, horseplay, and removal of safeguards; and mechanical or physical hazards, such as unguarded gears, unguarded point of operation, absence of rail guards, and insufficient light, result directly in accidents.
4	Accident	Events such as falls of persons, striking of persons by flying objects, etc., are typical accidents that cause injury.
5	Injury	Fractures, lacerations, etc., are injuries that result directly from accidents.

The occurrence of a preventable injury is the natural culmination of a series of events or circumstances, which invariably occur in a fixed and logical order. One is dependent on another and one follows because of another, thus constituting a sequence that may be compared with a row of dominoes placed on end and in such alignment in relation to one another that the fall of the first domino precipitates the fall of the entire row. An accident is merely one factor in the sequence. If this series is interrupted by the elimination of even one of the several factors that constitute it, the injury cannot possibly occur.

DEFINITION OF ACCIDENT

At this point it is advisable to consider what is meant by the term "accident."

An accident is an unplanned and uncontrolled event in which the action or reaction of an object, substance, person, or radiation results in personal injury or the probability thereof.

Industrial management and especially its staff members who specialize in *production efficiency* would do well to note that the foregoing definition applies to circumstances or conditions that create excess production cost, low volume, delay, and spoilage as well as to events causing personal injury; also that the causes and the remedies for both are alike.

It might be noted here that many today prefer to use the term "incident" rather than "accident." The reason for this is that in their definition of accident they eliminate the idea of "or the probability thereof," which is a definite and important part of our definition. An incident is defined as "an undesired event that could (or does) downgrade the efficiency of the business operation." The idea is to include in our work of safety management those things that occur that give us problems and cost us in time, effort, dollars, etc., that do not result in actual injury to a person. Since our definition above includes the notion of probability of happening as well as actually happening, we believe that our definition encompasses the idea of no-injury accidents as well as injury accidents. For any reader preferring the concept of incident over the concept of accident, everything that will be said in this text about accidents and accident prevention can literally be used in terms of incidents and incident prevention. The most important concept to zero in on is the fact of the event being undesired, and that it can (or could) result in injury.

AN UPDATED DOMINO SEQUENCE

The first update of the domino theory we might look at is the updated domino sequence originally presented by Frank Bird, Jr., and discussed in several of his publications. Following is his explanation:

> The five key loss control factors in the updated sequence will be explained in the order of the dominoes illustrated in Fig. 2-5.

LACK OF CONTROL – MANAGEMENT

> No factor in the sequence is more important for the safety/loss control manager to fully comprehend, in order that his program be based on widely-recognized and established tenets of professional management. The word "control" in this factor refers to the fourth function of professional management (PLANNING – ORGANIZING – LEADING – CONTROLLING). In its generic usage, as related to "loss control," the word "control" refers broadly to the general regulation, curbing, restraining or holding back of losses. CONTROL as a function of professional management is optimized through five established steps that systematically produce desired results.

Figure 2-5 An updated sequence. (Bird, Frank E., Jr., *Management Guide to Loss Control*, Institute Press, Atlanta.)

Simply stated, loss control management involves (1) the identification of the WORK activities in the program that management must engage in to produce desired results at an existing stage of the program's development (i.e., accident investigation, facility inspection, job analysis, personal communications, supervisory training, hiring and selection, design engineering, etc.), (2) the establishment of STANDARDS for management's performance in each work activity identified, (3) MEASURING management performance by the established standards in each of its work activities, and (4) CORRECTING performance by improving the existing program (see maintenance loop on chart) and/or expanding the existing program.

The American space endeavor has proven beyond doubt that, with unlimited resources, professional managerial and technological skills can be combined to produce a hardware system that is 99.9% reliable. Fully recognizing that most industrial systems are already operable and not 99.9% reliable, the application of professional loss control management will largely be concerned with upgrading the established system in a gradual developmental growth process as it is practical and economically feasible to do so. The loss control manager in general industry knows that resources are not unlimited, and therefore it is highly improbable that all accidents or downgrading incidents will be prevented without substantial program growth over a relatively long period of time. He therefore must utilize a system of professional management control to optimize values received for his management dollar and other resources invested. He must also recognize that, until such time as the system has sophisticated to the degree of reliability desired, he should expect that accidents and incidents will occur . . . and therefore his program must include organized loss control countermeasures directed at all factors in the accident/incident sequence.

BASIC CAUSE – ORIGINS (ETIOLOGY)

Since the path to achieving a highly-reliable safety/loss control system involves a developmental growth process, there will probably be management work activities that have not been identified at any given time, and, therefore, standards, measurement, evaluation and correction systems not yet established. In effect, the absence of a highly-reliable loss control system will permit the existence of personal and job-related factors referred to as the basic or underlying causes of accidents or downgrading incidents. Personal factors would include lack of knowledge or skill, improper motivation, and physical or mental problems. Job factors would include inadequate work standards, inadequate purchasing standards, normal wear-and-tear and abnormal usage. The professional manager realizes that only by identifying these basic causes can he really establish a system of effective control. ORIGINS (Etiology) refers to the sources, and its appropriate identification with the basic or underlying cause serves to reinforce our desire to achieve more effective control by getting at the root cause, rather than simply treating the symptoms of the problem.

IMMEDIATE CAUSES – SYMPTOMS

These are the factors in the accident/incident sequence that have historically been called the most important ones to attack. They are also the factors that receive the bulk of attention in governmental safety and health inspections around the world.

The safety manager most frequently refers to these causes as unsafe acts or conditions (i.e., operating without authority, nullifying safety devices, taking unsafe positions, inadequate guards, poor housekeeping, etc.).

The production or quality control manager could also refer to immediate causes of his problem (downgraded production) as substandard acts or conditions. In reality, the immediate cause is usually only a symptom of the deeper underlying problem. When we attack the symptom and do not identify the basic underlying problem, we have not really optimized the potential for permanent control. Again, the professional manager recognizes that, with an imperfect system, he can expect that the immediate causes of problems will be present, and he must design into his program a system to efficiently detect and classify these symptoms, so that appropriate countermeasures can be instituted systematically. At the same time, he will diligently strive to identify their basic causes, in order to institute more permanent control within the practical and economic parameters of his program's development and growth capabilities. With knowledge of his program's capabilities and limitations, he will then give proper professional perspective to the balance of his loss control countermeasures at the remaining stages of the sequence.

ACCIDENT – CONTACT

For practical purposes, the accident might be described as an undesired event that results in physical harm, injury, property damage. The word "accident" itself is purely descriptive, and has no real etiological connotation. While words such as "error" or "mistake" have been suggested as more indicative of an underlying management deficiency, they are also descriptive and would serve no more significant value in loss control than the word "accident". The writers use the word "accident" because of the probability of its wide continued use and application by government, legal bodies, and society as a whole. The word "contact" appears on the domino at this point in the sequence because an ever-increasing number of researchers and safety leaders around the world looks at the accident as a "contact" with a source of energy (electrical, chemical, kinetic, thermal, ionizing, radiation, etc.) above the threshold limit of the body or structure, or "contact" with a substance that interferes with normal body processes. This association gives more inference to control methodology, and is far more etiological in its connotation. As we consider the broader implications of associating "accident" with "contact," we see more clearly the important relationships of the safety, environmental health and fire loss control disciplines, and the increased import of interface between related specialists in a coordinated loss control program. It should be pointed out that the word "incident" could replace the word "accident," and might better represent a true loss control sequence in the domino illustrations provided. In using the word "accident," the writers are considering the developmental stage of loss control programming that currently exists in most organizations; we recognize, however, that substantial growth activities remain to be undertaken for most to achieve total accident control. We can look ahead to the day when all down-grading incidents will be included in our programs' realistic objectives. This point in the sequence is also referred to as the contact stage, and applications of the principles of deflection, dilution, reinforcement, surface modification, segregation, barricading, protection, absorption and shielding are examples of countermeasures frequently employed as loss control tools here.

INJURY – DAMAGE – LOSS

While the writers have used the word "injury" to broadly describe loss results that terminate in personal physical harm of a variety of types, it would seem appropriate to comment further on the meaning of this word. To the early safety practitioner, the terms "accident" and "traumatic injury" were almost synonymous. While occupational disease, fire and property damage were philosophically associated with industrial safety, actual accident prevention practices through the years have largely been devoid of these considerations and are quite injury-oriented. Thus, the word "injury" has been most frequently used to mean bodily damage or harm through traumatic accident. While occupational diseases have historically been counted in disabling injury rates, their cause and control have not until recently had their rightful inclusion in the accident prevention program.

Injury as used in this factor of the sequence, includes all personal physical harm, including both traumatic injury and disease, as well as adverse mental, neurological, or systemic effects resulting from workplace exposures.

The word "damage," as used in this factor of the sequence, is intended to broadly cover all types of property damage, including fire. While no major governmental or private safety organization publishes frequency or cost data on property damage accidents in America, there is enough related information from authoritative sources to indicate that our nation spends at least 20 billion dollars annually on this type of loss. This enormous loss justifies much greater attention in safety and loss control programming than it has received in the past. While the loss control manager who has achieved sufficient program sophistication can establish as his target the control of all "business interruptions," it is believed that the majority of these specialists have substantial control inroads yet to achieve in injury and damage reduction.

To optimize loss reduction, the professional will also direct substantial attention to control countermeasures at this last factor in the sequence, also referred to as the post-contact stage.

There is a tremendous reservoir of information to prove that the severity of losses involving physical harm and property damage can be minimized by the application of one or more countermeasures, even at this point. These could include education of personnel and property, prompt first aid and rehabilitation in cases of physical harm, prompt reparative action and salvage in cases of property damage, and the activation of fire devices including fire fighting by trained personnel.

ACCIDENT CAUSATION AND THE MANAGEMENT SYSTEM

In Frank Bird's updated accident sequence, he introduces the thought of managerial error into the sequence. Most accident cause theorists of today agree with this concept; most of the causation theories we shall discuss in this chapter include management in some way in their causation picture.

Edward Adams, director of loss prevention for Pet, Incorporated, of St. Louis, has presented an accident sequence similar to the updated domino sequence of Frank Bird. The sequence is shown in Fig. 2-6. He explains:

> Under this management theory, the 4th and 5th dominoes, the accident and the injury, remain essentially the same, although the refinements that Bird has made in these areas are recognized.
>
> The immediate cause of the accident has been retitled "Tactical Errors" to draw attention to the nature of unsafe acts and unsafe conditions within the management system. Essentially they are mistakes in employee behavior and in work conditions. The kind of mistakes made are not changed, nor are the problems they present altered by his new title. The new title is necessary, however, to draw attention to the nature of the decisions and work involved in their identification and correction.
>
> The major contribution of this management theory lies in the redefinition of the causes underlying the "Tactical Errors." The tactical errors in employee behavior and work conditions are seen as arising from "Operational Errors" made by managers and supervisors. These are administrative mistakes or omissions made by supervision and/or decisions made wrong or not made by managers in critical managerial areas. These errors are strategic, for they affect the very nature of the operation.
>
> These operational errors derive from the management structure; the objectives of the organization, how the management work is organized, and how operations are planned and carried out. These are the stable elements of the operating organization. They determine the "personality" of the organization. To a large degree the management structure is a reflection of the beliefs, objectives, and standards of the key decision-makers of the organization. It is here that the priorities, the standards, the guidelines for manager, and in turn supervisor, behavior are established.

SYMPTOMS OF OPERATIONAL ERROR

A third update of the domino theory has been provided by D. A. Weaver, and is shown in Fig. 2-7. He explains the concepts of operational error and symptoms:

> The operational errors which result in accidents and injuries also produce the endless array of other unplanned and undesired results which supervisory management contends with every day. The unplanned and undesired result is merely a symptom. The accident or injury is a

ACCIDENT CAUSATION WITHIN THE MANAGEMENT SYSTEM

MANAGEMENT STRUCTURE

Objectives

Organization

Operations

OPERATIONAL ERRORS

Manager Behavior

DECISIONS MADE WRONG OR NOT MADE IN THE AREAS OF:

POLICY
GOALS
AUTHORITY
RESPONSIBILITY
ACCOUNTABILITY
SPAN OF ATTENTION
DELEGATION

Cause Code

POLICY
Failure to assert a management will prior to situation at hand.

GOALS
Not made clear. Not projected as an "Action image."

AUTHORITY
Subordinates fail to exercise their power to decide. Decisions evaded or delayed. By-passing, too many bosses.

RESPONSIBILITY
Duties and tasks not made clear. Overlapping jurisdictions. Job descriptions inadequate. Conflicting goals established.

ACCOUNTABILITY
Failure to measure or appraise results. Appraisals inadequate or dwelt excessively on short-term results.

SPAN OF ATTENTION
Too many irons in the fire. Inadequate delegation. Inadequate development of subordinates.

DELEGATION
Failure to support and encourage subordinates to exercise their power to decide. Decisions too far above the problem.

Supervisor Behavior

ADMINISTRATIVE MISTAKES OR OMISSIONS IN THE AREAS OF:

CONDUCT
RESPONSIBILITY
AUTHORITY
RULES
COACHING
INITIATIVE
MORALE
OPERATIONS

Cause Code

CONDUCT
Sets a poor example

RESPONSIBILITY
Not Understood
Not Accepted
Allows pressure of immediate task to obscure full scope of responsibility

AUTHORITY
Fails to exercise power to decide. Orders fail to produce desired results. Orders not clear, not understood, not followed. Decisions exceed authority. Subordinates fail to exercise their authority to decide. Authority inadequate to cope with the situation.

RULES
Failure to make necessary rules or publicize them. Inadequate follow-up or enforcement of rules. Unfair or uneven enforcement. Weak discipline.

COACHING
Fails to tell why
Fails to listen.
Training not formulated, or need not foreseen
Failure to correct or to see need to correct
Failure to coach in new or unusual situation.
Instructions inadequate.
Instructions attempted but results show it didn't take.

INITIATIVE
Failure to see problems and exert an influence on them.

MORALE
Tension – subordinates lack faith in supervisor.
Insecurity about future of the job.
Team Spirit – subordinates fail to pull with supervisor.
Poor cooperation, failure to plan for coordination.
Compliance – work group sees no advantage to themselves.

OPERATIONS
Job placement – Hasty or improper selection or placement.
Lack – Absence of anything needed.
Clutter – Housekeeping
Job Procedure – Awkward, unsafe, inefficient, poorly planned.
Work Load – Pace too fast, too slow or erratic.
Work Flow – Inefficient, or hazardous stacking, piling, storing, routing, etc.

TACTICAL ERRORS

Employee Behavior

UNSAFE ACTS

Failure to use available personal protective equipment or to wear safe clothing.

Placing body or parts in unsafe or exposed position, or using poor lifting stance.

Using defective, unguarded, makeshift or otherwise unsafe equipment

Using equipment in unsafe manner, or for purpose not intended

Failure to use equipment provided, using hands instead of hand tools

Unsafe loading, placing, piling, arranging, carrying or moving

Distracting, teasing, horseplay, etc.

Operating, driving, working on, or otherwise moving self, machine, or materials at an unsafe speed

Making safety devices inoperative, failure to follow prescribed procedures, failure to heed posted warnings.

Failure to make secure, shut off, warn others

Poor teamwork

Working or operating on moving, electrically charged, or pressurized equipment.

Work Conditions

UNSAFE CONDITIONS

Improper design, construction or layout

Delayed, aged, worn, frayed or cracked

Unnecessarily slippery, rough, sharp-edged or sharp-cornered

Unsafely stored or piled tools or materials.

Poor housekeeping or congestion.

Unsafe established procedure, inadequate job planning, improper equipment provided

Inadequate aisle space, exits, etc.

Improperly guarded

Improper illumination

Improper ventilation

Personal protective equipment not adequate or not available

ACCIDENT INCIDENT

Injury Producing

The Near-Miss
No Injury Incident

The Property
Damage Incident

INJURY OR DAMAGE

To Persons

To Property

MANAGEMENT STRUCTURE

Objectives
GOALS
STANDARDS OF PERFORMANCE
APPRAISALS
MEASUREMENTS

Organization
CHAIN OF COMMAND
SPAN OF CONTROL
DELEGATION OF
  AUTHORITY
  RESPONSIBILITY
STAFF

Operations
LAYOUT
EQUIPMENT
PROCUREMENT
SCHEDULING
PROCEDURES
ENVIRONMENT

Figure 2-6 The Adams updated sequence.

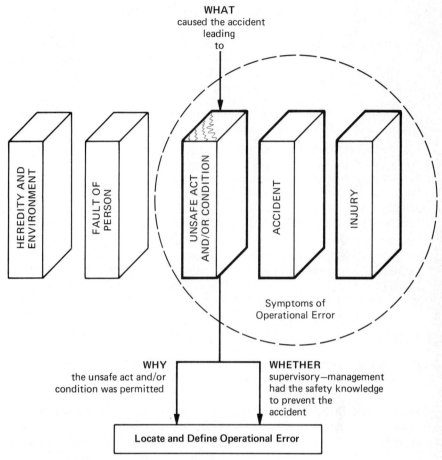

Figure 2-7 Weaver's updated dominoes.

symptom. So is the shipment that goes astray, the contaminated batch, the customer badly served, the snafus that snowball into major events. All are symptoms of the same underlying operational errors.

To seek cause and corrective action in supervisory-management practices, we mate two sets of ideas. We mate the idea of "locate and define operational error" to the inbred thinking of the domino format. The mating produces the principle that accidents and injuries as well as unsafe acts and conditions are all symptoms of operational error. Behind the unsafe scaffold, the unsafe act, the faulty tool, the defective machine or layout lie management practices. Behind any proximate cause (unsafe act and/or condition) ascribed to an accident lie management practices in policy, priorities, organization structure, decision-making, evaluation, control, and administration.

The input of safety technology and immediate correction are still achieved by identifying unsafe acts and conditions. We ask, "What unsafe act and/or condition," and receive a reply in terms of safety technology. But we expose operational error by asking two further questions: "Why the unsafe act and/or condition was permitted" and "Whether supervisory-management had the safety knowledge to prevent the accident." The what-why-whether process may be diagrammed as in Fig. 2-7.

The "whether" question asks whether the laws, codes, and standards applicable to the circumstances were known. Whether the safety director knew them. Whether the hazard had been identified by foresight. Whether the books, pamphlets, pass-outs, and knowledge needed were available. Whether the supervisor knew them. In short, did the organization possess knowledge of the safety technology available?

The "why" question asks why knowledge was not effectively sought or why it was not effectively applied. The question exposes operational error in the area of management policy, confusion in goals, staffing, housekeeping, responsibility, use of authority, line and staff relationships, accountability, rules, initiative, and much more.

UNPLANNED RELEASE OF ENERGY AS A CAUSE

Dr. Michael Zabetakis, director of MSHA's (Mine Safety and Health Administration) academy, provides a fourth update of our domino theory of accident causation in his *Safety Manual No. 4—Accident Prevention*, published by MSHA to assist mine management. This update is shown in Fig. 2-8. While similar to the others, it also introduces a new concept which we will see later in other causation models: the concept that the direct cause is an unplanned release of energy and/or hazardous material. This notion that accidents are caused by the unplanned transfer, or release, of energy is currently a part of a number of theories of causation.

Dr. Zabetakis explains:

> Most accidents are actually caused by the unplanned or unwanted release of excessive amounts of energy (mechanical, electrical, chemical, thermal, ionizing radiation) or of hazardous materials (such as carbon monoxide, carbon dioxide, hydrogen sulfide, methane and water). However, with few exceptions, these releases are in turn caused by unsafe acts and unsafe conditions. That is, an unsafe act or an unsafe condition may trigger the release of large amounts of energy or of hazardous materials which in turn causes the accident.

MANAGEMENT SAFETY POLICY AND DECISIONS

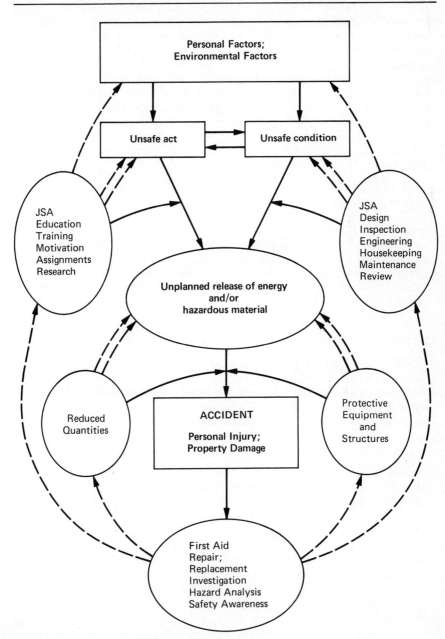

Figure 2-8 Zabetakis's theory of causation.

While we often think of hazardous acts and conditions as the basic causes of accidents, they are actually only symptoms of failure. The basic causes are usually traceable to poor management policies and decisions, and personal and environmental factors. Fortunately, most mine operators now realize that safety must be an integral part of the total operating system; these operators take great pains to prepare a written safety policy guide and to instill a safety awareness in their employees—from top management down to the individual worker. Further, the selection, training and placement of each employee and the purchase, inspection and maintenance of each piece of equipment and all supplies are considered as important to a successful accident prevention program, as are the maintenance of a safe and healthful environment and the establishment of adequate operating and emergency procedures.

DIRECT CAUSES

A detailed analysis of an accident must consider the direct cause (released energy and/or hazardous material) that produced it. Accident investigators are interested in the direct cause because it is often possible to redesign equipment, materials and facilities, to train miners to be aware of hazardous situations, and to provide personal protection in an effort to prevent injury.

INDIRECT CAUSES (SYMPTOMS)

As noted above, unsafe acts and unsafe conditions do not by themselves produce accidents. They are symptoms of poor management policy, inadequate controls, lack of knowledge, improper assessment of existent hazards, or other personal factors. A few of the more common unsafe acts and conditions found in mining and in our everyday activities include:

UNSAFE ACTS

1 Operating equipment at improper speeds
2 Operating equipment without authority
3 Using equipment improperly
4 Using defective equipment
5 Making safety devices inoperable
6 Failure to warn co-workers or to secure equipment
7 Failure to use personal protective equipment
8 Improper loading or placement of equipment or supplies
9 Taking an improper working position
10 Improper lifting
11 Servicing equipment in motion
12 Horseplay

13 Use of alcoholic beverages
14 Use of drugs

UNSAFE CONDITIONS

1 Inadequate supports or guards
2 Defective tools, equipment or supplies
3 Congestion of work place
4 Inadequate warning systems
5 Fire and explosion hazards
6 Poor housekeeping
7 Hazardous atmospheric conditions (gases, dust, fumes, vapors)
8 Excessive noise
9 Poor illumination
10 Poor ventilation
11 Radiation exposure

BASIC CAUSES

Many early accident prevention activities involved only the identification and correction of unsafe acts and conditions. While this is an important function, we now know that long-range improvements can best be made by identifying and correcting the basic causes. These can be loosely grouped into three interrelated categories:

Management Safety Policy and Decisions
Personal Factors
Environmental Factors

The first category, management safety policy and decisions, includes such items as management's intent (relative to safety); production and safety goals; staffing procedures; use of records; assignment of responsibility and authority, and accountability; employee selection, training, placement, direction and supervision; communications procedures; inspection procedures; equipment, supplies, and facilities design, purchase and maintenance; standard and emergency job procedures; and housekeeping.

The second category, personal factors, includes motivation; ability; knowledge; training; safety awareness; assignments; performance; physical and mental state; reaction time; personal care. The third category, environmental factors, includes temperature; pressure; humidity; dust; gases; vapors; air currents; noise; illumination; nature of surroundings (slippery surfaces, obstructions, inadequate supports, hazardous objects). Because these are so interrelated, much thought is given to the effects of a change of one factor (for example, employee selection) on each of the others (such as training, placement, equipment design, etc.).

MULTIPLE CAUSATION

While there are other similar adaptations of the domino theory, perhaps the above gives the reader the idea of the thinking of today stemming from the original accident sequence. Besides the models based on the domino theory, there have also been a number of other models of accident causation. One of these is the concept of multiple causation, described in *Techniques of Safety Management*:

> Today we know that behind every accident there lie many contributing factors, causes, and subcauses. The theory of multiple causation is that these factors combine together in random fashion, causing accidents. If this is true, our investigation of accidents ought to identify some of these factors—as many as possible—certainly more than one act and/or one condition.
>
> Let us briefly look at the contrast between the multiple causation theory and our too narrow interpretation of the domino theory. We shall look at a common accident: a man falls off a stepladder. If we investigate this accident under our present investigation forms we are asked to identify one act and/or one condition:
>
> *The unsafe act:* Climbing a defective ladder
> *The unsafe condition:* A defective ladder
> *The correction:* Getting rid of the defective ladder
>
> This would be typical of a supervisor's investigation of this accident under the domino theory.
>
> Let us look at the same accident in terms of multiple causation. Multiple causation asks what are some of the contributing factors surrounding this incident? We might ask:
>
> 1 Why was the defective ladder not found in normal inspections?
> 2 Why did the supervisor allow its use?
> 3 Didn't the injured employee know he shouldn't use it?
> 4 Was he properly trained?
> 4 Was he reminded?
> 6 Did supervision examine the job first?
>
> The answers to these and other questions would lead to these kinds of corrections:
>
> 1 An improved inspection procedure
> 2 Improved training
> 3 A better definition of responsibilities
> 4 Prejob planning by supervisors

Our narrow interpretation of the domino theory has put blinders on us and has severely limited us in finding and dealing with root causes in accidents.

One fact seems clear—when we look only deep enough to find the act or the condition, we deal only at the symptomatic level. This act or condition may be the "proximate cause," but invariably it is not the "root cause." To effect permanent improvement we must deal with root causes of accidents.

Root causes often relate to the management system. They may be due to management's policies, its procedures, supervision and its effectiveness, training, etc. In our example of the defective ladder, some root causes here could be the lack of inspection procedures, the lack of management's policy, poor definition of responsibilities (supervisors didn't know they were responsible for removing the defective ladder), the lack of supervisory or employee training.

Root causes are those which would effect permanent results when corrected. They are those weaknesses which not only affect the single accident being investigated, but also might affect many other future accidents and operational problems.

THE STAIR STEP MODEL

There is one additional management model that we might mention before moving on to some of the other categories of models. Hugh Douglas, safety director of the Imperial Oil Company and former president of the Industrial Accident Prevention Association of Ontario, Canada, in his book *Effective Loss Prevention*, proposed a different model of accident causation, which he has titled the Stair Step cause and effect sequence. He explains as follows:

> The *Stair Step cause and effect sequence* follows a logical series of steps. They are:
>
> Every activity, regardless of its nature, is undertaken to achieve a purpose, objective or goal. Purpose is defined as the general reason for the activity. The objective is the result expected. The goals are the intermediate, measurable achievement points along the route to accomplishing the end result. They are grouped because in common use they are used synonymously.
>
> Specific resources are required to accomplish the objective. These must be organized and administered in such a way that the desired objective is attained. The resources, their organization and administration, are symbolized as a rectangle.

STEP 1

(a)

STEP 2

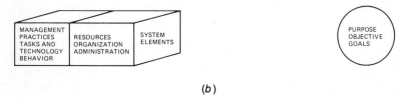

(b)

STEP 3

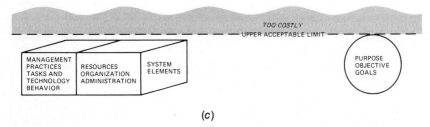

(c)

STEP 4

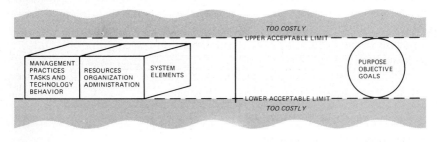

(d)

STEP 5

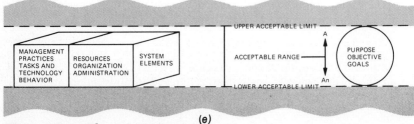

(e)

Figure 2-8a-g The stair step sequence.

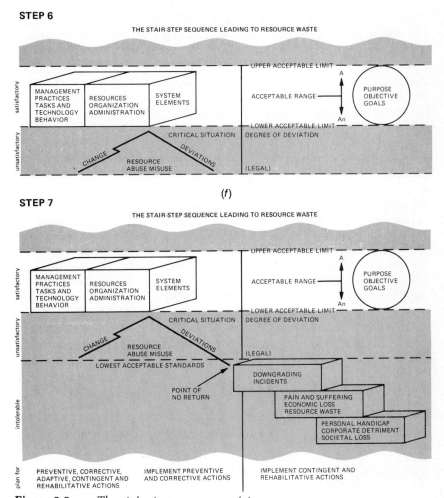

Figure 2-8a-g The stair step sequence. (g)

There is an upper level beyond which it is too costly to engage in the activity. It is indicated as a line joining the upper edges of the rectangle and the circle.

Likewise, there is a lower limit beyond which the costs would be too great.

The acceptable standard is not a single line. Rather, it is a range in which tasks and conditions can be blended together in a variety of ways, while meeting an acceptable level of performance. The upper and lower lines delineate the standard.

The range of performance or standard is shown in Step 5.

The activities and conditions required to achieve the objective in any situation may vary in number and variety, diagramatically illustrated as A - A_n in Step 5.

The width of the acceptable range and the level of the lower acceptable limit or standard will vary with the criticality of the inherent hazards, and the potential consequences for not setting and meeting a high enough standard.

A guideline for establishing the level and limit of standards is contained in *Rules of Reasoning* set forth by Sir Isaac Newton. He said, "Nature does nothing in vain and more is in vain when less will serve."

The lower acceptable limit is considered as the lowest standard individuals or business will accept as being satisfactory. In addition, there is another standard—the *lowest acceptable standard*. This is the standard set by law. In some cases, the lower and lowest acceptable limits are the same. Most organizations find it is economically to their advantage to keep the lower acceptable standard somewhat higher than the legal standard. This is because of the misuse of resources that occurs in the range between the lower acceptable limit and the legal standard.

The lower acceptable limit for the activities and conditions of any organization is determined by:

- Personal judgment
- Consensus opinion
- The demands of the market place
- The need to satisfy social and political values at the community interface
- Legal requirements

The quality, height, width or range of acceptability of a standard is determined by the personal values of whoever is responsible for setting the standards at an acceptable level and seeing that they are met.

The decision is influenced by—

- The potential for adverse consequences that could result from failure to set and meet an adequate standard for the situation, and,
- The social, legal, technological and economic influences that are present.

Setting the acceptable upper and lower limits may be difficult. An attitudinal guideline that is helpful for setting the height or quality and the upper acceptable limit is—*zeal for excellence*. The phrase "good enough" is *not* good enough for the lower acceptable limit.

TYPES OF CAUSATION MODELS

As we begin to look at some of the models that have evolved since the original domino theory of accident causation, we find a number of different approaches espoused and find that they begin to fall into various categories. The Bird model, the Adams model, and the Weaver model, as well as the Petersen model, all place a heavy emphasis on management as a prime causative agent in accidents. These might then be classified as management models. Some other categories might be behavior models, human factors models, systems models, epidemiological models, decision models, etc. While we can't do justice to all the models that have been devised, or even to all the categories of models, we will attempt to touch on a few of the most common models currently being discussed in safety management today.

SECTION 2   Additional Accident-Causation Models

In Sec. 1 we concentrated primarily on updates of the domino theory of causation, which led to a look at the so-called management models. In this section we'll discuss models that have not been derived from our original domino theory. We'll examine behavior models, human factors models, epidemiological models, systems models, and decision models. For the most part we'll simply describe the thinking without attempt at critique.

BEHAVIOR MODELS   These focus on tactical Error.

The first behavior theory historically was the notion of accident proneness. This theory assumes that there are certain somewhat permanent idiosyncracies which make a person more likely to have an accident. Probably the reason that this theory evolved and still exists is the simple fact that in examining any accident statistics one finds that the majority of people have no accidents, a relatively small percentage of people have one accident, and a very small percentage have multiple accidents. Thus it is obvious that since a small proportion of the population have most of the accidents, they must have some personal characteristics which increase their probability to be injured.

While historically the accident-proneness concept has been, and often still is, accepted, there are a number of arguments against it. The first argument is a statistical one. For any event having a low frequency but equal probability of occurrence to all members of the population, the distribution of occurrence will follow a Poisson distribution curve if plotted. This shows that many members of the population will have no occurrences, a smaller number will have one occurrence, and smaller numbers will have multiple occurrences. In short, accident data usually follow the Poisson distribution indicating that each person in the data has an equal probability to an infrequent event (the accident). This negates the idea of proneness (the unequal probability to the event).

A second argument against the proneness notion is the fact that numerous studies in both laboratory and clinical settings over a period of 40 years have failed to definitively identify any set of individual characteristics which are predictive of accidents across areas of activity and time. A third argument is that most studies show that those individuals experiencing a disproportionate number of accidents in one year tend not to continue that experience in succeeding years. Accident repeaters tend to be an ever-shifting group.

Today we think less in terms of proneness as a causative agent and look at other theories. One such theory attempts to explain the reason for accident repeaters.

The Life Change Unit Theory

A number of people have proposed and tested the idea that accident liability is somewhat situational—at times we are more liable to be involved in an accident than at other times. Schemes have been developed to measure and quantify situational factors which may precipitate an accident. These situational factors, or life events, may tax a person's capacity to cope, leaving the person more liable to accident. It has been shown that at such times the person is more susceptible to illness, and the research indicates he or she also may be more liable to be injured. Figure 2-9 shows one table of life change units. With this table it was found that of those persons who reported life change units (LCUs) totaling between 150 and 199, 37 percent had illnesses within 2 years. Of those with between 200 and 299 units, 51 percent reported illness, and with over 300 LCUs, 79 percent had injuries and illness to

Rank	Life Event	Mean Value
1	Death of spouse	100
2	Divorce	73
3	Marital separation	65
4	Jail term	63
5	Death of close family member	63
6	Personal injury or illness	53
7	Marriage	50
8	Fired at work	47
9	Marital reconciliation	45
10	Retirement	45
11	Changes in family member's health	44
12	Pregnancy	40
13	Sex difficulties	39
14	Gain of new family member	39
15	Business readjustment	39
16	Change in financial state	38
17	Death of close friend	37
18	Change to different line of work	36
19	Change in number of arguments with spouse	35
20	Mortgage over $xx,xxx	31
21	Foreclosure of mortgage or loan	30
22	Change in work responsibilities	29
23	Son or daughter leaving home	29
24	Trouble with in-laws	29
25	Outstanding personal achievement	28
26	Wife begin or stop work	26
27	Begin or end school	26
28	Change in living conditions	25
29	Revision of personal habits	24
30	Trouble with boss	23
31	Change in work hours, conditions	20
32	Change in residence	20
33	Change in schools	20
34	Change in recreation	19
35	Change in church activities	19
36	Change in social activities	18
37	Mortgage or loan under $xx,xxx	17
38	Change in sleeping habits	16
39	Change in number of family get-togethers	15
40	Change in eating habits	15
41	Vacation	13
42	Christmas	12
43	Minor violations of the law	11

Fig. 2-9 Table of life change units.

report. The literature has many similar studies from many researchers all tending to substantiate the concept. The concept might help to explain the accident repeater (the temporarily prone).

The Goals Freedom Alertness Theory

Dr. Willard Kerr has proposed a theory that states that great freedom to set reasonably attainable goals is accompanied typically by high-quality work performance. The theory regards an accident merely as a low-quality work behavior—a "scrappage" that happens to a person instead of to a thing. Raising the level of quality involves raising the level of alertness; and high alertness cannot be sustained except within a rewarding psychological climate. The more rich the climate in reward opportunities, the higher the level of alertness, the higher the level of work quality, and the lower the probability of accident. This theory is very much in harmony with current organizational psychological theory from all the current theorists in the field.

Motivation Reward Satisfaction Model

Starting from the goals freedom alertness theory, Petersen has developed a model of safety performance shown in Fig. 2-10. In the employee model (Fig. 2-10), safety performance of the employee is dependent upon his level of motivation and his ability to perform. Ability is a function of selection (can he do the task?) and of training (does he know how?). Motivation, however, is considerably more complex, being dependent on such things as the climate and style of the organization as he sees it (influenced primarily by his boss, but also by upper management and staff safety); by his own personality; by whether or not he is happy on the job he is in (is it any fun?); by the job motivational factors (for example, will it allow him to achieve, is there any responsibility to it, can he get promoted from it?); by his peer group (the norms established and enforced); and by the union.

Following performance there are all kinds of rewards (both positive and negative) which influence his level of satisfaction for the job performed. These rewards come from the boss (and the organization), from the peer group, from the union, and from his own feelings about the task accomplished (intrinsic rewards). Upon receiving these rewards he then compares the reward received to what he expected to

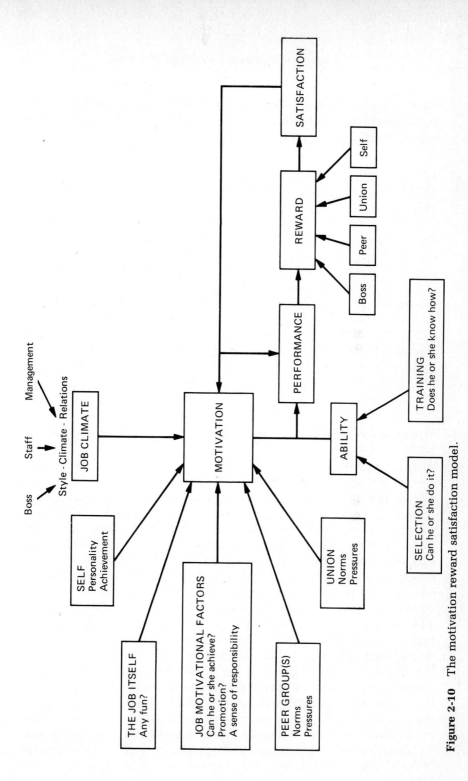

Figure 2-10 The motivation reward satisfaction model.

45

receive and is either satisfied or dissatisfied to a degree, which influences (feedback loop on the chart) whether or not he shall be motivated enough to perform again.

HUMAN FACTORS MODELS

The Adjustment Stress Theory

The adjustment stress theory of accident causation holds that unusual, negative, distracting stress upon the worker increases the worker's liability to accident or other low-quality behavior. This theory refers to distracting negative stress imposed upon the worker by either internal or external factors. This distracting stress is of a temporary, rather than permanent, nature. Some of the temporary stress factors which have been found to be significantly correlated with accidents are temperature at the workplace, illumination, mean-rated comfort of the workplace, degree of congestion, manual effort involved on the job, weight of parts handled, frequency of parts handled, alcohol consumption, and influence of disease on individuals.

This kind of a theory then states that the human being can be overloaded, and when this occurs, the person becomes more susceptible to an accident.

The Ferrell Theory

One theory of accident causation in this category comes from Dr. Russell Ferrell, professor of human factors at the University of Arizona. The theory is diagramed in Fig. 2-11. The theory states that accidents are the result of a causal chain (as in the multiple-causation theory), one or more of the causes being human error. Behind all initiating incidents of accidents is human error. Human error is in turn caused by one of three situations: (1) overload which is the mismatch of a human's capacity and the load to which he is subjected in a motivational and arousal state; (2) incorrect response by the person in the situation which is due to a basic incompatibility to which he is subjected; and (3) an improper activity that he performs either because he didn't know any better or because he deliberately took a risk. Since this is basically a human factors model, greater emphasis is then placed on the first two causes of human error, overload and incompatibility.

Overload can be examined in the model by looking at the sources of load: task load (physical requirements or information processing

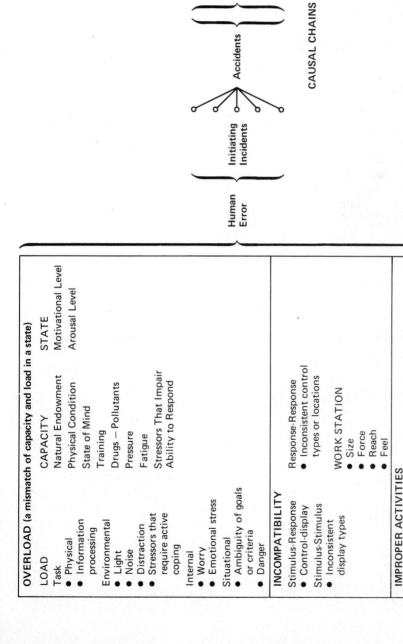

OVERLOAD (a mismatch of capacity and load in a state)

LOAD	CAPACITY	STATE
Task	Natural Endowment	Motivational Level
• Physical	Physical Condition	Arousal Level
• Information	State of Mind	
processing	Training	
Environmental	Drugs — Pollutants	
• Light	Pressure	
• Noise	Fatigue	
• Distraction	Stressors That Impair	
• Stressors that	Ability to Respond	
require active		
coping		
Internal		
• Worry		
• Emotional stress		
Situational		
• Ambiguity of goals		
or criteria		
• Danger		

INCOMPATIBILITY

Stimulus-Response
• Control-display
Stimulus-Stimulus
• Inconsistent
  display types

Response-Response
• Inconsistent control
  types or locations

WORK STATION
• Size
• Force
• Reach
• Feel

IMPROPER ACTIVITIES

Didn't Know
Deliberately Took Risk
• Low perceived probability of accident
• Low perceived cost of accident

Human Error   Initiating Incidents   Accidents   Outcomes

CAUSAL CHAINS

Figure 2-11 The Ferrell model.

requirements); environmental load (amount of illumination, noise, distraction, etc.); internal load (amount of worry, stress, etc.); and situational load (ambiguity of the situation, interpersonal problems, danger, etc.).

The sources of load can then be compared to the sources of capacity. These are a person's natural endowments, his physical condition, his state of mind, his training level, whether or not there is a drug or pollutant influence, the amount of pressure, and fatigue. And all this takes place when a person is in a certain arousal and motivational state.

Incompatibility can be examined in the model by looking at any basic incompatibilities that might exist between stimulus and response required, or by looking at incompatibilities in the work situation (is the wrong size, needs more force, can't reach, etc.).

Improper activities can be examined in terms of whether or not the person didn't know the correct activity (a training problem) or whether or not she deliberately took a chance. Such decisions on her part might be because she perceived the situation as having a relatively low probability of hazard, or because she perceived the potential cost of the accident as being relatively low. These then become either a personality- or attitudinal-problem kind of situation.

The Petersen Accident-Incident Causation Model

An adaptation of the Ferrell human factors model of causation is the Petersen model shown in Fig. 2-12. This model differs from the Ferrell model in that it allows two possible causes for accidents much as the original Heinrich domino theory did: human error and/or systems failure. Causes of accidents and/or incidents can be from either or both.

This model suggests that behind human error are three broad categories: overload, traps, and decision to err. Overload approximates the Ferrell model very closely, and is again defined as a mismatch of capacity with load in a state. Items under each are slightly different, however, with the load category including the concept of LCUs, the job hazard situation, etc., and with the state category including four classes: motivational, arousal, attitudinal, and biorhythmic.

The major difference, however, is in the third category, called "decision to err." This category suggests that employees often commit human error through conscious decision (or unconscious decision). There are times, many times, when workers will choose to perform a task unsafely because it simply is much more logical in their situation to perform it unsafely than it is to perform it safely, due to peer

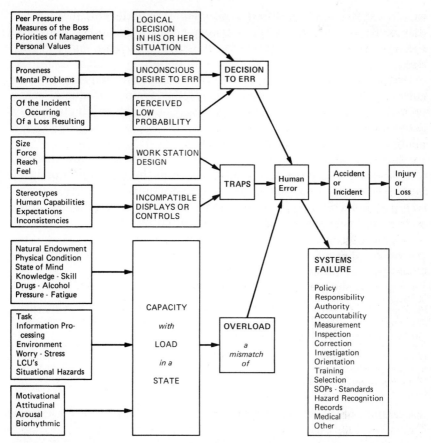

Figure 2-12   The Petersen accident-incident causation model.

pressure, the priority system they live under, the pressure for production, etc. The model also incorporates the persons who are truly accident-prone. While rare, they do exist, and will be involved in accidents through the unconscious desire they have to be injured.

And finally, the model suggests that many workers do perform unsafely simply because they perceive a low probability of an accident happening to them or because they perceive a low potential cost to them of the accident.

EPIDEMIOLOGICAL MODELS

The purpose of epidemiology has been described as the search for causal association between diseases or other biologic processes (perhaps accidents) and specific environmental experiences. It has been only in recent years that accidents have been considered an epidemic and thus subject to epidemiology. It is very much in that category today, and a number of studies are underway which could be considered epidemiological in nature. Perhaps the largest and best known is the current national survey of consumer products, being conducted under the auspices of the Consumer Products Safety Commission. Other smaller studies are also being made in products safety, traffic safety, and industrial safety.

Suchman stated that the accident phenomenon is "the unexpected, unavoidable unintentional act resulting from the interaction of host, agent, and environmental factors within situations which involve risk taking and perception of danger." This definition is almost a definition of the approach of epidemiology as the study of host, agent, and environmental interactions causing disease.

Suchman has proposed the model shown below based on epidemiology:

Predisposition characteristics	→	Situational characteristics	→	Accident conditions	→	Accident efiects
Susceptible host		Risk-taking		Unexpected		Injury
Hazardous environment		Appraisal of		Unavoidable		Damage
Injury-producing agent		margin of error		Unintentional		

According to this approach, injuries and damage are the measurable indices of the accident, but the accident itself is the unexpected, unavoidable, and unintentional act resulting from the interaction of the victims of the injury or damage deliverer and environmental factors within situations which involve risk taking and perceptions of danger. This model is analogous to that used for the study of disease. In applying this approach one seeks an explanation for the occurrence of accidents within the host (accident victim), the agent (injury or damage deliverer), and environmental factors (physical, social, and psychological characteristics of a particular accident setting).

SYSTEMS MODELS

Perhaps the largest single category of accident-causation models is the systems category. With the development of systems safety, which will be covered in some detail in Chap. 5, many people have suggested new causation theories in a systems format. Similarly when we look at the control of accidents later in this chapter, we'll see a number of systematic approaches. We'll touch on only a few in this section.

Like the epidemiological model, a systems model recognizes the inseparable ties between individuals, their tools and machines, and their general work environment. Bob Firenze illustrates this with his model depicted in Fig. 2-13. He explains:

Each worker, whether he be a machinist, chemist, foundry worker, etc., performs his job as part of a network called a man-machine system.

Such a system is composed of the *physical equipment*, the *man* or *men* who perform functions with the equipment, and the *environment* where the process takes place. This integrated group of components is designed to bring about a desired end—a product or task within the limits of its environment, and within an acceptable period of time.

Providing that the system functions as planned, the expected result is usually obtained. However, a failure—whether it be human, equipment, or environmental—will invariably detract from the efficient accomplishment of the operation.

In some instances, it wipes out the operation, resulting in a total failure or catastrophic situation. The match of human and equipment functions in a given environment are critical to a system's effectiveness and its capacity to perform as it was designed to.

As shown in Fig. 2-13, there exists a void between a man-machine system and its task. The process that takes place in this space holds one of the keys to understanding the man-failure-accident problem.

In order for the system in Fig. 2-13 to move towards its objective, a series of sequential processes must take place.

First, the human element must make decisions. Based on these decisions, the human component will take certain risks in an effort to reach his objective.

In each case, the man needs information upon which to base his decisions. The better his information, the better his decision, and, subsequently, the more calculable the risk becomes. The poorer his information, the greater the chance for bad decisions, bad risks, and failures that may be responsible for accidents.

Secondly, the equipment in the system must function effectively without failure. Poorly designed equipment or tools—or those poorly maintained—may, in themselves, be the trigger mechanism that leads to an accident.

Thirdly, the environment plays a significant role in the system. A failure in the environment may affect either the man, his machine, or both, thus setting up an accident situation (e.g. toxic atmospheres, glare, etc.)

Before decisions are made, man seeks information that serves to remove some or all of the uncertainty involving his task. The uncertainty centers around two major areas: the requirements of the task; the nature of the harmful consequences. If his "information bank" is adequate, the ultimate

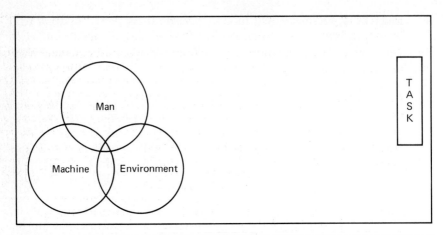

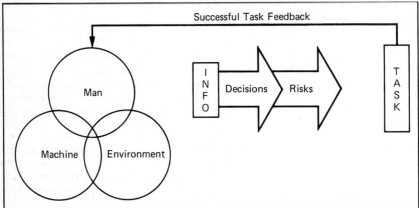

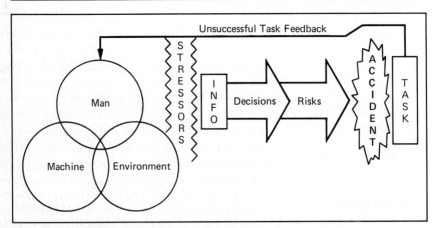

Figure 2-13 The Firenze systems model.

risk resulting from his decisions will be within calculable limits of risk taking, and the chance of his failing will be lessened.

For that reason, one of the primary efforts must be to equip workers with as much information as possible through training, so that their decision-making is facilitated, their actions are effective, and their chance for human failure is minimized.

There is, however, an exception to the rule that a man with full knowledge of his job will always make wise, calculated decisions.

Variables known as "stressors," or blocks to a man's decision-making capability, often appear in the decision process and cloud his ability to make sound, rational decisions (see Fig. 2-13). These stressors can be of psychological, physiological, or physical origin.

They act to distort and, sometimes, prevent the decision-making process from taking place. Narcotics and alcohol are examples of physiological stressors that have an effect upon the organism.

Anxiety, aggressiveness, and fatigue are examples of psychological stressors. Glare, temperature extremes, and low levels of illumination are examples of the environmental type.

Each type of stressor has the capacity by itself or in combination with other factors to cause otherwise safe behavior to be faulty.

It is often during the period when the man is distracted that he makes an error. The error oftentimes is primarily responsible for an accident.

Accident causation involves consideration of as many of the variables that affect the system as possible. The design of tools and equipment must be considered as must the elements in the environment that may detract from the successful completion of the operation.

Lastly, but ever so important, it must be considered that the human variable in the system is a non-perfect entity subject to many forces in his environment. With this in mind, it must be understood that no matter how intelligent the man is, how much he is trained, or how much information he possesses, he will still—under certain circumstances—make errors. Some of these errors will lead him into an accident situation.

This is not to imply that trying to improve the man's knowledge of his tasks should be abandoned. Chances are that if his decision-making ability is sufficiently developed, along with his comprehension of the hazards connected with his job, and his ability to anticipate and counter accident situations, he stands a better chance of surviving without injuries than if he had no comprehension of the problem at all.

The Ball Model

Earlier we introduced the concept of energy release as a necessary part of the accident-causation process. Dr. Leslie Ball, former director of safety for NASA, has introduced a causation model based upon this concept. It is shown in Fig. 2-14. His thesis is that all accidents are caused by hazard, and all hazards involve energy, either due to involvement with destructive energy sources or due to a lack of critical energy needs. This model is most useful in the identification of hazards, and will be discussed in more detail in Chap. 5 on systems safety.

DECISION MODELS

The final category of causation models is decision models. In this section we'll present only one, the Surry model; however, in the later section on control of accidents, we'll look at other decision models.

The Surry Model

Jean Surry has developed a model of accident causation stemming from the epidemiological model of Suchman. The Surry model has three principal stages with two similar cycles linking them. First a dangerous situation is built out of a secure situation, and then the danger is released causing injury or damage. She separates the behavior of a person during these two phases. The model is shown in Fig. 2-15. The

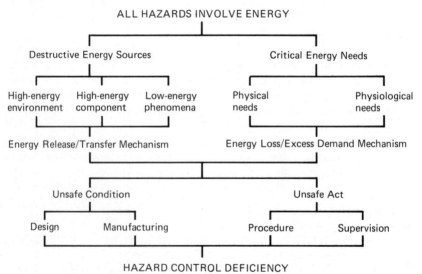

Figure 2-14 The Ball model.

Principles of Accident Prevention

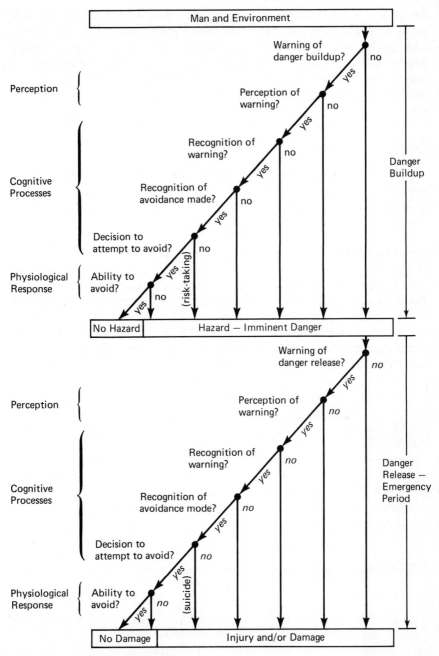

Figure 2-15 The Surry model.

first stage is the total environment, both spatial and temporal, of a person. This stage is similar to Suchman's predisposing characteristics. The model assumes that by a person's action or nonaction danger occurs to the person. If any negative responses to the question are shown during the danger build-up cycle, the danger becomes imminent. If all replies are positive, the danger will not grow and no injury will ensue. Similar questions are posed in the danger release cycle, and a negative response to one of the questions will lead to inevitable injury.

An accident can be the result of many different routes through the model (20 routes). There are somewhat fewer routes leading to no-injury situations.

AXIOM 1 AND NEWER MODELS

Axiom 1 in Fig. 2-1 gives the first model of accident causation. It is still the most used model. Most inspection and investigation systems in industrial safety even today are built on the domino theory. The theory has been updated by many. We have presented four updates, all leading to management models of causation.

Starting from different points, other theories have also evolved in recent years. While we've only presented a few, they are typical of the thinking currently being expressed. As the reader can see, there is a tremendous variety in approach.

In this section we've not attempted to present the one single best model of accident causation. We believe there is no single best model. Each presented in this section seems to have value and validity. Readers may wish to select one to guide their thinking, or perhaps select several which seem to be particularly descriptive of reality.

In the next sections we'll look at other axioms from Fig. 2-1, the updates to those beliefs, and other current thinking in that area.

SECTION 3   Humans versus Machines

The most ardent supporters of the belief that worker-failure accident causes are predominant are, nevertheless, firmly convinced that mechanical guarding and correction of mechanical and physical hazards are fundamental and first requirements of a complete safety

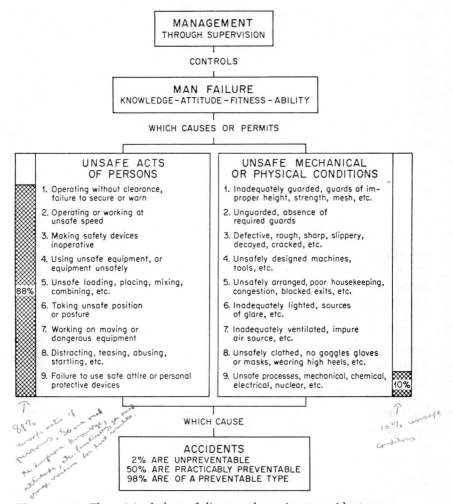

Figure 2-16 The original chart of direct and proximate accident causes.

program. They believe, and act on the belief, that safety begins with safe tools, safe machines, safe processes, and safe environment. This attitude is not at all inconsistent with the emphasis placed herein on the importance of worker failure as a causative factor, and is more readily visualized when one considers "corrective action." In the same breath it can be truthfully said that although worker failure *causes* the most accidents, mechanical guarding and engineering revision are nevertheless important factors in *preventing* the most accidents.

Figure 2-16 shows our original chart of direct and proximate accident causes and indicates the results of early research on the relationship between worker and machine causes. Twelve thousand cases were taken at random from closed-claim-file insurance records. They covered a wide range of territory and a great variety of industrial classifications. Sixty-three thousand other cases were taken from the records of plant owners. Through analysis of these 75,000 cases, through study of actuarial records and engineering reports, and with the cooperation of employers, it was found that 98 percent of industrial accidents are of a preventable kind.

It was discovered that 25 percent of all accidents would, according to usual but improper methods of analysis, be charged to defective or dangerous physical or mechanical conditions, but that in reality the causes of many accidents of this group were either wholly or chiefly worker failure and only partly physical or mechanical. This group, therefore, was found actually to be 10 percent instead of 25 percent. This difference (15 percent) added to the 73 percent of causes that are obviously of a worker-failure nature gives a total of 88 percent of all industrial accidents that are caused primarily by the unsafe acts of persons. Check analyses, made subsequently on a smaller scale, produce approximately the same ratios.

In this research major responsibility for each accident was assigned *either* to the unsafe act of a person or to an unsafe mechanical condition, but *in no case* were *both* personal and mechanical causes charged.

It is at this point that most critics of this original research begin to argue with its validity. Their argument is that in most, if not all, accident situations both worker and machine causes are present. To eliminate accidents where there is a combination of causes is to eliminate most accidents. Many believe that there must be such a combination in every accident.

To be sure, extraordinary hazards exist and many serious and perplexing problems, social and economic, arise because of mechanization. These problems necessitate adjustments that are often costly. In the final analysis, however, it must be admitted that people invented the machine, built it, and put it to work. They alone give it life and motion. It moves when and where they direct, at the speed they desire, and stops when they stop it or when the energy they give it expires. People devise intricate mechanical processes and procedures together with such safeguards, rules, regulations, and instructions as they alone decide are necessary. This being so, it is fair to conclude that even injuries resulting wholly from mechanical fault, with no personal unsafe action whatever, are nevertheless basically chargeable to humans and not machines. However, the discussion in this section deals not with basic, or sub-, causes but with *direct* causes. The point is being established that worker failure *directly* as well as indirectly causes the great majority of accidents, as portrayed graphically in Fig. 2-16.

The machine is dangerous as a person makes it so. It is a person's use of the machine—more correctly, abuse of it—that creates danger.

Nor is the appalling degree of worker failure fully portrayed by its direct results in loss of life, limb, and dollars. Misunderstanding of instructions, recklessness, violent temper, and lack of knowledge or training result in unsafe acts, for many of which no penalties whatever are exacted in the form of personal injuries or property damage. For example, automobiles left unsafely parked on grades run wild through busy streets and crash into buildings, and no one is hurt. Steelworkers "ride the ball" and are hoisted hundreds of feet into the air in the erection of skyscrapers; yet they do not always fall. Guards on heavy machines are removed, people stand under suspended loads, get on and off moving vehicles, leave obstructions in walkway areas, pile material insecurely, refuse to wear goggles, gloves, masks, and other protective equipment, and yet injuries from such unsafe personal actions occur but once in several hundred cases.

The situation reminds one of an honestly conducted lottery where most of the tickets fail to draw anything, a few draw small prizes, and a still smaller number draw big ones. In the case of the lottery, however, the prizes are hoped for and desired, whereas in the accident situation the reverse is the case. It is hard to understand why those who have been in the habit of playing lotteries are so confident that the occasional event certainly will come to pass once in so often, whereas in the accident field, where the mathematical principles involved are so similar, workers seem to feel an equal degree of confidence that the occasional event never will come to pass.

It would appear that the laws of chance provide a substantial safeguard. No individuals, however, know whether they will pay the penalty for the first or the last or some intermediate unsafe act. The moral is to profit by the opportunity to learn, to realize that repeated violations of commonsense safe practice eventually and invariably lead to injury, and to avoid or prevent repetition of them.

If knowledge is power and accidents arise primarily out of the unsafe acts of people, then power in the conservation of life against the ravages of accidents must come from adequately aroused interest and then from knowledge of specific unsafe acts of persons and the reasons why those acts are committed. With this background of knowledge the reasonable and effective approach can only be the elimination or modification of such unsafe mechanical conditions as exist, even though these are in the minority, *plus* simultaneous action, both direct and continuing, toward control of the all-important factor, unsafe performance of persons.

It seems almost unbelievable that with the knowledge that people cause most accidents, knowledge that has been available since the early 1930s, so much time and effort since that time have been spent by industry with primary, often total, attention on physical conditions. It is even more unbelievable that in 1970, some 38 years after this knowledge was available, the United States would turn to a national approach based almost entirely upon the control of physical conditions: the Occupational Safety and Health Act (OSHA).

Unbelievable or not, this is precisely what transpired. With almost universal belief in the principle that safety is primarily determined by people, the principle was almost totally rejected by the Congress, who chose to legislate a law based upon a totally opposite principle: that accidents are caused by conditions—by things.

SECTION 4   Foundation of a Major Injury

Analysis shows that, in the average case, for every mishap resulting in an injury there are many other similar accidents that cause no injuries whatever. From review of data available concerning the frequency of potential-injury accidents, it is estimated that in a unit group of 330 accidents of the same kind *and involving the same person*, 300 result in no injuries, 29 in minor injuries, and 1 in a major lost-time injury. The determination of this no-injury accident frequency followed a study of over 5,000 cases.

In the accident group (330 cases), shown in Fig. 2-17, a major injury is any case that is reported to insurance carriers or to the state compensation commissioner. A minor injury is a scratch, bruise, or laceration such as is commonly termed a first-aid case. A no-injury accident is an unplanned event involving the movement of a person or an object, ray, or substance (slip, fall, flying object, inhalation, etc.), having the probability of causing personal injury or property damage. The great majority of reported or major injuries are not fatalities or fractures or dismemberments; they are not all lost-time cases, and even those that are do not necessarily involve payment of compensation.

There has been much confusion about this original ratio in industrial accident prevention. The 300-29-1 ratio was initially viewed as an aid in accident prevention, as an opportunity. When an

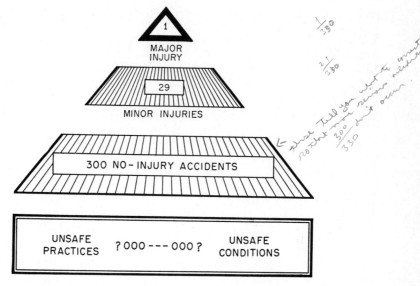

00.3 per cent of all accidents produce major injuries
08.8 per cent of all accidents produce minor injuries
90.9 per cent of all accidents produce no injuries

The ratios above—1-29-300—show that in a unit group of 330 similar accidents occurring to the same person, 300 will result in no injury, 29 will produce minor injuries, and 1 will cause a serious injury.

These ratios apply only to the *average* case. The major injury may result from the very first accident or from any other accident in the group.

Underlying and causing all accidents, including those resulting in no injury or in either minor or major injury, there is an unknown number of unsafe practices or conditions, often running into the thousands.

Moral 1. Prevent the accidents and there can be no injuries.

Moral 2. Prevent the unsafe practices and unsafe conditions and there can be neither accidents nor injuries.

Figure 2-17   Foundation of a major injury.

employee, either because of her repeated unsafe action or repeated exposure to an unsafe mechanical condition, suffers 300 no-injury accidents (actual events, such as slips and falls but fortunately not causing injury) *before* she sustains even a minor injury, surely there can be no lack of opportunity in preventive effort. They should and can be controlled long before one of the 300 no-injury accidents ultimately causes an injury. Another significant point in this ratio is sometimes overlooked. It indicates still greater opportunity. The fact is that the base of 300 no-injury accidents includes *only those events where there was a narrow escape from injury.*

ACCIDENTS—NOT INJURIES—THE POINT OF ATTACK

Accident prevention is too frequently based upon an analysis of the causes leading to a major injury. This situation exists, for the most part, because of a misunderstanding of what an accident really is. Precise terminology is of much value when its lack indicates misdirection in both thought and action. Then the matter of accuracy in the use of words or phrases becomes decidedly important.

Throughout industry, reference is made to so-called major and minor accidents. No-accident contests and campaigns are usually based upon lost-time or major-accident frequency. Tables and statistics feature lost-time accidents (or others in the so-called major group) that involve fatalities, fractures, dismemberments, and other serious injuries; in general, attention is centered upon these more spectacular occurrences to the exclusion, in part at least, of adequate consideration of minor accidents. Not only is this true with regard to cause-and-type analysis and tabulation, but subsequent action in prevention work follows along the same line.

The expression "major or minor accidents" is misleading. In one sense of the word there is no such thing as a major or minor *accident*. There are major and minor *injuries*. In any case, in prevention work, the importance of any individual accident lies in its potentiality for creating injury and not wholly in the fact that it actually does, or does not, so result. When the causes of lost-time or so-called major accidents only are selected for study as a basis for records and for guidance in prevention work, efforts are often misdirected, valuable data are ignored, and statistical exposure is unnecessarily limited.

An injury is merely the result of an accident. The accident itself is controllable. The severity or cost of an injury that results when an accident occurs is difficult to control. It depends upon many uncertain and largely unregulated factors—such as the physical or mental condition of the injured person, the weight, size, shape, or material of the object causing the injury, the portion of the body injured, etc. Therefore, attention should be directed to accidents as properly defined, rather than to the injuries that they cause.

Further, in the length of time over which experience is analyzed (usually from 1 month to 1 year), the average plant, or department of a plant, does not develop sufficient exposure to justify the use of the comparatively small number of serious injuries, either as an indication of progress in accident-prevention work or as a safe guide to the real causes of the predominating types of accident.

In basing accident-prevention work upon the cause analysis of major injuries alone, therefore, not only is the importance of the accidents that produced them overestimated, and the field of research thus limited, but the resulting data also are seriously misleading when used to determine the proper corrective action to be taken.

The foregoing statements and figures justify the conclusion that in the largest injury group—the minor injuries—lie the more valuable clues to accident causes. Equally evident is the conclusion that the unsafe practices and conditions, which can cause accidents and result in either major or minor injuries, should be corrected *before* injuries result.

In making a survey of 100 typical manufacturing plants, it was found that, in the majority of them, the causes of the serious injury accidents, over a limited period, did not fairly picture the unsafe practices and conditions needing first attention. Accident-prevention work in these plants was misdirected, since it was based largely upon the investigation of major injuries, and many other serious injuries of a slightly different nature later occurred.

Other similar studies along this line have also been made. Frank E. Bird, Jr., reports that in 1969 he made an analysis of 1,753,498 accidents reported by 297 cooperating companies while at the Insurance Company of North America. The companies represented 21 different industrial groups, employing 1,750,000 employees who worked over 3 billion work-hours during the period analyzed. The results are shown in Fig. 2-18. Frank E. Bird, Jr., concludes:

> The 1-10-30-600 relationships in the ratio would seem to indicate quite clearly how foolish it is to direct our total effort at the relatively few events terminating in serious or disabling injury when there are 630 property damage or no-loss incidents occurring that provide a much larger basis for more effective control of total accident losses.

At the beginning we indicated that there had been considerable confusion surrounding the triangle concept, the relationship between

Figure 2-18 The Bird accident ratio study.

severe- and minor-injury causes. And there has been. The difficulty stems from the use of statistics. Our original data of 1-29-300 were based on "accidents of the same kind and involving the same person." The figures are averages of masses of people and all kinds of different accident causes and types. It does not mean that these ratios apply to all situations. It does not mean, for instance, that there would be the same ratio for an office worker and for a steel erector. It might mean that they could be averaged to this (or a similar) ratio; but certainly neither of these extremes would fit the ratio.

It also does not mean, as we have too often interpreted it to mean, that the causes of frequency are the same as the causes of severe injuries.

Our ratios and figures in this area have confused us. We have typically believed a 1-29-300 ratio, believed it might apply to all kinds of accident types and causes, and then seen national figures as in Fig. 2-19 that show that different things cause severe injuries than the things that cause minor injuries. Obviously then there are different ratios for different accident types, for different jobs, for different people, etc. The triangle for the accident type "electricity" is a different looking triangle than the one for "handling materials."

Common sense dictates totally different relationships in different types of work. For instance, the steel erector would no doubt have a different ratio from the office worker.

This very difference might lead us to a new conclusion. Perhaps circumstances which produce the severe accident are different from those that produce the minor accident.

Safety workers for years have been attacking frequency in the belief that severity would be reduced as a by-product. As a result, our frequency rates nationwide have been reduced much more than have our severity rates. One state reports a 33 percent reduction in all accidents in the last 10 years, while during the same period the number of permanent partial disability injuries has actually increased. This state is typical of others, and its figures are typical of our national figures. In the period 1926 to 1967, the national frequency rate improved 80 percent while the permanent partial disability rate improved only 63 percent. (This could be, of course, partly due to the changing definition of a partial disability as laws vary.)

If we study any mass data, we can readily see that the types of accidents that result in temporary total disabilities are different from the types of accidents that result in permanent partial disabilities, or in permanent total disabilities or fatalities. For instance, the National Safety Council (Fig. 2-19) shows that handling materials accounts for 25 percent of all temporary total disabilities and 21 percent of all

Type of Accident	Temp. Total, %	Perm. Partial, %	Perm. Total, %
Handling materials	24.3	20.9	5.6
Falls	18.1	16.2	15.9
Falling objects	10.4	8.4	18.1
Machines	11.9	25.0	9.1
Vehicles	8.5	8.4	23.0
Hand tools	8.1	7.8	1.1
Electricity	3.5	2.5	13.4
Other	15.2	10.8	13.8

Figure 2-19 Accident types and severity.

permanent partial injuries, but only 6 percent of all permanent total injuries and fatalities. Electricity accounts for 13 percent of all permanent totals and fatalities but accounts for a negligible percentage of temporary totals and permanent partials. These percentages would not differ if the causes of frequency and severity were the same. They are not the same. There are different sets of circumstances surrounding severity. Thus, if we want to control serious injuries, we should try to predict where they will happen. Today we can often do just that.

Statistics show that we have been only partially successful in reducing severity by attacking frequency. In the last 40 years National Safety Council figures show an 80 percent reduction in the frequency rate. During that period the same source shows only a 72 percent reduction in the severity rate, a 67 percent reduction in the fatal and permanent total rate, and a 63 percent reduction in the permanent partial disability rate.

Studies in recent years by a number of different people suggest that severe injuries are fairly predictable in certain situations. Some of these situations involve:

1 Unusual, nonroutine work. This is the job that pops up only occasionally, or the one-of-a-kind type of situation. These situations may arise in production or in nonproduction departments. The normal controls that apply to routine work have little effect in the nonroutine situation.
2 Nonproduction activities. Much of our safety effort has been directed to production work. But there is tremendous potential exposure to loss in nonproduction activities such as maintenance and research and development facilities. In these types of activities most work tends to be nonroutine. As it is nonproduction, it often does not get the attention from safety, and it is not usually procedurized. Severity is predictable here.

3 Sources of high energy. We can associate high energy with severity usually. Electricity, steam, compressed gases, and flammable liquids are examples.
4 Certain construction situations. High-rise erection, tunneling, working over water, etc. (Actually, construction severity is an amalgam of the previously described high-severity situations.)

These are just a beginning point. A long list could be made which more extensively specifies the areas where severity is predictable.

SECTION 5   Reasons for Unsafe Habits

Originally these four reasons or causes for the unsafe acts of persons were identified:

1 Improper attitude
2 Lack of knowledge or skill
3 Physical unsuitability
4 Improper mechanical or physical environment

And from these causes, these categories of remedial action were proposed:

1 Engineering revision	Guarding, redesign, relocation, etc.
2 Persuasion and appeal	Instruction and reinstruction in safe practices, providing proof and illustration, inspiring enthusiasm, persuading, convincing, and appealing to motivating characteristics. Application of psychology.
3 Personnel adjustment	Assignment of workers to relatively less hazardous work. Medical attention and psychology.
4 Discipline	Enforcement of rules, mild reproval, admonition, militaristic methods, penalties, etc. Applicable in rare cases and as a last resort only.

These four remedies might look somewhat familiar to anyone who has any familiarity with industrial safety for the past 20 or 30 years, for they are the basis of the "3 E's of safety." Many a safety program has been built on the three E's of safety: engineering, education, and enforcement.

Most safety professionals today no longer believe in this approach. It is overly simplistic—it may not even lead us to solid safety programs.

To attempt to summarize the whole of human behavior into four simple categories and to attempt to summarize the whole of management into three or four simple categories seem today to be much too simplistic an approach.

Nonetheless the reasons and remedies still provide us with some help and direction. Perhaps, however, we might do well to look beyond these reasons and remedies for help. In fact in looking at the reasons behind the performance of unsafe acts, we should be looking way beyond the field of safety management, into the behavioral sciences; for we must begin to understand the human being a little better if we are to understand human behavior on the job.

To a certain extent, of course, consideration has always been given to the need for finding why persons do unsafe things. But such consideration too often has been intermittent and casual—seldom has it been conscious and continuous. This line of investigation has not been planned and carried on as a separate step—rather, when done at all, it has been instinctive and incidental, although actually it is a most important single factor in accident analysis and in the selection and application of helpful remedies.

A few illustrations may serve to prove that in many instances there is a need for finding the reasons for unsafe acts, and that when these are discovered they lead to the proper selection and application of effective measures in accident prevention.

Example 1
An accurate and thorough analysis has been made of public and property damage accidents in which a commercial fleet of automotive vehicles had been involved over a period of years. Mechanical equipment was found to be safe. The drivers were at fault. Specifically, they followed too closely and at too great speed for the conditions of traffic. Question—what should be the remedy?

Previously, in this case, remedial action, which had been selected at random, consisted in issuing instructions (and attempting to enforce them) to the drivers to follow cars at a safe distance. This had not served to reduce the accident frequency, however. There were several groups of drivers and several departments of work. Finally, the decision was reached to find out the "why" or the reason for the violations, preparatory to planning a better remedy. The results were startling.

In one group of drivers it was found that the few who were responsible for the poor record were inherently reckless. The remedy became a matter of managerial personnel procedure.

The drivers in group 2 were not reckless at all, but they were wholly unconvinced that it was unsafe to follow closely. The remedy, obviously, was a special form of education.

In a third group there were many drivers with defective eyesight who were unable to judge distances accurately. The remedy was found to lie in ocular examination and the provision of glasses with prescription lenses.

Group 4 drivers found it *impossible to follow at a safe distance* because of time-schedule requirements. In heavy city traffic the creation of a gap of much more than a car length or two provided a temptation for another driver to pass and pull in ahead. This necessitated dropping back again, thus creating another gap. In short, any attempt to live up to the rule for following safely destroyed all chance of meeting the time schedule. The remedy lay in rerouting and in revision of schedules.

The remedial actions indicated by the reasons enumerated in this example were made effective, and, as may readily be anticipated, the accident frequency dropped immediately and substantially.

It is of interest to note that although four *distinctly different* methods of correction were used, they were applied effectively to *exactly the same unsafe practice.*

Example 2

In this case employees violated the company safe-practice rule relating to washing before eating luncheon. All efforts in the way of instruction, supervision, and education had failed to produce results. Finally when the reason for the unsafe act was found, it developed that the washrooms were uncomfortably drafty and that the water was cold and hard. In short, the unsafe act was committed largely because it was *inconvenient* and *uncomfortable* to follow safe practice. When ventilation was improved, more suitable free-lathering soap provided, and warm water made available, the workers readily made proper use of washroom facilities. This is one of the many instances where engineering revision is found to be an answer to worker-failure accident-and-health problems. The moral, moreover, is that success came about only when cause analysis was carried to the point where the reason for the unsafe act was ascertained.

Example 3

In this instance, extreme difficulty was encountered in getting shipping department employees to stop the practice of jumping off loading platforms. Here it was found that the workers *were by no means convinced* that jumping approximately three feet was at all unsafe. They were not reckless, it was quite convenient to use the stairs at either end of the platform, and no other probable reason, except the one selected, was applicable.

The safety engineer, being now adequately informed, proceeded to develop an effective educational campaign designed to remove all possible doubt as to the hazards of jumping. He obtained the services of a surgeon-physician who gave talks illustrated by charts. The charts showed the bone and muscular structure of the human as compared to that of four-footed jumping animals such as the dog and the antelope. It was conclusively demonstrated that humans are poorly equipped, indeed, when it comes to jumping, by comparison with such animals. A dog lands on well-padded toes, and the shock is absorbed by several more joints and by less rigid anatomical structure than is the case with a human being.

It was pointed out, further, that the abdominal support of the human was designed for horizontal (four-footed) carriage, and that in walking erect the original support was necessarily ineffective and was replaced by a thin membrane. The very first lecture was so impressive that an employee who was the chief offender tightened his belt as he left the room and "held himself in" as he stepped carefully over the threshold of the doorway, saying at the same time, "I'm sold."

Suffering heavily by contrast with these examples is the slipshod, hit-or-miss guesswork which presupposes that a single form of prevention work—education or guarding, for example—will suit any and all occasions.

Surely, it would be a waste of time to preach the doctrine of safety to a group of drivers who already believed in it and who drove unsafely only because they could not see well enough to judge distance properly. Especially would it be wasteful if the defects in vision *were not known* and consequently if *nothing were done about it*. How much simpler and more direct and effective is the procedure finally adopted in Example 1, namely, *finding the reason for* the unsafe act and *basing the remedy on that reason*.

Surely, too, it would be useless to fit glasses to drivers, such as those in group 4 of Example 1, or to educate or discipline them. Those drivers were physically sound, they were fully aware of the danger in following the car ahead too closely, and they were not at all reckless— they were forced to drive unsafely by improper routing and timing.

The method whereby unsafe acts are to be corrected is most effective when it suits the *reason* for the occurrence of an unsafe act.

How many kinds of these reasons and remedies are there? Is the question involved and complicated? Does it enter the field of psychology or psychiatry? Is it one that the average safety engineer can solve?

Fortunately, from the viewpoint of progress, the answers to these questions are all cheerfully simple and encouraging. Indeed, psychology of a sort is involved, but at the supervisory level it need only be of an elemental and understandable nature which the average safety engineer or industrial supervisor can readily apply. The matter therefore is not impractical or too complicated.

The above examples are perhaps too simple. The answers to the problems fall back into the traditional areas of engineering revision, of training, of fitting the person to the job, etc. However, all safety problems today are not that simply solved. It's true many of our solutions can be traditional. If a worker does not know what to do, we can teach the worker and solve some of his or her safety problems. If a worker has inadequate or unsafe equipment, we can change that and thereby cure his or her safety problems.

There are, however, today problems that are quite more complex than these—problems stemming from job disenchantment, from boredom, from alienation, from lack of involvement, etc. We cannot solve these with training, with discipline, or with engineering revision. We must look beyond. Chapter 10 attempts to look at some of these areas.

SECTION 6   Analogy between Methods of Controlling Accident Causes and Production Faults

Controlling the quality and quantity of product and controlling the frequency and severity of accident occurrence have much in common. In many cases the same faulty practice is involved, and the reason for existence of the fault is similar, both for accident occurrence and for unsatisfactory production.

If it is known as a result of a correct fact-finding job that a particular unsafe practice is chiefly responsible for accident occurrence, it can safely be assumed that the methods best suited to correct that particular practice are identical with managerial and supervisory methods such as would be used if the practice were not unsafe but were one that resulted in impaired or high-cost production.

Suppose, for example, that in stevedoring work analysis discloses that workers are being injured because they "work under open hatches when drafts are being hoisted." If it is also known that this practice is being carried on in willful disregard of instruction, an executive with ordinary supervisory ability should be able to select and apply a remedy successfully. To make the point clear it is necessary only to visualize what such an executive would do if, instead of violating that safe-practice rule, the same employees willfully disregarded rules on the overloading of drafts so that materials spilled out and were smashed, creating a property damage loss which the employer must pay.

The employees are paid to perform their work in a prescribed way—that is, to stand clear of the hatches while material is being hoisted. The supervisors are paid to see that employees carry on their work as instructed. The employer has a right to expect wage-earning employees to obey commonsense rules. The employer wants the workers to stand clear, and she wants the supervisor to see that they do so. Both groups of wage earners fail to do as she asks. What further is there to say, except that *the responsibility lies first of all with the employer*? If she has an earnest desire to reduce the frequency and cost of accidents, if she recognizes her responsibility for the safety of her workers, and if she is aware of the fact that methods of achieving safety are analogous with methods of controlling production, she will exercise her prerogative and obtain compliance with instructions. Moreover, she will see that unsafe mechanical conditions as well as unsafe employee practices are eliminated, just as she would do if they resulted not in accidents but in delays, spoilage, breakage, defective merchandise, or low production.

In a drop-forge plant an inspector was definitely instructed to gauge roughed-out forgings in three dimensions. He disregarded the instruction, however, and gauged them for length and width only. As a result, a large order was canceled when the purchaser found that several forgings of the first batch received from the shop were of less than the specified thickness.

This situation is analogous to many accident-prevention problems, and its manner of correction is identical with that herein advocated for accident occurrence, as the following comparison shows:

EXAMPLE 1	*EXAMPLE 2*
Defective-Product Problem	*Accident Control Problem*
1 Improper practice—gauging forgings for length and width, but not for thickness.	1 Unsafe practice—throwing hot billets across aisle space between furnace and base of drop press.
2 Cause or reason for existence—improper attitude, disregard of instruction.	2 Cause or reason for existence—improper attitude, disregard of instruction.
3 Remedy—supervisory enforcement of instruction to gauge for three dimensions.	3 Remedy—supervisory enforcement of instruction to carry billets to drop press with tongs.

The executives in charge of the drop-forge plant referred to in both examples experienced no difficulty whatever in applying the remedy for defective product in Example 1. They would hardly have been satisfied to deal with such elemental and definitely known faults only by means of general education through bulletins, meetings, and talks, or by other indirect methods. They lost no time in identifying the improper practice and in correcting it by eliminating the reason for its existence through appropriate supervisory procedure.

For two years, however, they struggled with the accident problem (Example 2), though it could have been disposed of in a very short time by employing methods similar to those used in the first case.

The foregoing is what is meant by the analogy between supervisory control of *production* and supervisory control of *unsafe performance*. If a supervisor can do the one, he can do the other. By means of the analogy a *short cut* may be found to accident prevention. However, it presupposes that a supervisor is *already capable of enforcing production rules*, in which case he has merely to apply to accident prevention the very same techniques he already applies to production. It by no means nullifies the assertion elsewhere in the text that industry will profit tremendously by training its practicing and prospective supervisors in the art of worker-performance control measures.

Two primary thoughts have been stated: that the causes of accidents are also the causes of other operational errors, and that the same methods that control production and other managerial problems will also control accidents. On the first notion, that of operational error, D. A. Weaver defines the problem:

> Operational error has occurred whenever unplanned and undesired results stem from the acts or decisions of supervisory management, or the failure to act or decide. The term "supervisory management" encompasses the entire management structure from chief executive to the lowest level of front-line leadership.

> What are "unplanned and undesired results?" Examples are endless, including occasional accidents and injuries. If the customer ordered green and we send him pink, operational error has occurred. If 300 gallons of product go down the drain (memorable since the product was beer), operational error has occurred. If the crew goes to one location, the power equipment to a second, and an angry supervisor waits at the actual job site, we have an unplanned and undesired result stemming from operational errors. The examples of waste, "ball-ups," and snafus are indeed more common and more costly than the occasional accident or injury, but such incidents have no name. We recognize, define, and name the particular incident we call an accident; but all such unplanned and undesired results stem from operational error.

In searching modern safety management literature we find general agreement with our original two thoughts. In fact, we believe today that it might well be that those two ideas expressed the best thinking in the original editions of this text, but for some reason, they are ideas that safety people have not seemed to live by in the past.

Consider the difference in how we handle safety compared to quality, cost, and quantity of production. How does management get other things that it wants—production for instance? When management officials decide they want a certain level of production, they first of all tell somebody what they want, or they set a policy and definite goals. Then they say to some one, "You do it." They define responsibility. They say, "You have my permission to do whatever is necessary to get this job done." They grant authority. And finally they say, "I'll measure you to see if you are doing it." They fix accountability. This is the way management motivates its employees to do what it wants in production, in cost control, in quality control, and in all other things except safety.

In safety, industry has seemed to take a quite different tack. Management officials have not effectively used the above tools of communication, responsibility, authority, and accountability. Rather they have chosen committees, safety posters, literature, contests, gimmicks, and a raft of other things that they would not consider in quality, cost, or production. In those other areas management has not worried too much about motivating people. Management there has decided what it wants and then made sure that it gets exactly that.

In safety, we have gotten into the ludicrous position of pleading for management support, instead of advising how management can better direct the safety effort to attain its specified goals.

Buying the concept of utilizing normal managerial principles and techniques to achieve safety results, we then must begin to look at these principles and techniques. Section 7 begins this by looking at responsibility and accountability.

SECTION 7   Responsibility for Occurrence and Prevention of Accidents

Good for people to be in a safe workplace, and there are legal responsibilities

Existing laws, rules, and ordinances require that workplaces, machinery, and equipment be maintained in safe condition, that adequate first-aid facilities be provided, that accidents be reported, and that the employee be compensated when accidentally injured. The burden of responsibility for compliance with these requirements is fixed and defined by law. The moral obligation of employers to their employees and to society requires that a reasonably safe working environment be maintained. It likewise demands that consideration be given to the physical and mental fitness of employees to perform safely the tasks to which they are assigned, that adequate training and instruction of employees in safe methods be provided, and that a systematic effort, suited to the individual circumstances and conditions, be made to eliminate, minimize, and control the physical and mechanical hazards and the unsafe actions of persons.

OSHA requires an employer to comply with a rather large pile of individual standards which specify in detail the conditions that must exist at the workplace. The responsibilities are quite clear. Penalties can be severe, including fines and imprisonment for management in certain instances.

Dependence must be placed largely on the recognition by management of its moral rather than its legal responsibility for preventing accidents, not only because existing legal requirements are limited in scope, but also because it is a most difficult task to be practical and fair in any attempt by law to achieve wholly safe employee working conditions. This is due in part to the fact that workers themselves are often negligent and in part to the complexity and tremendous variation of industrial processes.

As herein used, the term "management" applies broadly to the entire managerial and supervisory staff of an industrial organization. In the case of a "one-person" company, where the establishment is so small that the owner is also the superintendent and the supervisor, managerial responsibility clearly rests on this one person. Where the organization consists of several executives, they must all share the obligations of management.

The line of demarcation, therefore, between management and employee lies in the authority to issue orders or to direct work. A supervisor, a leader, or even a "straw boss" is a representative of management, and because he is authorized to direct the work of employees, he is a part of management.

MANAGEMENT CONTROLS MECHANICAL OR PHYSICAL CAUSES

Mechanical or physical causes of accidents include such typical hazards as unguarded or inadequately guarded machines, tools, and equipment; machines and tools or other devices that are worn, frayed, broken, of insufficient strength, or otherwise defective; and inadequate light or ventilation in workplaces.

These conditions are obviously within the control of management. To begin with, management selects, purchases, installs, and makes use of the equipment. It may in certain cases actually design and build the equipment. Management is the owner of such equipment and may be said to be the sole authority in final decisions as to its handling, operation, maintenance, placing, and guarding. Management pays the bills when, as a result of accident, equipment is damaged or destroyed or when accidents interfere with the continuity of production. The persons who are charged with the task of building safety into mechanical equipment, planning safe and efficient manufacturing processes and procedures, repairing and replacing defective equipment, and otherwise making and maintaining safe working conditions are directed by management. Where control is so placed and opportunity so great, it cannot fail to be evident that moral responsibility also exists. In every sense of the word, therefore, management is logically and properly responsible for safe mechanical and physical conditions in the workplaces of which it has charge.

MANAGEMENT CONTROLS PERSONAL CAUSES

Management's responsibility for controlling the unsafe acts of employees exists chiefly because these unsafe acts occur in the course of employment that management creates and then directs. Management selects the persons upon whom it depends to carry on industrial work. It may, if it so elects, choose persons who are experienced, capable of and willing to do this work, not only well but also safely. Management may also train and instruct its employees, acquaint them with safe methods, and provide competent supervision. In following the principles of delegated authority, management, through its representatives in the supervisory staff, may set a safe example, establish standards for safe performance, and issue and enforce safe-practice rules.

It is tacitly understood, by a person who accepts wages in return for her services, that her employer has a logical right to ask that she do the work for which she is remunerated in the way her employer wants it done, provided always that the request is reasonable. Surely it is reasonable to expect that safe-practice rules will be obeyed. The employer, therefore, has an opportunity that he is morally obligated to fulfill, to obtain compliance with safe-practice rules just as he obtains compliance with rules and instructions relating to the quality or tolerances of manufactured product. In the case of unsafe personal performance, therefore, no other proper conclusion can be reached than that because of its ability and its opportunity, management is responsible for prevention.

COST OF ACCIDENTS AN INCENTIVE FOR MANAGEMENT

In the cost of industrial accidents there lies another incentive for management to prevent accidents—one that indicates moral responsibility for taking action. The responsibility rests in this case not directly with the employee or society so much as with the industrial establishment, its stockholders, directors, or company members whose time and capital are invested and at stake.

The direct cost of accidents, as represented by compensation payments, first aid, medical and surgical expense, plus legal fees and overhead, may be portrayed approximately by compensation insurance rates or premiums. In the high-hazard industrial classifications these compensation insurance rates may be as much as 20 to 30 percent or more of the total payroll. Direct accident cost in itself is thus shown to be a material factor in efficient management, but it is nevertheless tremendously overshadowed by the vastly greater cost referred to herein as "hidden" or "incidental."

RESPONSIBILITY OF SAFETY ENGINEER

The safety engineer referred to in this discussion is the industrial safety engineer who is employed by management. She is representative of and a part of the managerial and supervisory staff that directs the work of employees, and she therefore shares the responsibility of management for accident prevention as already described.

Being specially qualified in safety work and often in direct charge of it, the safety engineer has the opportunity to impart her knowledge to others. Her functions were shown in Fig. 1-7 and 1-8. Further discussion of her legal responsibilities is presented in Chap. 15.

RESPONSIBILITY OF SUPERVISOR

The statements heretofore made with regard to the responsibility of management are applicable to the supervisor also.

In other than so-called one-person organizations, the principle of delegated authority must obviously be followed, and this places responsibility on the supervisor as management's representative to issue and enforce orders.

The first-line supervisor, moreover, is in a peculiarly strategic and tremendously important position so far as attaining results in accident prevention is concerned. He is the person who actually gives orders and instructions to the employees. He explains, instructs, interprets, and enforces the orders. From the employees' point of view the supervisor's word is authoritative.

The supervisor is closely associated with his workers and may work side by side with them. He is often as skilled or more so than the employees. He knows his workers personally and has a splendid opportunity to become acquainted with their habits, grievances, attitudes, and personal as well as business qualities. His influence and example, as well as his authority, provide him with a degree of control over his workers that is of the greatest importance in safety work.

No safety program, in view of the foregoing circumstances, can hope to be wholly successful without the sympathetic and intelligent support of able supervisors.

Industry, while fully cognizant of these facts, has still a long way to go in making effective use of its supervisors. It has assumed erroneously that the art of supervision is difficult to express and to teach as one consisting of specific factors and methods. Therefore, supervisory training methods too frequently depend on discussion of the generalities of human engineering and leadership, psychology, hints, and tips. Practicing and prospective supervisors are instructed to listen, lead but not drive, explain changes in advance, ask for suggestions, praise wherever possible, give credit when due, explain the reason for an order, display an interest in employees and their home circumstances, make an inventory of employee characteristics, also of his own traits and characteristics, get his workers to like him, be fair, kindly, patient, firm, impartial, and do literally hundreds of similar things.

There is nothing wrong with this teaching except that it has no pattern or basic structure. These admonitions, instructions, hints, and tips are all good. They are definitely individual parts of the art of supervision. But their relation to one another and to the art as a whole must be made clear. This can be and has been done (see Chap. 12).

RESPONSIBILITY OF EMPLOYEES

The responsibility of employees for accident prevention is primarily and naturally a selfish one. They are of course responsible for their own safety, but they are also under an obligation to their dependents and to society for keeping themselves physically sound enough to maintain their efficiency as wage earners.

However, individual employees do not issue or enforce orders; they lack authority to instruct fellow employees and cannot well be held responsible for the mechanical and physical hazards of the establishment in general.

In the sphere of their own specific duties, employees have certain responsibilities to their employers for the safe conduct of their work and the maintenance of equipment. They should report unsafe conditions to their supervisors when these are not within their authority to control. They should see that other employees do not interfere with their work in such a way as to create accident possibilities. They may have opportunities to impart their knowledge and experience in safety matters to their fellow employees and may take an active and helpful part in safety meetings. As a member of an organized labor group they play an increasing part, directly and through the shop steward or other union representative, in inspecting plant safety, in formulating safety rules, and in recommending safe work environment.

In general, it would seem that although the opportunities and responsibilities of the individual employees are limited, they are nevertheless of great cumulative importance.

Nonemployees, with rare exceptions, have no part in managerial responsibility, nor have they any obligations for the safety of any part of the industrial work of a given establishment.

Their responsibility is chiefly that of avoiding interference with persons or procedures which might create unsafe conditions. Nonemployee safety engineers and other persons who are invited by management to take part in industrial safety work and specialists and experts who are authorized to cooperate with the managerial or supervisory personnel are not included in the foregoing reference to nonemployee responsibility.

JOINT RESPONSIBILITY—LABOR UNION AND MANAGEMENT COOPERATION

Obviously the prevention of accidental injury is of importance to both union and management. While in the past there have been conflicts arising out of safety issues, it seems that there is now more cooperation in safety areas than in the past. A current trend in union contracts is the establishment of joint union-management safety committees. While these are only as effective as the members that serve on them, it seems a distinct possibility that the approach could be of genuine help in the future.

DEGREE AND CLASSIFICATION OF RESPONSIBILITY

The legal responsibility of management is fixed and defined by law. Moral responsibility cannot well be classified nor can its extent or degree be definitely established as it applies to individuals or groups included in "management."

An illustration may serve to make this point clear.

An employee, while filling a drum with a highly corrosive acid from the swinging spout of an overhead gravity tank, allowed his attention to wander. The drum overflowed and the employee was seriously burned.

Gloves, apron, goggles, and mask had been provided, and instructions to wear them had repeatedly been given. They were not worn. The operation was one that was commonly conducted in other plants. No previous trouble had been experienced.

The employee undoubtedly was negligent, first in paying insufficient attention to his work, and second, in failing to wear the equipment that had been provided.

Supervision, however, was also at fault. It was inadequate and most certainly ineffective in that it had not impressed the injured man sufficiently with the hazards and with the need for care, nor in this case did it observe and correct the unsafe practice that had been carried on for some time before the accident.

Higher executives were likewise responsible in several ways. The process could have been, and in fact was after the accident, more safely designed, employee and supervisor selection and training could have been better, and there was lack of system and foresight in the failure of management to keep closely informed concerning the probable hazards of its manufacturing processes.

Accident prevention is a cooperative task of vital importance in our social and economic life, in which management assumes the major responsibility, the employees must do their share, and supervisors bear most of the burden of detail.

There has been little controversy or disagreement on the notion that management has the key responsibility for accident prevention. Most safety professionals agree with this wholeheartedly today. Much of our effort in the last 10 or 15 years has been on how to better build management into our programs. We've incorporated management statements of policy into most programs, and attempted other techniques to better build them in, realizing the importance of the safety program being "their program." It probably hasn't been enough, for many executives today are still not involved, not active in safety. But there is no doubt that they are responsible for safety; for OSHA has made that clear. But being responsible for it and being active in it are two different things.

There has been some rethinking of the key person concept. We have always, it seems, thought of the supervisor as the key person. We have utilized course after course preaching this. The concept has been axiomatic in our thinking. Supervisors are those persons between management and the workers who translate management's policy into action. They have eyeball contact with the workers. Are supervisors the key persons? In a way, yes they are. However, although supervisors are the key to safety, management has a firm hold on the key chain. It is only when management takes the key in hand and does something with it that the key becomes useful.

In short, yes the supervisors are the key people for getting things done. But we also know today that they will get things done only when their bosses want them to get things done in safety. And this leads us away from the concept of responsibility for safety and toward the concept of accountability for safety.

The key to effective line safety performance is management procedures that fix accountability. Any line manager will achieve results in those areas in which management is measuring him. The concept of "accountability" is important for this measurement. The lack of procedures for fixing accountability is safety's greatest failing. We have preached line responsibility for many years. If we had spent this time devising measurements for fixing accountability on line management, we would still be achieving a reduction in our accident record.

When people are held accountable, they will accept the given responsibility. If they are not held accountable, they will not in most cases—they will place their efforts on those things that management is measuring: on production, on quality, on cost, or on wherever the current management pressure is.

Frank E. Bird, Jr., discusses this in his book *Loss Control Management*, coauthored by Bob Loftus. He first presents his principle of point of control: "The greatest potential for control tends to exist at the point where the action takes place." This, of course, is a restatement of the key person concept, but he then goes on to state:

> With this important statement made, it may seem contradictory to indicate that while the front-line supervisor is a key man in the program, he is not *the* key man. In order of importance to the success of the program, he

could be the least important of all management levels. This "point of control" supervisor sees the safety and loss control program exactly as his "boss" wants him to see it. He is merely a mirror reflection of upper management attitude toward the subject. The "key" man in any program (whether it be loss control, quality, production or costs) is the highest level of operating management.

And D. A. Weaver gives us this principle:

Accountability should be fixed near the point of control. The point of control lies in the line organization. Therefore, safety management must devise procedures to fix accountability at the point of control. This means something counted or measured with sufficient reliability and validity that line management accepts it for appraisal, praise, blame, correction, and reward. Correction of supervisory safety performance (with suitable input of safety expertise) should be the task of immediate superiors at every echelon, *because it matters to them*, because they also are being held accountable by the same procedures.

SECTION 8   Incidental Costs of Accidents

Compensation laws place the burden of industrial accident responsibility upon the employer of labor, and educational methods reach the employee mainly (and often only) through the employer. The guarding of machinery must be accomplished by the employer or with the employer's full cooperation. The elimination of unsafe methods and processes is under the control of the employer. And in fact the entire structure of the accident-prevention program rests upon employer-executive participation and action. It is fortunate for the cause of safety that the action of these employer executives is defensible from a purely business viewpoint as well as from that of humanity.

HIDDEN COSTS

To industrial executives is offered the incidental or, more correctly, the hidden or indirect, but nonetheless real, employer cost of accidents as indisputable evidence of the need for recognizing accident prevention as an essential of sound business management.

The many and varying estimates of the annual cost of industrial accidents are stated in terms of millions of dollars and are usually based upon the lost time of the injured worker and medical expense. This is largely an employer loss, inasmuch as the employee is partially compensated; but it is far from being all of the cost to the employer.

In addition there are the very real so-called incidental costs of accidents. These incidental costs of accidents have been estimated to be 4 times as great as the actual costs. There have been innumerable studies made, discussions held, articles written, and arguments presented about that figure. No one has disputed the concept, the fact that there are indirect, or incidental, or hidden costs surrounding accidents. All tend to agree with that. There is tremendous disagreement, however, about how much those costs might amount to. We might first briefly look at what the costs involved are, and then look at some of the controversy over how much they really amount to.

FACTORS AND EXAMPLES OF HIDDEN COST

The calculations from which the 4 to 1 ratio was derived were based on the factors that follow. Compensation and liability claims, medical and hospital cost, insurance premiums, and cost of lost time except when actually paid by the employer without reimbursement are excluded:

Factors of Hidden Accident Cost

1 Cost of lost time of injured employee
2 Cost of time lost by other employees who stop work:
   a Out of curiosity
   b Out of sympathy
   c To assist injured employee
   d For other reasons
3 Cost of time lost by foremen, supervisors or other executives as follows:
   a Assisting injured employee
   b Investigating the cause of the accident
   c Arranging for the injured employee's production to be continued by some other employee
   d Selecting, training, or breaking in a new employee to replace the injured employee
   e Preparing state accident reports, or attending hearings before state officials
4 Cost of time spent on the case by first-aid attendant and hospital department staff when not paid for by the insurance carrier Co. nursing station.
5 Cost due to damage to the machine, tools, or other property or to the spoilage of material
6 Incidental cost due to interference with production, failure to fill orders on time, loss of bonuses, payment of forfeits, and other similar causes
7 Cost to employer under employee welfare and benefit systems
8 Cost to employer in continuing the wages of the injured employee in full,

after his or her return—even though the services of the employee (who is not yet fully recovered) may for a time be worth only about half of their normal value

9 Cost due to the loss of profit on the injured employee's productivity and on idle machines

10 Cost that occurs in consequence of the excitement or weakened morale due to the accident

11 Overhead cost per injured employee—the expense of light, heat, rent, and other such items, which continues while the injured employee is a nonproducer

This list does not include all the points that might well receive consideration, although it clearly outlines the vicious and seemingly endless cycle of events that follow in the train of accidents.

To illustrate how the 4 to 1 ratio was determined, following are some examples:

Case 1

Total cost of compensation and medical aid	$209
Total additional incidental cost, paid directly by the employer	937

The following accidents occurred on a building erection job:

Number of Accidents	Description		Compensation and Medical Cost
3	Fractures and contusions	Material hoist	$106
18	Rivet burns, cuts, bruises	Miscellaneous operations	76
21	Falling materials	Miscellaneous operations	15
30	Slips and falls	Miscellaneous operations	12

The incidental cost was computed as follows:

Time lost by injured employees, paid directly by the employer	$116
Time lost by other employees	310
Time lost by foremen and superintendent	78
Property damage	158
Payment of forfeits (2 days) for failure to complete the job on time	200
Portion of overhead-cost loss during delay	75

An interesting point developed by this analysis is the relatively high cost to the employer on account of the time lost by employees other than those who were injured. This was due chiefly to one of the material-hoist accidents. Shaftway enclosures were not maintained as good practice demands, and one employee, who was working in the vicinity of the hoist, was injured when he allowed a heavy plank to project into the path of the ascending car. The hoisting cables were torn from their fastenings, causing the car to drop. Labor on the upper tiers was necessarily suspended pending the completion of repairs.

Another feature worthy of note is the loss due to forfeits. The contractor herself estimated that delays caused by the accidents interfered with the completion of the job at the time agreed upon, thereby penalizing her to the extent of $200.

Case 2

Total cost of compensation and medical aid	$ 66
Total additional incidental cost, paid directly by the employer	277

This example was obtained from the records of a hardware manufacturing plant and covers an experience of 6 months.

The following accidents occurred:

Number of Accidents		Description	Compensation and Medical Cost
1	Lost nail of index finger	Punch press	$61.50
1	Lacerated forearm	Baling press	4.50
9	Cuts on hands	Handling sheared metal	0.00
10	Slips and falls, dropping objects	Miscellaneous operations	0.00

The incidental cost was computed as follows:

Value of labor and material in connection with canceled order	$107
Time lost by injured employees (paid by employer)	36
Time lost by other employees	34
Cost of repairs to stamping dies	33
Unearned wages (the employer paid a slightly injured skilled employee 10 days' full wages—$12 a day—receiving in return unskilled labor worth $8 a day)	40
Time lost by foreman and superintendent	27

The visual direct cost in this example lies in the relatively infrequent serious accidents, whereas close analysis shows that a large part of the incidental cost is due to the trivial injuries.

One of the accidents listed on the employer's records under "slips and falls—dropping objects" occurred to a toolmaker who, on account of laxity in supervision, was indulging in a bit of gossip while operating an engine lathe. His attention being temporarily diverted from his work, he allowed the lathe tool to feed into a shoulder of a jig that was mounted on the face plate of the lathe, tearing the work loose from the face-plate clamps and causing it to drop upon his fingers and jam them against the lathe bed. The jig (original value $48) was ruined. The cost of the jig, however, is not included, since it was merely being repaired for stock, and its actual value at the time of the accident was difficult to ascertain.

What is of more interest, however, is the fact, that, because of his bruised and bandaged fingers, this skilled toolmaker was unable to undertake the building of a set of blanking dies that were ordered the same day that the accident occurred. Another toolmaker was employed for that purpose, and the injured man was transferred to work requiring less skill. Thus, for a time, the employer paid the wages of two skilled workers and received in return little more than the labor of one.

The punch-press injury is also of interest from the point of view of incidental cost. Orders had been issued to the effect that the pressroom supervisor must be called in case stock became jammed in the dies. These orders were not rigidly enforced, and one of the press operators, observing that a blank had been pulled up by the punch so that it obstructed further feeding, tried to remove it with a metal rod and accidentally stepped on the clutch pedal at the same time. The dies were thrown out of alignment and seriously sheared, and at the same time the employee's finger was nipped under the spring stripper plate. The delay incidental to repairs in this case was sufficient to cause cancellation of the order on which the operator was working, thus creating a labor and material loss of $107, according to the manufacturer's statement.

Case 3

Total cost of compensation and medical aid	$ 0
Total additional incidental cost, paid directly by the employer	154

The incidental cost was computed as follows:

Cost of wasted materials	$93
Cost of time lost by injured employees	37
Cost of time lost by employees other than those injured	24

For a period of 60 days, following the inauguration of a no-accident campaign in this plant, there was not a single compensable injury or one that required professional medical treatment. There were 66 minor cases, however—chiefly burns due to the unsafe acts of employees when handling caustic or corrosive materials.

One of the workers who was not seriously injured but who was nevertheless in no condition to carry on his work was a skilled chemist upon whom the concern was depending to provide an analysis prior to the fulfillment of a large rush order. In his absence the work was done by a less skilled employee, who bungled the job. The work of filling the order was begun before the error in the analysis was discovered, resulting in waste of materials and dissatisfaction on the part of the customer on account of delay in the delivery of goods.

Case 4

Total cost of compensation and medical aid	$ 59
Total additional incidental cost, paid directly by the employer	262

Here is the cost record of 36 injuries in an average woodworking plant manufacturing interior trim and doing some cabinetwork.

The following accidents occurred:

Number of Accidents		Description	Compensation and Medical Cost
1	Hand severely cut	Jointer	$51
15	Cuts and bruises	Handling material	5
10	Slips and falls, falling objects	Miscellaneous operations	3
1	Finger slightly lacerated	Band saw	0
1	Bruised forehead	Struck against machine	0
8	Miscellaneous cuts and bruises	Hand tools	0

The incidental cost was computed as follows:

Time lost by injured employees (paid by the employer)	$ 48.00
Cost of time lost by other employees	116.00
Time lost by foremen and superintendents	79.00
Spoilage of material	11.40
Broken and damaged tools	7.60

Here, again, a significant feature is that the incidental cost of the time lost by employees other than those who were injured was greater than the incidental cost of the time lost by the injured employees themselves. A part of this time was lost because fellow workers crowded about in sympathy or curiosity or to give assistance to the injured employees. The major time loss, however, was due to the fact that a certain supervisor was not readily available for consultation for a period of 1 week, because during that time she was obliged to operate a production machine.

This situation was brought about by a preventable accident on a partially guarded jointer that was being used temporarily as a molder. It was essential that production be continued, and the supervisor happened to be the only remaining available skilled operator. While she was engaged in running the jointer, considerable delay and confusion existed in her department because employees who looked to her for instruction and advice were obliged to wait. Loss of production is not included in the estimate of incidental cost, but the lost time of uninjured employees in itself constitutes an item worthy of consideration.

Case 5

Total cost of compensation and medical aid	$11
Total additional incidental cost, paid directly by the employer	49

An employee in a machine shop was injured while reaming a casting on an engine lathe. She attempted to grasp the dog, which had started to revolve when the reamer pulled away from the tail-stock center. Three fingers were lacerated.

The incidental cost was computed as follows:

Injured employee, upon returning to work with her hand bandaged, was engaged for 2 weeks at work ordinarily performed by unskilled employees at a low wage rate. The employer, while paying *full* wages for 2 weeks, received but *part* value	$33
Time spent by supervisor and assistant superintendent in investigating damage to the tools and to the casting and in planning the replacement of the ruined casting	8
Lost time of several employees who left their work to assist or sympathize with the injured worker and to discuss the accident	6
Cost of new reamer, to replace the one broken in the accident	2

The cost of a new casting (estimated at $50) is not included in this example, nor is the lost time (4 days) of the injured employee, because there may have been some salvage on the casting, and the employee received no wages while away from the shop.

The specific point of value in this example lies in the first item of incidental cost—the wages of the convalescent employee, which continued at 100 percent, while her services, being rendered on unimportant work, were greatly reduced in value.

Case 6

Total cost of compensation and medical aid	$ 22
Total additional incidental cost, paid directly by the employer	107

An operator of a hot drop-forge press received an eye injury caused by flying hot scale. He was a skilled workman, and the assistant supervisor was obliged to substitute for him for a period of three days, so that an important order could be filled on time.

The incidental cost was computed as follows:

Cost of time lost by employees who depended upon the assistant supervisor for advice and instruction	$50
Cost of time lost by the assistant supervisor, in addition to that which he would have spent in supervision as indicated by foregoing item	30
Cost of time spent by the safety committee in the analysis and investigation of the accident	12
Cost of time lost by the injured worker (paid by the employer)	10
Cost of time lost by other employees when the accident occurred	5

This example illustrates another point in incidental cost—namely, that in filling the gap caused by an injured employee's absence, and in maintaining production, other important work is often neglected.

Accident cost in the aggregate is made up of the greater volume of low-cost minor injuries, rather than the lesser volume of high-cost fatalities and serious injuries.

SPECTACULAR EXAMPLES

The examples of hidden accident cost given in this chapter include no fatalities, major dismemberments, or major permanent injuries, nor do they feature spectacular costs that result from trivial injuries.

1 A premature quarry blast, set off inadvertently by an employee who became startled by a slight injury, resulted in serious damage to six drill rigs and a steam shovel. This equipment was buried under tons of rock, and the total loss was over $10,000.
2 The flooding of a tunnel occurred as the direct result of a minor injury. The loss amounted to over $25,000.
3 A slight injury to a lineman caused him to drop a coil of wire, which resulted in a short circuit. This led to claims against the power company and a loss of several thousand dollars.
4 An injury to an engineer who was adjusting the stuffing box of a large steam engine while the engine was in operation caused him to drop a wrench into the path of the moving crosshead. The cylinder head was broken, and the engine was thrown out of alignment and badly wrecked.
5 An explosion of an oven (loss $17,000) was the result of an error on the part of a worker whose attention was diverted by a minor injury.

Although 11 hidden accident-cost factors have been specifically listed in this chapter, it should be emphasized that, in calculating the ratio of 4 to 1 between hidden and direct costs, only a few of the factors were employed, these being the ones for which data were readily available. This fact, coupled with the omission of calculations for spectacular accidents that resulted in huge hidden costs, fortifies the main argument. Circumstances such as those specifically described in the examples might not prevail again, but others similar to them are likely to arise with considerable frequency.

The original research resulting in the 4 to 1 ratio was made in 1926.

Since 1926 obviously there has been much discussion and some research around the idea of hidden accident costs. Frank E. Bird, Jr., accepts the iceberg principle of hidden costs, dividing them into two categories: uninsured costs of property damage which can be easily quantified and uninsured miscellaneous costs which are harder to quantify. His estimates of each are higher than our original 4 to 1 estimate (see Fig. 2-20).

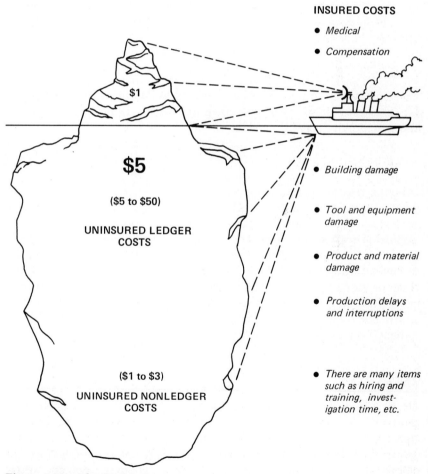

Figure 2-20   The Bird iceberg.

Simonds and Grimaldi provide a method of quantifying uninsured costs in their book *Safety Management*. Their method is based on the formula:

Uninsured cost = A times number of lost-time cases+
  B times number of doctor's cases+
  C times number of first-aid cases+
  D times number of no-injury accidents

After thoroughly defining each of the above four categories, they provide these cost estimates:

Lost-time cases	$220
Doctor's cases	55
First-aid cases	12
No-injury cases	400

These figures are based on wage levels as of February 1974.

This approach is also subscribed to by the National Safety Council in its *Accident Prevention Manual*.

While a number of current safety writers still are studying and discussing the concept of and quantification of hidden costs, a few are tending today to downplay their use. There are reasons for this, for although hidden costs are very real costs, they are extremely difficult to demonstrate in any meaningful way. To arbitrarily pick a ratio—any ratio of hidden to insured costs—and use it to sell management is often asking for trouble, for the costs are unproven and often seem unreal. To go through the formula above, while easy enough, also could be somewhat difficult to prove to management unless it accepts the concept arbitrarily from the beginning. And to attempt to actually quantify hidden costs is an almost impossible task, and probably not worth the effort. If management believes in the concept, it is often unnecessary to have to quantify.

SAFETY AND EFFICIENT PRODUCTION

The first part of axiom 10 states that the safe establishment is efficient productively and the unsafe establishment is inefficient. It is a statement that is not universally endorsed.

Is the axiom true? The question is probably unanswerable until we define terms and look at specific situations. There are specific situations in industry where it is obvious that to have a safe operation might cost so much that a company could be forced from business in a competitive situation with a company not taking the safeguards. A job shop press shop might be an example. This is not to say the axiom is untrue. Until terms are defined and situations are examined we will not know. And for all practical purposes the entire question is somewhat academic. We must, in safety management, do all that is possible to achieve safety within the confines we must live with on the job. As Grimaldi points out in his book *Safety Management*, we could achieve highway safety in this country by setting a 6 mile per hour speed limit—but this is a price that this country will not pay to achieve safety. Similar thinking is real world in industry.

We can state the fact that the arbitrary preaching of the thought in the first part of the axiom by safety people to line people and management people has done considerably more harm to our goal than good.

REFERENCES

Section 1

Adams, E., "Accident Causation and the Management System," *Professional Safety*, October 1976.

Bird, F., *Management Guide to Loss Control*, Institute Press, Atlanta, 1974.

Bird, F., and H. O'Shell, *Principles of Loss Control*, International Safety Academy, Houston, 1972.

Petersen, D., *Techniques of Safety Management*, McGraw-Hill, New York, 1971.

Weaver, D., "Symptoms of Operational Error," *Professional Safety*, October 1971.

Zabetakis, M., *Safety Manual No. 4, Accident Prevention*, MSHA, Washington, 1975.

Section 2

Alkov, D., "The Life Change Unit and Accident Behavior," *Lifeline*, September–October 1972.

Ball, L., in R. Stein, *Recording and Cataloging Hazards Information*, NASA, Huntsville, Ala., 1973.

Firenze, R., "Hazard Control," *National Safety News*, August 1971.

Kerr, W., "Complementary Theories of Safety Psychology," *Journal of Social Psychology*, vol. 45, 1957, pp. 3–9.

Johnson, W., *Management Oversight and Risk Tree—MORT*, U.S. Atomic Energy Commission, Washington, 1973.

Petersen, D., *Safety Management—A Human Approach*, Aloray, Englewood, N.J., 1975.

Suchman, E., "A Conceptual Analysis of the Accident Phenomenon," *Behavioral Approaches to Accident Research*, Association for the Aid of Crippled Children, New York, 1961.

Surry, J., *Industrial Accident Research*, University of Toronto, Toronto, 1974.

Section 4

Bird, F., *Management Guide to Loss Control*, Institute Press, Atlanta, 1974.

Petersen, D., *Techniques of Safety Management*, McGraw-Hill, New York, 1971.

Section 6

Bird, F., and R. Loftus, *Loss Control Management*, Institute Press, Atlanta, 1976.

Petersen, D., *Techniques of Safety Management*, McGraw-Hill, New York, 1971.

Petersen, D., *Safety Supervision*, American Management Association, New York, 1976.

Weaver, D., *Strengthening Supervisory Skills*, Employers Insurance of Wausau, Wausau, Wis., 1964.

Weaver, D., "Symptoms of Operational Error," *Professional Safety*, October 1971.

Section 7

Bird, F., and R. Loftus, *Loss Control Management*, Institute Press, Atlanta, 1975.

Petersen, D., *Techniques of Safety Management*, McGraw-Hill, New York, 1971.

Weaver, D., "Symptoms of Operational Error," *Professional Safety*, October 1971.

Section 8

Accident Prevention Manual for Industrial Operations, 7th ed. National Safety Council, Chicago, 1974.

Bird, F., *Management Guide to Loss Control*, Institute Press, Atlanta, 1974.

Grimaldi, J., and R. Simonds, *Safety Management*, 3d ed., Irwin, Homewood, Ill., 1975.

PART TWO

Accident-Prevention Method

THREE

A Flowchart

Basically simple as accident prevention is, success cannot be achieved when corrective action is not well planned, its steps are not well organized, and there are not sufficient personnel to carry on the work. Organization constitutes the vehicle—the systematic procedure—by means of which interest is created and maintained and safety activities are correlated and directed. Since, however, the detail required to describe organization fully may interrupt continuity of thought, there is definite advantage to the reader in describing, at this point, at least the framework of complete and practical preventive method, while reserving the detail for subsequent chapters.

The framework of action referred to in the foregoing is illustrated graphically in Fig. 3-1.

In line with the theme of the entire text, accident prevention, as shown in Fig. 3-1, is built on and about the core of hazards or causes, their finding and analysis, and the selection and application of appropriate remedies. If hazards are removed, or better still, if they are *anticipated* and *prevented from originating*, the job is done, period.

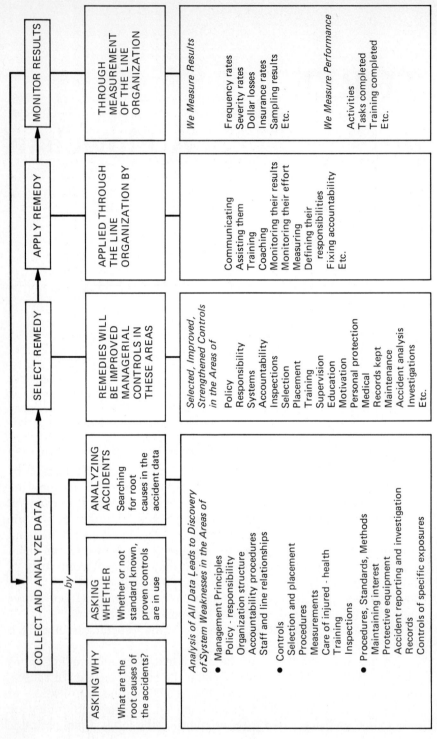

Figure 3-1 A flowchart of the safety management process.

COLLECTING AND ANALYZING DATA

The first step in this systematic approach to safety management is to collect information, to find out where you are in terms of the current activities and results in accident prevention. The figure in 3-1 indicates there are three basic information-gathering approaches: to ask why certain accidents and incidents are happening (investigation for causes); to ask whether or not certain known proven accident controls are in effect and are effective in the management system; and to analyze the accidents en masse, their trends, and the circumstances that surround their happening.

The methods of data collection and analysis are covered in some detail in Chaps. 4 to 6 of this part of the text. Chapter 4 deals with the basic approaches of asking why and whether. This information ties in with the fifth principle of safety management discussed in *Techniques of Safety Management*:

> *The function of safety is to locate and define the operational errors that allow accidents to occur. This function can be carried out in two ways: (1) by asking why—searching for root causes of accidents, and (2) by asking whether or not certain known effective controls are being utilized.* The first part of this principle is borrowed from W. C. Pope and Thomas J. Cresswell in their article, "Safety Programs Management," in the August 1965 issue of the *Journal of the American Society of Safety Engineers*. This article defines safety's function as locating and defining operational errors involving (1) incomplete decision making, (2) faulty judgments, (3) administrative miscalculations, and (4) just plain poor management practices.
>
> They suggest that to accomplish our purposes we in safety would do well to search out not what is wrong with people but what is wrong with the management system that allows accidents to occur.
>
> This thinking is borne out in the ASSE publication, "Scope and Functions of the Professional Safety Position." In this the position of safety is diagramed as in Fig. 1-7.

In further describing the four major areas under identification and appraisal of the problem, the publication states that it is the function of safety to:

> Review the entire system in detail to define likely modes of failure, including human error, and their effects on the safety of the system. . . . Identify errors involving incomplete decision making, faulty judgment, administrative miscalculation and poor practices. Designate potential weaknesses found in existing policies, directives, objectives, or practices.

This new concept directs the safety professional to look at the management system, not at acts and conditions.

The second part of principle 5 suggests that a two-pronged attack is open to us: (1) tracing the symptom (the act, the condition, the accident) back to see why it was allowed to occur, or (2) looking at the system (the procedures) that our company has and asking whether or not certain things are being done in a predetermined manner that is known to be successful.

Chapter 5 deals with a newer approach to hazard identification and safety management, system safety. Popular in the aerospace industry since the 1960s the techniques of system safety are now finding their way into many other industries. System safety is a new and somewhat unknown discipline to industrial safety people. Although at times they may feel that it is of little or no value to them, there are, no doubt, concepts in system safety that could be usefully applied to industrial safety.

Industrial safety and system safety start from a common base—the desire to save lives and property. Yet system safety is completely oriented toward analysis and improvement of hardware, that is, of systems; and industrial safety directs itself to people more and more each year. Both start from a common base, but they start with different fields of endeavor. Industrial safety strives primarily to control accidents to employees on the job. So far, system safety has worked primarily in the area of product safety in the aerospace and automotive fields.

Industrial safety engineers operate in a fixed manufacturing situation. They work in the midst of hazards which have often been there for a long time, many of which are accepted by production as a necessary component of their way of operating. They must work within that framework, perhaps teaching the employees how to work around those hazards instead of removing them.

The first concern of system safety engineers is that a given system should work as it has been designed to, that is, that the design should be foolproof and it should be impossible for anyone to get hurt.

Chapter 6 deals with the third block shown under data collection in Fig. 3-1, accident analysis.

SELECTING REMEDIES

There is a wide choice of remedies, and the one selected should fit the case to which it is applied. Machine guarding and other forms of engineering revision should always be considered first, however, even when a worker-failure cause is plainly evident. An excellent slogan is "Foolproof the task mechanically when possible and then proceed with control of personal unsafe action."

Illustrating the value of choosing a remedy wisely is the case of a crane operator having defective vision, but no other physical or attitudinal fault, who swings her crane load perilously near to other workers. An engineering-revision remedy is not directly indicated, nor would instruction or discipline be helpful—she should be fitted with glasses.

Remedies often fall into the areas indicated on the chart in Fig. 3-1. Most are managerial in nature.

APPLYING REMEDIES

Here is what everyone wants to know—what to do first and how to go about the job. There is little value in knowing how and why accidents occur, in finding the causes and selecting appropriate remedies, *if the remedy isn't applied*. Fortunately, application is simple, even if sometimes laborious and long drawn out. In adverse circumstances— when, for example, there is lack of authority, personnel, funds, equipment, or support of management—the people in charge of safety just can't apply effectively a proved remedy, *but they can try*. In such cases, as staff rather than line employees, they must begin further back—they must convince management, and they may have to sell the idea, that it should provide personnel, equipment, and funds as necessary or grant proper authority. In other cases where they cannot apply the remedy themselves, they may rely on the supervisors or on existing facilities for indoctrination and training.

Invariably the remedy must be applied through the line organization, and usually through the methods shown in the chart on Fig. 3-1.

MONITORING RESULTS

The final step in the flowchart is to monitor results. Without this step there is no way we can know whether or not our prescribed remedy is effective. Some possible measures are shown in the chart on Fig. 3-1, and detailed measures are discussed in Chap. 9.

SUPERVISORY CONTROL OF EFFICIENT PRODUCTION

This approach would not be complete without special emphasis on certain facts of the greatest value to industry in considering the need and importance of accident prevention, namely:

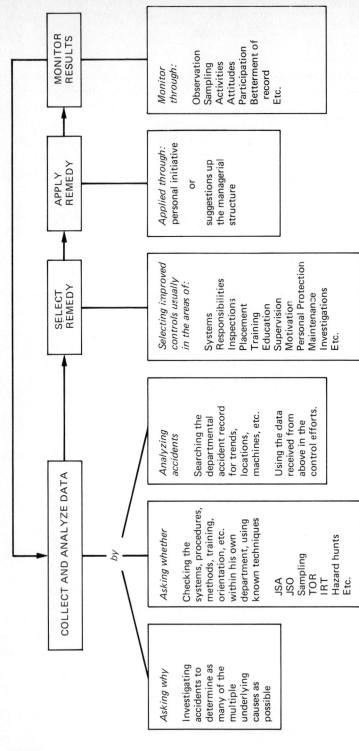

Figure 3-2 A flowchart for the supervisor.

1 The causes of both faulty production and accidental injury are of the same general nature.
2 The available forms of correction for both faulty production and accidental injury are identical.
3 Direct controls for both faulty production and accidental injury lie in the hands of the same person—the supervisor.

Industry in the United States has well earned the reputation for know-how in the supervisory control of quality and volume of production. It needs only to apply to accident prevention the exact same methods it already applies so well to production in order to reduce the sanguinary annual record of deaths and injuries.

Figure 3-2 attempts to convert the flowchart in Fig. 3-1 to the supervisory level. This approach to accident control is basically the same as the approach used on the corporate level.

REFERENCES

Petersen, D., *Techniques of Safety Management*, McGraw-Hill, New York, 1971.
Pope, W., and T. Cresswell, "Safety Programs Management," *Journal of the ASSE*, August 1965.

FOUR

Collecting and Analyzing Data

Assuming that a sufficient degree of executive and employee interest has been created to support an attack upon the accident problem and that a suitable organization has been established, it becomes necessary to develop such pertinent facts as are required to formulate practical effective procedures in accident prevention. These facts should relate not only to accidents that have already occurred but also to circumstances and conditions that have the probability of producing accidents. Thus, the sources of information lie in past accident history, the data obtained from inspection and survey, and the knowledge and imagination of informed persons.

In safety traditionally, the emphasis of data collection has been on the "direct and proximate" causes of accidents, on information relating to the unsafe acts and unsafe mechanical and or physical conditions (the third domino in that theory of causation).

The selected unsafe act or mechanical hazard, however, is only the first, direct, or proximate cause of an accident. There is also a subcause of the *accident* which is synonymous with the direct cause of the *unsafe act* or *mechanical hazard*. Here again the illustration showing the row of dominoes is helpful. It will be seen from the illustration (Fig. 2-2) that a "fault of person" not only is the *direct* cause of the unsafe act or unsafe condition but also may be termed the *subcause* of the *accident*.

All the factors in the chain of events culminating in accidental personal injury, therefore, are "causes." For example, environment causes faults of persons. These in turn cause unsafe acts of persons. The unsafe acts cause accidents, and the accidents cause injuries. To go a step further—one that is not necessary to discuss in a text on *accident* prevention—it is clear that the injury also is a cause, inasmuch as injuries result in cost and suffering.

Here we are using the term "cause" to mean the unsafe act and condition, "subcause" to mean the specific fault of person, and "underlying cause" to mean the managerial and supervisory faults and the social and environmental conditions outside the workplace. In fact, most causation theory recently has tended to emphasize the supervisory, personal, and managerial causes as considerably more important than the unsafe act or unsafe condition. As Weaver emphasizes, the act and condition are mere symptoms of something wrong. Our task is to attempt to diagnose and find out what is wrong. Concentrating on the act and condition is stopping at the symptomatic level.

Perhaps of all the models discussed in Chap. 2, the Weaver model, Fig. 2-7, best describes our process in the text. We shall view the act, the condition, the accident, and the injury all as symptoms of something wrong in the management system, and our collecting and analyzing of data are all in an effort to find out what's wrong.

To start we list these as examples of unsafe acts and unsafe conditions (symptoms):

Inadequately guarded
Unguarded
Defective (rough, sharp, slippery, decayed, corroded, frayed, cracked, etc.)
Unsafe design or construction
Hazardous arrangement, etc. (piling, storage, aisle space, exits, layout, overload, misalignment)
Unsafe illumination (inadequate or unsuitable)
Unsafe ventilation (inadequate or improperly distributed)
Unsafe dress or apparel
Unsafe methods, processes, procedures, planning, etc.
Operating without authority, failure to secure or warn
Operating or working at unsafe speed
Making safety devices inoperative

Using unsafe equipment, hands instead of equipment, or equipment unsafely
Unsafe loading, placing, mixing, combining, etc.
Taking unsafe position or posture
Working on moving or dangerous equipment
Distracting, teasing, abusing, startling, etc.
Failure to use safe attire or personal protective devices

These unsafe acts of persons are merely general heads or categories under which more specific acts may be grouped. For the individual plant or industrial operation it is more informative to specify the particular unsafe act that is committed. For example, the general head Making Safety Devices Inoperative can be subitemized as

1 Removing guards
2 Tampering with adjustment of guard
3 "Beating" or "cheating" the guard
4 Failing to report defects

And these reasons behind the act and condition can be listed:

1 *Improper attitude.* Willful disregard, reckless, lazy, disloyal, uncooperative, fearful, oversensitive, egotistical, jealous, impatient, absent-minded, excitable, obsessive, phobic, inconsiderate, intolerant, mentally unsuited in general.
2 *Lack of knowledge or skill.* Insufficiently informed, misunderstands, not convinced of need, indecisive, inexperienced, etc.
3 *Physically unsuited.* Hearing, sight, age, sex, height, illness, allergy, slow reaction, disability, intoxication, physical handicap in general.
4 *Improper mechanical or physical environment.* Space, light, heat, arrangement, ventilation, materials, tools, equipment, procedures, company policy, routing, etc., make it awkward, difficult, inconvenient, embarrassing, or impossible to follow safe-practice rules.

These are typical examples of managerial and supervisory fault:

1 Lack of organized safety procedures, including plant safety inspector, committees, accident investigation, forms, etc.
2 Inadequate or ineffective safety work
3 Lack of executive direction of and participation in safety work
4 Failure to guard machines and to provide adequate light, ventilation, first-aid, hospital, and sanitary facilities, personal protection, safe tools, and safe working environment in general
5 Lack of suitable procedures for examining new employees for physical fitness and work experience

6 Lack of suitable procedures for assignment of employees to work that they can do safely
7 Poor morale of employees
8 Lack of suitable training and instruction of employees in safety and of supervisors in the art of supervision
9 Lack of enforcement of safe-practice rules
10 Failure to place responsibility for accident occurrence

Although all three categories are discussed, the primary emphasis traditionally has been on methods to identify and eliminate the first category, the unsafe act and condition. It is that emphasis that is changed here. We have found that to concentrate at this symptomatic level is to institute only temporary cures. To throw away the broken ladder does not ensure that there will be no more broken ladder incidents. To inform a worker that what he or she has done is unsafe does not ensure that the worker will not do it again. To deal at the next level at least gives us some insight into why conditions are allowed and acts are performed. We begin to look at whether or not persons know they are doing something unsafe, whether or not we have attitudinal problems, etc. But even at this level we find it difficult to zero in on improved controls.

At the next level, so often ignored in the past, we begin to find an area where we can operate very meaningfully: at the managerial and supervisory level. For here we can analyze the situation, spot weaknesses, install improvements, and measure results. And at this level we can institute procedural changes that will be more permanent in nature. We can institute ladder inspection programs that find unsafe ladders before the accident. We can look at our training needs and at attitudinal problems, and we are at a level where we can change things.

As indicated in our flowchart in Fig. 3-1, we have three basic approaches to collect data to determine what and where our accident problems are. We can wait for a large number of accidents to occur and then summarize and analyze them to determine trends and areas to concentrate on. This will be discussed in Chap. 6. A second approach is to wait for one accident to occur and then start asking why that accident occurred. We ask what went wrong, where the system broke down, what is missing that would have stopped the causal chain. In this approach we can look at all levels, the employees (their needs, training, attitudes, etc.), the supervisors (their knowledge, time, priorities, etc.), management (its systems, procedures, policies, priorities). A third method is not to wait until the accident occurs, but rather to immediately begin to ask whether or not we are in fact doing all the things we might do to control conditions and behaviors so that accidents do not happen.

As indicated in our flowchart, we use all three approaches to accomplish results in safety. As shown in Fig. 3-2, a supervisor uses all three approaches to get a handle on his safety problems. The approach ties in with all the updated causation theories in Chap. 2. In short, it is an accepted way to achieve results in accident prevention. In this chapter we'll look at the latter in two ways mentioned above.

ASKING WHY

The first approach is to wait until an accident occurs and then begin to ask *why* it happened, to investigate.

The primary accident investigation function has always been the supervisor's. Usually management provides the supervisor with a

---

SUPERVISOR'S REPORT OF INJURY

Name of injured _____

Injury date _____ Time _____ A.M.–P.M.

Did injured return to work ? _____ Time _____ A.M.–P.M.

Witnesses _____
_____

Nature of injury _____
_____

Where and how did the accident occur ?_____
_____
_____

Unsafe act or condition _____
_____
_____

Measures taken in preventing a similar type of accident _____
_____
_____
_____
_____
_____

Figure 4-1 Typical accident investigation report.

simple form on which to record the results of the investigation (see Fig. 4-1).

Here we ask the supervisor to investigate thoroughly and determine only one "cause" for the accident. This violates the multiple-causation principle. Asking the supervisor to identify only one act or condition violates the principle that the act and condition are merely symptoms of true accident causes. Figure 4-2 shows a form that might be more in line with our current thinking. The key to supervisory investigation is that the supervisor be instructed to also look beyond symptom to determine cause. This holds true also with the additional

```
                  SUPERVISOR'S REPORT OF INJURY

    Name of injured _____

    Injury date _____ Time _____ A.M.-P.M.

    Did injured return to work ?_____ Time _____ A.M.-P.M.

    Witnesses _____
    _____

    Nature of injury _____ __
    _____

    Where and how did the accident occur ?_____
    _____
    _____

    Identify:
```

Acts and conditions	Possible causes

```
    Measures taken in preventing a similar type of accident
                    ( List on the reverse side)

    Supervisor's signature _____ Department_____
```

Figure 4-2 Suggested accident investigation report.

more detailed investigation by the safety professional. Usually investigators are asked to identify a number of facts surrounding an accident:

1 The time of injury (hour, day, month, and year)
2 The place (town, plant or place of work, department or specific location)
3 The person injured (name, number, shift, or crew)
4 The nature and severity of injury
5 The witnesses and participants
6 The kind of accident (fall of person, struck by, etc.)
7 The agency—meaning the machine, tool (also parts), process, radiation, substance, or person most closely associated with the accident and injury usually because of its hazardous condition or action

Corollary Facts

In addition to the foregoing identifying facts and the major facts of causation, other data are often of value in determining the best practical remedy, as follows:

1 The age and experience of the injured person
2 The previous accident record of the injured person
3 The outside or "home" environment of the injured person
4 The foreman or supervisor and his or her supervisory abilities and procedures
5 The operation or process involved in the accident
6 The extent of lost time and the cost of the injury
7 The existence of safe-practice rules covering the individual case
8 The name of the attending physician or surgeon and the record of treatments
9 The previous accident record of the machine or other agency, and the place and department of work where injury occurred
10 Description and cost of property damage or spoilage

While most investigation forms require this kind of information, it should be kept in mind that the primary reason for the investigation is to unearth some basic root causes, not just circumstances surrounding the accident. And root causes invariably relate to the management system.

There are additional methods other than accident investigations that ask why. One system that forces us to look at the management system for root causes is a tracing system known as TOR (technique of operations review). In this system, invented by D. A. Weaver, starting from the proximate cause of the accident (act or condition) or from any other cause which seems to be paramount in the accident, the investigator traces through other possible related causes using the table shown in Fig. 4-3. As a result a number of other causes are looked at and either are discarded or are listed as additional causes of the incident that need managerial improvement. TOR leads us to root causes, regardless of where we start in our analysis.

Another method is the incident recall technique. In this employees are systematically encouraged to recall situations and incidents that could well have led to accidents and injuries but that did not. Incident recall assists in locating some of the whys behind accidents.

The whole idea of asking why leads also to the area of understanding people's behavior. This will be discussed in Chap. 11.

ASKING WHETHER

The second broad approach to collecting data is to ask *whether* or not certain known workable controls are in effect. There are a number of ways to do this, and a number of levels that this approach can be used on. First, at the companywide level we can begin to look in some detail at what we do: our policy, our procedures, our systems, etc. These tend to fall into certain areas of operation. Often a checklist is helpful as an aid to thinking. One such checklist is shown in Fig. 4-4.

Another method of asking whether is to compare what you are doing to currently accepted standards. The entire OSHA approach is a whether approach. Each company in the country is compared to a broad range of standards which state how things should be. Certainly at the very least each company should use both the whether approach indicated by the Loss Control Analysis Guide in the figure and the comparison to standards approach dictated by OSHA. In this way the important whethers will be asked for both managerial and physical condition problems and accident causes. Inspection is another commonly used whether approach. We physically inspect to see whether we have adequately controlled those physical hazards that we should have.

Inspection was and is one of the primary tools of the safety specialist. Before 1931 it was virtually the only tool, and from 1931 to 1945 it was still the one most used. Until 1960 it was the primary tool of many outside service agencies, and even today it remains the primary (and sometimes the only) tool of some safety people.

It is generally agreed that the responsibility for conditions and for people is the line supervisor's. Thus responsibility for the primary safety inspection must also be assigned to the line supervisor. By

1 COACHING

10	Unusual situation, failure to coach (new man, tool, equipment, process, material, etc.)	44,24,62
11	No instruction. No instruction available for particular situation	44,22,24,80
12	Training not formulated or need not foreseen	24,34,86
13	Correction. Failure to correct or failure to see need to correct	42,20,30
14	Instruction inadequate. Instruction was attempted but result shows it didn't take	15,16,42
15	Supervisor failed to tell why	44,24,83
16	Supervisor failed to listen	11,81
17		
18		
19		

2 RESPONSIBILITY

20	Duties and tasks not clear	44,34,14,53
21	Conflicting goals	80
22	Responsibility, not clear or failure to accept	26,14,54,82
23	Dual responsibility	47,34,13
24	Pressure of immediate tasks obscures full scope of responsibilities	36,12,51
25	Buck passing, responsibility not tied down	80,86
26	Job descriptions inadequate	80,86

3 AUTHORITY
(Power to decide)

30	Bypassing, conflicting orders, too many bosses	44,13
31	Decision too far above the problem	36,83,85
33	Authority inadequate to cope with the situation	81,83
33	Decision exceeded authority	20,26,14
34	Decision evaded, problem dumped on the boss	36,14,85
35	Orders failed to produce desired result. Not clear, not understood, or not followed	40,46,13,15
36	Subordinates fail to exercise their power to decide	26,12,83,85
37		
38		
39		

4 SUPERVISION

40	Morale. Tension, insecurity, lack of faith in the supervisor and the future of the job	15,56,64,80
41	Conduct. Supervisor sets poor example	13,84
42	Unsafe Acts. Failure to observe and correct	24,11,52
43	Rules. Failure to make necessary rules, or to publicize them. Inadequate follow-up and enforcement. Unfair enforcement or weak discipline	25,36,12,52

27		
28		
29		

5 DISORDER

51	Work flow. Inefficient or hazardous layout, scheduling, arrangement, stacking, piling, routing, storing, etc.	41,24,31,80
52	Conditions. Inefficient or unsafe due to faulty inspection, supervisory action, or maintenance	21,32,14,86
53	Property loss. Accidental breakage or damage due to faulty procedure, inspection, supervision, or maintenance	43,20,80
54	Clutter. Anything unnecessary in the work area. (Excess materials, defective tools and equipment, excess due to faulty work flow, etc.)	44,36,80
55	Lack. Absence of anything needed. (Proper tools, protective equipment, guards, fire equipment, bins, scrap barrels, janitorial service, etc.)	44,36,80
56	Voluntary compliance. Work group sees no advantage to themselves	40,15,41
57		
58		
59		

44	Initiative. Failure to see problems and exert an influence on them	22,34,30
45	Honest error. Failure to act, or action turned out to be wrong	10,12,15,81
46	Team spirit. Men are not pulling with the supervisor	40,21,56
47	Cooperation. Poor cooperation. Failure to plan for coordination	23,25,15,66
48		
49		

6 OPERATIONAL

60	Job procedure. Awkward, unsafe, inefficient, poorly planned	44,32
61	Work load. Pace too fast, too slow, or erratic	44,51,63
62	New procedure. New or unusual tasks or hazards not yet understood	43,44
63	Short handed. High turnover or absenteeism	80,40,61
64	Unattractive jobs. Job conditions or rewards are not competitive	81,46
65	Job placement. Hasty or improper job selection and placement	80,86
66	Co-ordination. Departments inadvertently create problems for each other (production, maintenance, purchasing, personnel, sales, etc.)	45,35,13
67		
68		
69		

Figure 4-3 Table TOR review cause code

111

7 PERSONAL TRAITS
(When accident occurs)

70	Physical condition—strength, agility, poor	44,26,65
71	Health—sick, tired, taking medicine	44,24,65
72	Impairment—amputee, vision, hearing, heart, diabetic, epileptic, hernia, etc.	44,24,65
73	Alcohol—(if definite facts are known)	80
74	Personality—excitable, lazy, goof-off, unhappy, easily distracted, impulsive, anxious, irritable, complacent, etc.	44,13
75	Adjustment—aggressive, show off, stubborn, insolent, scorns advice and instruction, defies authority, antisocial, argues, timid, etc.	44,13
76	Work habits—sloppy. Confusion and disorder in work area. Careless of tools, equipment and procedure	44,13
77	Work assignment—unsuited for this particular individual	42,65
78		
79		

8 MANAGEMENT

80	Policy. Failure to assert a management will prior to the situation at hand	24,81,83
81	Goals. Not clear, or not projected as an "action image"	83,86
82	Accountability. Failure to measure or appraise results	36
83	Span of attention. Too many irons in the fire. Inadequate delegation. Inadequate development of subordinates	12,86
84	Performance appraisals. Inadequate or dwell excessively on short range performance	20,65
85	Mistakes. Failure to support and encourage subordinates to exercise their power to decide	36
86	Staffing. Assign full or part-time responsibility for related functions	66
87		
88		
89		

Figure 4-3 (Continued)

A LOSS CONTROL ANALYSIS GUIDE

I Management Organization

A Does the company have a written policy on safety?

B Draw an organizational chart, and determine the line and staff relationships.

C To what extent does executive management accept its responsibility for safety?
1 To what extent does it participate in the effort?
2 To what extent does it assist in administering?

D To what extent does executive management delegate safety responsibility? How is this accepted by:
1 The superintendent or top production people?
2 The foremen or supervisors?
3 The staff safety people?
4 The employees?

E How is the company organized?
1 Is there staff safety personnel? If so, are the duties clear? Are responsibilities and authorities clear? Where is staff safety located? What can it reach? What influence does it have? To whom does it report?
2 Are there safety committees?
   a What is the makeup of the committees?
   b Are the duties clearly defined?
   c Do they seem to be effective?
3 What type of responsibility is delegated to the employees?

F Does the company have written operating rules or procedures?
1 Is safety covered in these rules?
   a Is it built into each rule, or are there separate safety rules?

II Accountability for Safety

A Does management hold line personnel accountable for accident prevention?

B What techniques are used to fix accountability?
1 Accountability for results:
   a Are accidents charged against departments?
   b Are claim costs charged?
   c Are premiums prorated by losses?
   d Does supervisory appraisal of supervisors include looking at their accident records? Are bonuses influenced by accident records?
2 Accountability for activities:
   a How does management ensure that supervisors conduct toolbox meetings, inspections, accident investigations, regular safety supervision and coaching?
   b Other?

III Systems to Identify Problems—Hazards

A Are routine inspections accomplished?
1 Who is responsible for inspection functions?
2 By whom are inspections made?

Figure 4-4 An analysis checklist

3 How often are they made?
4 What types are they?
5 To whom are the results reported?
6 What type of follow-up action is taken?
7 By whom?

B Are any special inspections made?
1 Boilers, elevators, hoists, overhead cranes, chains and slings, ropes, hooks, electrical insulation and grounding, special machinery such as punch presses, X-ray, emery wheels, ladders, scaffolding and planks, lighting, ventilation, plant trucks and vehicles, materials-handling equipment, fire and other catastrophe hazards, noise and toxic controls.

C Are any special systems set up?
1 Job safety analysis
2 Critical-incident technique
3 High-potential accident analysis
4 Fault-tree analysis
5 Safety sampling

D What procedure is followed to ensure the safety of new equipment, materials, processes, or operations?

E Is safety considered by the purchasing department in its transactions?

F When corrective action is needed, how is it initiated and followed up?

G When faced with special or unusual jobs, how does the company ensure safe accomplishment?
1 Is there adequate job and equipment planning?
2 Is safety a part of the overall consideration?

H What are the normal exposures for which protective equipment is needed?
1 What are the special or unusual exposures for which personal protective equipment is needed?
2 What personal protective equipment is provided?
3 How is personal protective equipment initially fitted?
4 What type of care maintenance program is instituted for personal protective equipment?
5 Who enforces the wearing of such equipment?

IV Selection and Placement of Employees

A Is an application blank filled out by prospective employees?
1 Does it ask the right questions?

B What type of interview and screening process is the prospective employee subjected to before being hired?

C How are employee references and past history checked?

D Who actually does the final hiring?

E Is the physical condition of the employee checked before hiring?
1 If a physical exam is given, how complete an examination is it?
2 How is the information used?

F Are any skill, knowledge, or psychological tests given?

G Are job physical requirements specified from job analysis?
1 Are these requirements considered in new hires?
2 In job transfers?

Figure 4-4 (Continued)

V Training and Supervision

A Is there safety indoctrination for new employees?
   1 Who conducts it?
   2 What does it consist of?
B What is the usual procedure followed in training a new employee for a job?
   1 Who does the training?
   2 How is it done?
   3 Are written job instructions based on job analysis used?
   4 Do they include safety?
C What training is given to an older employee who has transferred to a new job?
D What methods are used for training the supervisory staff?
   1 New supervisors?
   2 Continuous training for the entire supervisory force?
   3 Who does the training?
   4 Is safety a part of it?
E After an employee has completed the training phases of the job, what then is his or her status?
   1 What is the quality of the supervision?
   2 What use is made of the probation period?

VI Motivation

A What ongoing activities are aimed at motivation?
   1 Group meetings, literature distribution, contests, film showings, posters, bulletin boards, letters from management, incentives, house organs, accident facts on plant operations, other gimmicks and gadgets, and activities in off-the-job safety.
B What special-emphasis campaigns have been used?

VII Accident Records and Analysis

A What injury records are kept? By whom?
B Are standard methods of frequency and severity recording used?
C Who sees and uses the records?
D What type of analysis is applied to the records?
   1 Daily analysis
   2 Weekly
   3 Monthly
   4 Annual
   5 By department
   6 Cost
   7 Other
E What is the accident investigation procedure?
   1 What circumstances and conditions determine which accidents will be investigated?
   2 Who does the investigating?
   3 When is it done?
   4 What types of reports are submitted?
   5 To whom do they go?

Figure 4-4 (Continued)

6 What follow-up action is taken?
7 By whom?
F Are any special techniques used?
   1 Estimated costs
   2 Safe-T-Scores
   3 Statistical control charts

VIII   *Medical Program*

A What first-aid facilities, equipment, supplies, and personnel are available to all shifts?
B What are the qualifications of the people responsible for the first-aid program?
C Is there medical direction to the first-aid program?
D What is the procedure followed in obtaining first-aid assistance?
E What emergency first-aid training and facilities are provided when normal first-aid people are not available?
F Are there any catastrophe or disaster plans?
G What facilities are available for transportation of the injured to hospital?
H Is a directory of qualified physicians, hospitals, ambulances, etc., available? ·
I Does the company have any special preventive medicine program?
J Does the company engage in any activities in the health education field?

Figure 4-4   *(Continued)*

"primary safety inspection" we mean the inspection intended to locate hazards. Any inspections performed by staff specialists should be only for the purpose of auditing the supervisor's effectiveness. Hence the results of our inspection become a direct measurement of the line supervisor's safety performance—or effectiveness.

One other whether approach that can be effectively used by the supervisor is the job safety analysis approach. Job (or methods) safety analysis is a procedure that identifies the hazards associated with each step of a job. It is a simple but effective approach commonly used at the supervisory level.

Methods Safety Analysis

Methods safety analysis is primarily concerned with investigating the mechanical hazards of the methods of an operation.

The fundamentals of the method analysis for safety are the same as for any operations analysis. They are as follows:

1　Break down the job or operation into its elementary steps.
2　List them in their proper order.
3　Then examine them critically.

The only difference is that while operations analysts investigate the steps of a job with the aim of eliminating those that are unnecessary, or substituting for those that are inefficient, safety engineers examine each step for its possibility of causing an accident. Sometimes they achieve the same results as operations analysts who are principally concerned with the efficiency of the job, for safety analysts will find the opportunity to eliminate hazardous steps rather than guard them. At other times, they may be able to reduce the number of steps as a whole and thereby decrease the hazardous aspects of an operation, for the fewer the steps to be performed, the less will be the exposure to accidents for any one cycle of an operation.

There are four units to be considered when analyzing a job operation for its possible hazards; these are as follows:

1　*Worker:* A representative word including one or all of the following: The operator, the supervisor, any other person or individual responsible for safety on the job, the injured person.
2　*Method:* The working procedures of the process being investigated.
3　*Machine:* The machine tools employed; there may be one or more.
4　*Material:* The substances and articles, other than machine tools, employed in the process.

"Worker" is generally the unit it is desirable to protect from accidental injury caused by faults in the other three units of the group. Therefore the investigator is concerned first with the potential hazards to personnel, involving the method, machine, or material. These are represented by the usual types of accident as falls of persons, struck by, caught in or between, etc.

Making the Analysis Methods safety analysts examine each step of a job or process from its very beginning with respect to the method, machine, or material involved to see if any or all of these three can be responsible for the occurrence of accidents. If they find a step that is potentially hazardous, they notate (alongside the step description) the type of accident apt to be caused. They have a record, therefore, of (1) the steps of the job, (2) those steps which might be hazardous, and (3) a means of reference for additional study of these steps, if necessary, in order to apply the proper corrective.

Example

Two employees in an inspection department, at separate times, had received smashed toes when a stack of 30 pieces of armor plate, approximately 36 × 24 × ⅜ inches, standing on end against a workbench, lost its support and slid to the floor. It was reported the pieces of armor plate were stacked on end because the department was crowded and there was not sufficient room to leave them on the skid pallet on which they were delivered. (The pallet containing the armor plate had been towed from the receiving department to the inspection crib, a distance of approximately 50 feet.) The pallet was approximately 6 × 4 feet in size. Inasmuch as each piece had to be Rockwell-tested, the workers piled the pieces on end as they removed them from the pallet. This was apparently easier than finding space to place them flat on the floor, then later lifting them to the Rockwell machine workbench. The supervisor's report showed that similar accidents had happened several times but only in these two instances had an injury occurred.

After the usual accident investigation, it was recommended that employees working in this inspection crib wear safety-toe shoes, and the hazard was believed to be eliminated. When the accident was analyzed according to the methods safety analysis, a more efficient elimination of the accident-causing factors was achieved. The analysis is outlined in the following table.

It will be noticed that the number of steps of the operation have been reduced from seven to five, and at the same time the safety conditions have been distinctly improved.

Comparison of Old and New Methods of Handling Armor Plate

	Old Method		New Method	
Step	Description	Accident Possibility	Description	Accident Possibility
1	Armor plate delivered (30 pcs/day), on pallet from component plant	Struck by sliding material	Armor plate delivered on steel table	
2	Hydraulic lift picks up pallet		Tow to inspection department	
3	Tow lift to inspection department		Armor plate remains stored on table until ready for test	

Comparison of Old and New Methods of Handling Armor Plate

Step	Old Method Description	Accident Possibility	New Method Description	Accident Possibility
4	Remove armor plate from pallet and stack against bench	Overexertion	Slide armor plate off table to Rockwell bench as needed	Struck by sliding material
5	Armor plate remains here until worked on	Struck by sliding material	Send table back to component plant	
6	Remove pallet			
7	Lift armor plate, piece at a time, to Rockwell table	Overexertion		

Recommendation. Provide a welded tubular steel table 40 inches × 30 inches, the height of a workbench. Equip table with 6-inch casters and a wheel lock for two of the wheels. Make two so that one is always at component plant waiting to be loaded.

Note. Apparently the hazard in step 4, permitting injury to the fingers of the worker who is sliding the armor plate, still exists. However, this is overcome by using a ¾-inch × ¾-inch × 2-foot piece of wood as a wedge placed under the armor plate as it is slid from the pile to the workbench. The wedge is pulled out when the piece of armor plate is in position and is used in the same manner for the next piece.

Advantages of Methods Safety Analysis The advantages of such an investigation procedure as the methods safety analysis are many, some of which may be cited as follows:

1. It maps out all the details of an operation so that they can be studied and restudied.
2. It is quick, simple, factual, and objective, so that its results can be used easily and with confidence.
3. It permits a ready comparison of the undesirable method and the proposed new method, demonstrating the types of hazards and number of steps involved in each.
4. It presents a picture of the effect on production which a safety improvement will have.
5. It is an aid in showing management, when necessary, that a properly designed safety feature does not hinder productive output and, therefore, can assist in convincing management of the benefits to be gained by improving the safety of a method.
6. It permits the engineer to discern and improve ways for increasing the productivity of an operation.
7. It facilitates the analysis of the safety potential of a job before an accident has occurred.

8. It assists in the thorough investigation of those methods which have involved accidents, but in which the causes are obscure.

Other Whether Approaches

There are obviously any number of other whether approaches. It is our basic method of collecting information. We can only begin to make improvements, to select and apply remedies, when we have solid, factual information on what we are currently doing.

In the following chapter we'll look at two specific techniques for asking whether. First we'll discuss a concept of hazard recognition, a way of asking whether or not we are currently handling our physical problems. Then we'll examine the most complex and sophisticated whether we currently have in accident prevention work, system safety.

REFERENCES

Grimaldi, J., Paper at the ASME Standing Committee on Safety, Atlantic City, N.J., 1947.

Petersen, D., *Techniques of Safety Management*, McGraw-Hill, New York, 1971.

Weaver, D., "Symptoms of Operational Error," *Professional Safety*, October 1971.

Weaver, D., "TOR Analysis: A Diagnostic Training Tool," *Journal of the ASSE*, June 1973.

FIVE

System Safety

System safety technology is a well-established and formally recognized segment of modern system engineering. Much of its methodology was developed to help prevent accidents in United States government-sponsored or -controlled systems, such as missile, aircraft, space, and nuclear power systems. However, today, understanding of and application of system safety methods to industrial accident prevention is a moral, professional, and often a legal necessity.

The term "system safety" was chosen by the pioneers, in the early 1960s, because they were engaged in safety studies of complex systems. In order to perform these studies successfully, it was necessary to perform both "hazard identification analyses" and "hazard control evaluation analyses" in a very formal, step-by-step, or "systematic," manner. It is these systematic hazard identification and systematic hazard control evaluation features of system safety technology that can and often must be applied to industrial accident prevention. Consequently discussion of these features is the main substance of this chapter.

SYSTEMATIC HAZARD IDENTIFICATION

The Accident-Hazard Relationship

System safety terminology makes a very clear distinction between the terms "accident" and "hazard." An accident is an event in which damage to property or injury to personnel is occurring or has occurred. A hazard is any real or potential condition or act that could cause damage to property or injury to personnel, but has not yet done so. For example, the term "explosion" describes a type of accident, while a crack in a pressure vessel weld describes a type of hazard.

In spite of this clear, logical, and important distinction, the safety literature contains many lists of hazards that contain confusing mixtures of terms, some of which define types of accident and some of which define hazards.

Types of Accident

It is recommended that step 1 in systematic hazard identification be the preparation of a checklist of the types of accident that could occur for the product, equipment, system, or operations area that is being studied.

Note that Fig. 2–14 is divided into two subdivisions, namely, "energy transformation" and "energy deficiency" types of accident. This use of scientific language may initially offend or discourage the industrial safety practitioner who must, at least on the factory floor, use more earthy terms such as "unsafe conditions" and "unsafe acts." However, it is very important and helpful for all safety practitioners to realize that (1) energy is involved in all accidents and (2) study of energy factors is the key to success in systematically identifying and controlling all hazards, including a wide variety of obscure but potentially catastrophic hazards.

An energy transformation type of accident occurs whenever a stable or controlled form of energy is transformed in a way that damages property or injures people. For example, gasoline in an automobile tank is a normally stable form of chemical energy. When the engine is running, the chemical energy of the gasoline is transformed into the kinetic energy of the moving automobile in a controlled manner. By contrast, a rear-end collision can transform the chemical energy of the gasoline into the thermal energy of a catastrophic fire. Thus "fire" is an energy transformation type of accident.

An energy deficiency type of accident occurs whenever the energy needed to perform a vital function is not available and damage to property or injury to personnel results.

An energy deficiency may be either a direct cause or an indirect cause of an accident. For example, when a child is suffocated in an abandoned refrigerator, chemical energy deficiency in the form of lack of oxygen is the direct cause of death. When an airplane runs out of fuel, chemical energy deficiency causes loss of power, followed by a crash, and thus it is the indirect cause of death.

Energy Sources and Transformation Mechanisms

For an energy-transformation-type accident to occur, there must be both an *energy source* and an associated *energy transformation mechanism*. Both the sources and the mechanisms should be treated as hazards. Thus the next steps in systematic hazard identification should be (2) preparation of a list of all the potentially harmful energy sources and (3) for each such source, preparation of a list of potential energy transformation mechanisms. Preparation of a specific list for a particular product, equipment, system, or operating area can be aided greatly by using such a list.

Lists should be divided into three subdivisions, namely (1) high-energy environments, (2) high-energy components, and (3) low-energy and aesthetic.

High-energy natural environments, such as lightning and tornadoes, have threatened people since the dawn of history. The threats of man-made environments, such as the heat from a blast furnace or the noise from a forge hammer, have become targets for government control. Consequently, every safety practitioner is familiar with a number of potentially harmful environments. The reason for writing them down in checklist form is to help assure that none is overlooked during a systematic safety analysis.

High-energy sources that are "components" of a product, equipment, system, or operating area may be classified into groups in several ways. For example, mechanical, electrical, pneumatic, and hydraulic groups could be listed.

The low-energy and aesthetics subdivision is somewhat of a catchall for a wide variety of hazards. The physiological group covers accidents in which a low-energy source is combined with a specific human sensitivity. For example, poisoning involves small amounts of chemical energy combined with a sensitive physiological reaction. Aesthetic hazards are included for completeness, even though doing so pushes the energy concept a little far. However, even an accident in which a piece of furniture is scratched or coffee is spilled on clothing does involve a small expenditure of energy.

Energy transformation mechanisms are of two types. This duality was recognized in our earlier editions and is represented by the familiar pair of terms "unsafe conditions" and "unsafe acts."

Some energy transformation accidents are caused by a single mechanism in the form of one unsafe condition or one unsafe act. For example, oil spilled on a concrete floor is a single unsafe condition that can cause a fall that transforms the potential energy of an upright person into a broken hip. Striking a match to look for gasoline in a tank is a single unsafe act that can cause an explosion.

Unfortunately, many actual and potential accidents involve sequences or combinations of unsafe conditions and unsafe acts that result in quite complex energy transformation mechanisms. A major objective of system safety technology is first to predict such potential sequences and combinations and then to evaluate the means that are provided for preventing their occurrence. To achieve this objective, three major types of analysis have been developed. These analytical methods will be outlined in the section Complex Hazard Analyses.

Energy Needs and Deficiency Mechanisms

For an energy-deficiency-type accident to occur there must be both a need for some type of energy to perform a function that is essential for safety and an occurrence that makes the available energy insufficient to perform the safety critical function.

Safety critical functions may be subdivided into two types, namely, physiological and physical. For predicting and describing both types of accident, considerable imagination and latitude in the use of the words "accident" and "energy" are needed.

For example, if a person enters an enclosure where air has been displaced by nitrogen gas and dies, there is no question that a physiological energy-deficiency-type accident has occurred. If a person entered the enclosure and was rescued so quickly that no injury occurred, one person may say "there was an accident" while another says there was a "near miss."

In the manned space program, energy in the form of small rocket jets is needed to orientate the manned modules so that they will not be incinerated during the atmosphere reentry. Complete loss of this energy by leakage of the rocket fluid does not of itself cause "damage to property or injury to personnel." However, it places the astronauts in a position where they must choose between attempting reentry and being incinerated or staying in space until death occurs by starvation or anoxia.

The semantic question arises: After the rocket energy is lost but before damage or injury occurs, has an "accident" happened? To get around this semantic problem, system safety technology uses the term "undesired event." All accidents that produce damage or injury are undesired events. However, every loss of ability to perform a safety critical function, such as spacecraft attitude control, is defined as an undesired event, even in the absence of damage or injury. Thus, for example, if an automobile loses its brake fluid and thereby loses the safety critical braking function, an undesired event has occurred irrespective of whether damage or injury results. It is recommended that industrial safety practitioners adopt the system safety practice of including in their safety studies undesired events as well as damage- and injury-producing accidents.

In the previous section, it was mentioned that many actual and potential accidents involve sequences or combinations of unsafe conditions and unsafe acts. An energy deficiency undesirable event often creates an unsafe condition that in turn contributes to the probability of occurrence or the severity of a damage- or injury-producing accident. For example, if the energy needed to pump water into a fire-fighting system is deficient, then a controllable fire may become a catastrophic accident. It follows that system-safety-type studies of undesired events should be included in all industrial safety programs.

The When Aspects of Safety Analysis

So far we have discussed only *what* could happen to cause damage to property, injury to people, or, at least, an undesired event. Another important aspect of both traditional industrial safety studies and modern system safety technology is concerned with *when* an accident or undesired event could occur.

In industrial safety practice, when-type studies may be focused on single hazardous activities, such as welding, or on a step-by-step analysis of an operation such as crane handling or molten metal.

System safety is concerned with a product, system, or operation from its earliest concept until its final demise. Consequently the when aspects are concerned with (1) when the analysis is done and (2) when an accident or undesired event might take place. For example, MIL-STD-882, *System Safety Program for Systems & Associated Subsystems & Equipment*, includes an Appendix B, "System Life Cycle-Safety Activities." This appendix defines when specific types of safety analysis will be performed during each of five phases in the development, production, and use of military equipment. The system safety concern with when an accident would occur focuses first on broad functions such as during testing, manufacturing, transportation installation, operation, and decommission. However, it does eventually focus on specific hazardous activities, such as the landing of an airplane.

COMPLEX HAZARD ANALYSES

The principal purposes of system safety analysis are:

1. To identify hazards which singly or in combination could cause an accident or an undesired event
2. To evaluate the adequacy of hazard controls
3. To reconstruct accidents that have happened

In general, the system safety use of the term "hazard" covers the same physical situations and human actions as the industrial safety terms "unsafe condition" and "unsafe act." All three terms should be interpreted as covering potential as well as actual conditions and actions.

When analysis of what could happen or what has happened involves only a few hazards or unsafe conditions or acts, the analytical procedure can well consist of nothing more than commonsense thinking supported by safety checklists. As mentioned earlier, these checklists should include:

1. Types of accident checklists
2. Potentially harmful energy sources checklists
3. Energy transformation mechanisms checklists
4. Safety critical functions (energy needs) checklists
5. Energy deficiency mechanism checklists

Items 3 and 5 are expressed in scientific terms that may not be familiar to the practical industrial safety engineer. However, most of the checklist items will be recognized immediately as familiar unsafe conditions or unsafe acts.

When analysis of what could happen or what has happened involves complex sequences or combinations of hazards or unsafe conditions or acts, formal, documented analytical procedures are necessary. Attempts to predict and then prevent such complex accidents are somewhat like the proverbial search for a needle in a haystack. A random search may, by luck, find the needle or identify the way in which hazards could combine to cause an accident, but only a methodical search or analysis will give high confidence in finding the needle or in recognizing the potential accident.

The system safety literature includes a bewildering variety of names for formal hazard identification and hazard control evaluation analysis methods. This variety is due in part to the use of different names for the same type of thinking depending on when it is performed during a development program, in part to the use of more different names depending on the broad function that is being studied, and in part to the newness and consequent lack of standardization of system safety terminology.

For existing products, equipment, systems, or operations, it is natural to perform (A) hazard identification analysis and (B) hazard control evaluation analysis simultaneously. However, it should be recognized clearly that these are two very different analytical processes.

In developing a new product, equipment, system, or operation it is a basic management principle to require clear separation of (1) hazard identification from (2) hazard control evaluation. The primary reason for doing so is that after hazards have been identified, product designers and operations planners must be given ample opportunity to provide controls over both (1) the probability of occurrence and (2) the severity of the effects of each potential accident. Consequently, the first step in applying complex hazard analysis methods to industrial accident prevention is to establish the two analysis categories:

A Hazard identification analysis methods
B Hazard control evaluation analysis methods

Hazard Identification Analysis Methods

There are three basic thought processes by which hazards may be identified:

A1 Deductive, based on a list of undesired events
A2 Inductive based on physical part failure modes
A3 Inductive based on human failure modes

No consensus exists on the names for these three thought processes or analytical methods, but the following titles are logical.

1 Accident–undesired event cause analysis
2 Physical failure modes and safety effects analysis
3 Human failure modes and safety effects analysis

In practice, combinations of deductive and inductive thought processes may occur. They do so when an analyst focuses first on component failures and then thinks deductively about what could cause them and inductively about their effect on the safety of a system. Also, combinations of methods A2 and A3 occur when an analyst thinks simultaneously about the effects of both physical part failures and human errors.

Method A1 may be performed with or without graphical representation of the system that is being studied and of the potential hazard sequences and combinations. When graphical representation is used, the method is known as "fault tree analysis." Figure 5-1 illustrates the basic features.

Responsibility for Hazard Identification and Control

Good management practices require identification of five distinct functions:

1. Safety engineering
2. Design engineering
3. Operations planning
4. Specialist consultation
5. Operations supervision

Systematic hazard identification and listing of identified hazards are a responsibility of the safety engineering function. Building hazard controls into designs and operating procedures is the responsibility of the design engineering and operations planning functions. Evaluation of the adequacy of these controls is a responsibility of safety engineering supported by specialist consultation. Preserving the effectiveness of built-in hazard controls by strick adherence to training, operating, and

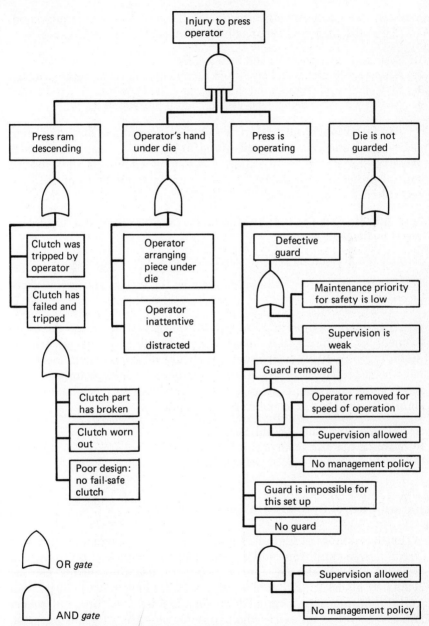

Figure 5-1 Fault tree analysis. (Petersen, *Techniques of Safety Management*, McGraw-Hill, New York, 1978)

maintenance procedures is a responsibility of operations supervision. Thus the principal responsibilities of safety engineering are:

1. Systematic identification and listing of hazards
2. Assistance to design and operations personnel in devising hazard controls
3. Systematic evaluation of the adequacy of hazard controls
4. Auditing of compliance with hazard control procedures

An important contribution by safety engineering to helping design engineering and operations planning to establish hazard controls is teaching the system safety principle that is known as the "hazard control order of precedence." This principle states that control over each identified hazard should be provided by:

First, minimizing the probability of occurrence and the potential severity of the effects by the basic design

Second, adding warning and protection devices to the basic design

Third, writing hazard controls into training, operating, and maintenance procedures

The safety engineering function should greatly increase the effectiveness of teaching the hazard control order of precedence by providing supporting hazard control checklists. Many such checklists have been prepared. However, their usefulness could be improved by much more systematic arrangement. As a start, all available items should be arranged into the following segments:

A Design for minimum hazard checklists
B Warning and protection devices checklists
C Hazard control by procedures checklists

Hazard Control Evaluation Analyses

When the number of identified hazards is small, or when the ways in which unsafe conditions and unsafe acts could combine to produce an accident are all simple, the adequacy of hazard controls can be evaluated by assuring that all checklist items have been incorporated into the design or into procedural documents. The "order of precedence" should be followed when applying checklists for this purpose.

For more complex situations, it is necessary to repeat the same three types of analysis that were used for systematic hazard identification. For brevity, the analyses may be called "fault trees" or, even more simply, "safety analyses."

There is an important difference between the original and the repeat performances of safety analyses. The original performance was done for the purpose of "guiding" design decisions and writing procedures by identifying unsafe conditions and unsafe acts which could combine to cause an accident. The repeat performance is done for the purpose of "evaluating" the degree to which the design decisions that have been made and the procedures that have been written are adequate for controlling the identified hazards.

In general, the "guide" analysis will have been done several times, first on the basis of design concepts and functional block diagrams, then on the basis of preliminary design sketches and equipment block diagrams, and then on the basis of tentative drawings and circuit diagrams which may or may not include warning and protection devices. By contrast, the evaluation analysis must be done on the basis of the completely finished and signed design drawings and associated procedural documents.

The guide analysis may include some rough quantitative estimates of the probability of occurrence of each type of potential accident. The evaluation analysis may be limited to answering the question, Have the designers and procedure writers done everything that should have been done to control hazards? Whenever it is feasible to do so, quantitative predictions of accident probabilities should be made. The fault tree technique provides for such quantitative predictions.

PRODUCT LIABILITY PREVENTION

For a long time, compensation for the victims of industrial accidents was limited to the money and medical treatment provided by the workers' compensation law. Recently, the law courts have reflected changing social attitudes by finding manufacturers of industrial equipment liable for any injury to a worker in which deficiency in the design or manufacturing of the equipment or in an associated manual contributed to the accident. Consequently, all system safety engineers employed by a manufacturer of industrial equipment should know what they can do to help prevent liability suits against their employer.

More specifically, system safety engineers should know what a prudent and safety-conscious equipment manufacturer does to assure adequate consideration of safety in these six areas:

A Product requirements specification
B Development program specification
C Product design
D Product conformance with design

E Product support documentation
F Response to user experience

Area A includes recognition and documentation of items, such as (1) customer requirements, (2) environment of use, (3) predictable misuses, and (4) known problems with similar equipment.

Area B includes requiring, funding, performing, and auditing of (1) analyses, such as fault tree, (2) safety development tests, and (3) design reviews.

Area C covers the actual design of the equipment, including (1) compliance with standards, (2) compliance with safety checklists, (3) adequate safety margins, and (4) control of hazards identified by analysis or test.

Area D corresponds with an adequate quality control program, including (1) inspection reports, (2) test reports, and (3) problem disposition reports.

Area E covers (1) owners or operating manuals, (2) training manuals, and (3) maintenance manuals.

Area F covers adequate (1) surveillance over performance of the equipment, (2) feedback of reports of field problems to the manufacturer's home plant, (3) action on recognized problems.

In 1976, it was reported that over half a million product liability suits were pending in the courts of the United States. If the average award was $100,000, this would mean a loss to equipment manufacturers of $50 billion. Obviously the time has come when prudent industrial managers must insist on excellence in the system safety segment of their total industrial accident prevention programs.

REFERENCES

Brown, D., "Systems Engineering in the Design of a Safety System," *Journal of the ASSE*, February 1973.

Hammer, W., *Handbook of System and Product Safety*, Prentice-Hall, Englewood Cliffs, N.J., 1972.

Johnson, W., *The Management Oversight and Risk Tree—MORT*, U.S. Atomic Energy Commission, Washington, 1973.

MacCollum, D., "Reliability as a Quantitative Safety Factor," *Journal of the ASSE*, May 1969.

Requirements for System Safety Program for Systems and Associated Subsystems and Equipment, Military Standard 882, July 1969.

Rodgers, W., *Introduction to System Safety Engineering*, Wiley, New York, 1971.

SIX

Accident Analysis

W hen an accident and its resultant injury have occurred, the event becomes a matter of history. Singly or in groups, accident records provide a fertile field for analysis. If the work of investigation and record keeping has been done well, the facts of each past accident will already have been recorded in such manner that analysis covering a considerable period of wide exposure will require merely that facts already available be *assembled* and that conclusions be drawn.

VALUE OF FACTS IN SELECTION OF REMEDY

The creation and maintenance of executive and employee interest and of the will to achieve, the adoption of proper methods of analysis for determination of cause, and the development of causal facts would all be of little avail in accident prevention if the correct remedy were not selected, or if when selected it were not effectively applied. It is necessary, therefore, when the pertinent facts of accident occurrence have been determined and recorded, to *draw conclusions* as to conditions and circumstances that are in need of first attention and as to practical methods whereby corrective action can be taken. Cause analysis will permit:

Grouping Without Regard to Cause

1 Group accidents by:
   a Plant
   b Department or operation
   c Name of supervisor
   d Type of accident
   e Nature and severity of injury
   f Day of month and time of day
   g Agency and part of agency
   h Age, sex, experience, and length of employment of worker
   i Miscellaneous identifying and corollary facts

Grouping for Causal Facts

2 Group accidents by:
   a Unsafe act
   b Reason for unsafe act
   c Mechanical or physical hazard
   d Underlying causes (managerial and supervisory)

Selection of What and Where to Attack

3 Select place, operation, or circumstance according to:
   a Highest frequency and severity
   b Highest frequency alone
   c Highest severity alone
   d Operation or procedure most in need of improvement
   e Predominating unsafe acts
   f Predominating hazardous agencies
   g Predominating accident types and injury types

In addition, cause analysis is a necessary prerequisite to the selection of an effective remedy.

Further, particularly when considering large groups of accidents wherein there is little similarity of type, it is highly important to record the causes before selecting a cure, since the cause and the remedy may be alike for a number of accidents which, from the point of view of type alone, are dissimilar and seemingly difficult to prevent.

A case in point is that of a cotton-spinning-and-weaving plant where a high accident frequency presented a troublesome condition. Workers received slivers in their fingers and hands from handling spools; other employees slipped on floors; still others were hurt on carding machines and pickers or while carrying sheath knives. There was no general run or trend of accident types. Even when several accidents of apparently the same kind did occur, investigation showed that the manner of occurrence varied. Cause analysis provided immediate relief. It indicated that the same fault was an underlying factor in over 75 percent of all the accidents, notwithstanding that these accidents were of different types.

Sliver injuries occurred, for example, because of the unsafe practice "failure to inspect, remove, and replace defective spools." Likewise, there were slips and falls because instructions with regard to limiting the area of floor to be wet down at any one time were not followed. In the carding room, employees, contrary to rules, made the card-cover locks inoperative; and unsafe practices occurred also in the operation of pickers and from the use of knives.

In this example the unsafe acts were so varied in nature as to make the job of correction an apparently difficult one. However, when the *cause* was found, the remedy immediately became clear. It was determined by investigation that in each case the employees violated safe-practice rules *because they had not been properly instructed and were not fully aware of safe procedure.*

Thus, cause analysis uncovered a predominant fault in supervision and an effective and definite place and method of attack.

Situations of the kind just described are the ones that present a most mystifying appearance to safety engineers and that prove most difficult to solve—until they try true cause analysis.

In a rubber-tire-manufacturing plant, notwithstanding a frequency of several thousand accidents annually, there seemed to be no target type of accident or of injury. One employee would be hurt while handling a pipe wrench, another while using a pair of shears. In other departments one worker would collide with a fixed object, whereas another would be struck by a moving conveyor. Other employees slipped and fell, or were struck by flying objects, or were hurt by falling material. A bewildering maze existed, and the selection of a practicable remedy was difficult until, here again, analysis and the recording of true causes pointed clearly to congestion of working space (direct mechanical or physical cause) as the cause of more than 60 percent of all accidents.

It is necessary only to emphasize the fact, therefore, that selection of a remedy without cause analysis is difficult and can be attained only by guess or instinct, both of which may be unreliable.

SELECTING PREDOMINANT CAUSES FOR ATTACK

There is no need to point out that the most important faults should be attacked first, but it may be in order to substantiate the bare assertion. No effort that is spread over too great an area can be effective. For this reason, predominant causes—those that result in a high frequency of accidents, coupled with actual or probable high severity—should receive first attention.

It is seldom, if ever, advantageous to attempt the prevention of all types of accident from all causes at one and the same time. Even in a small plant or in one that has a low accident frequency or, in rare cases, in a plant where true cause analysis reveals that no one cause predominates and where all causes have equal significance, it is inadvisable to attempt the immediate correction of more than a few causes at one time. In this respect, accident prevention again may be compared with other phases of business activity. When the red "rush" tag for important correspondence, for example, is clipped on all correspondence, it immediately loses its value.

SIMILARITY OF CAUSES IN DIFFERENT INDUSTRIES

A New England granite quarry, having employed the cause analysis method, found that employees were *not convinced of the need* of following safe-practice rules and that this resulted in the following specific unsafe practices:

Employees stood under suspended loads.
Others failed to get out of the zone of danger when blocks of stone were being hauled to the cars and hoists.
Workers were in the habit of spilling rock, with little regard for the safety of other workers who might be in line with flying pieces.
Debris was not kept cleared back from the outer edges of the working levels.
Frayed cable was used for guys and hoist cables.

Lack of conviction as to safe performance is one of the most important of the causes of accidents in all industrial operations as shown by the following cases.

A Pennsylvania Coal Mine

Jumping on and off moving trips
Riding front end of cars instead of in empties
Standing between bumpers while coupling cars
Failing to sprag standing cars properly
Pushing cars by hand with hands on top of coal or corners where clearance was small

A Bakeshop

Cutting dough out of mixers without waiting for revolving paddles to slow down
Cleaning dough mixer while paddles were revolving
Lifting sacks and barrels of flour with back in a bent position
Cleaning rolls of dough brakes from the inrunning side
Piling barrels and boxes too high

Erection of a Skyscraper

Riding material hoists
Riding crane loads
Laying tools on girders
Walking across narrow partially suspended beams
Throwing boards and other debris on walks and stairways
Removing protective barriers from hoistways

Erection of a Small Dwelling

Throwing concrete form boards with protruding nails in working spaces
Failure to clean up premises
Use of ladders that are too short
Failure to shore trenches
Standing on insecure working platforms

A Steel Rolling Mill

Failure to stand clear as crane loads are picked up
Crane workers climbing up columns instead of using permanent ladders
Crane workers walking on girders instead of on catwalk
Improper balancing of loads
Employees walking across tables containing rolls, in front of mills, instead of using runways
People working in pairs while chipping billets, working face to face
Climbing over and between railroad cars—jumping on and off rapidly moving trains

The above examples are cited to show that although the extent and the nature of industrial operations must affect the type, severity, and frequency of accidents, the problems of prevention may be solved by applying the same scientific principles in each case. In fact, the same causes of accidents may apply in many lines of work. In each of the foregoing cases lack of complete appreciation of the real need of following safe-practice rules was a common factor.

The student of accident prevention may ask: Why select the cause "lack of conviction" in the foregoing examples; why not "recklessness" or "lack of skill"?

In each of the cases referred to, the supervisor had repeatedly issued instructions but was mistaken in believing that he had explained them clearly. The employees did not fully appreciate the dangers and were not *sufficiently impressed* with the necessity for following the instructions for safe procedure. Instead of having the supervisor issue more instructions only to have them disregarded, instead of taking the long-way-round path that rests on the assumption that the employees are reckless, it was found highly and immediately effective to select a remedy based on the knowledge that disregard of instruction was chiefly the result of misunderstanding and lack of conviction. Such a remedy would take the form of more explicit instruction coupled with explanation and "reasons why."

Under other circumstances these same practices might well be charged to "unaware of safe practice," as, for example, when the supervisor failed altogether to instruct untrained employees in the safe methods of doing their work.

Incorrect analysis would have failed in all the instances cited. It would have led to no such clarity of conditions and would have produced no appreciable success. Referring to the steel mill again, prior to true cause analysis and with regard to the accidents that occurred when crane loads were being lifted, it was known only that employees were injured by falling objects, that they were hurt while engaged in moving or hoisting or handling material, or that a crane load slipped, etc. This directed attention merely to better methods of securing loads, warnings to crane workers, inspection of slings and chains, and other activities that were undoubtedly good procedure but did not correct the real condition that was chiefly at fault, namely, that employees did not appreciate the need or importance of standing clear of the load when the crane took hold because they were not convinced that it really was necessary.

These unsafe practices were brought to light only through true cause analysis. They remained uncorrected when ordinary accident-prevention methods only were used, notwithstanding their obviousness and the fact that they were improper. They were not the only unsafe practices in the plant. Every single operation—literally thousands of varied kinds of work—was subject to criticism from the point of view of safety. Without analysis for *predominating* cause, therefore, the accident-prevention engineer had no measure of the importance of any one practice and was obliged to attempt the correction of all unsafe practices, with the result that her efforts covered too great an area and were so thinly spread as to be ineffective. As an alternative, she might by guesswork or judgment have selected certain causes for attention only to find later that she had spent time on matters that had little bearing upon the occurrence of accidents in her particular plant.

Most large organizations have begun to utilize the computer in their injury record keeping. Most often companies utilize an approach to data collection which is based primarily on the American National Standards Institute's standard (ANSI Z16.2). The purpose of this standard is to provide a method of recording the essential facts relating to injuries experienced in the course of employment and to the accidents which produced those injuries, for use in accident prevention. The standard provides for a complete recording for any one injury case consisting of the identification of one item in each of the following categories to describe the accident:

1 Nature of injury
2 Part of body affected
3 Source of injury
4 Accident type
5 Hazardous condition
6 Agency of accident
7 Agency of accident part
8 Unsafe act

One of the problems with the utilization of the standard is the fact that it is very much based upon our old domino theory of accident causation and thus deals primarily at the symptomatic level. Computer programs based on this standard then tend to be summaries and analyses of large numbers of symptoms. Following are examples of how the standard is utilized.

1 A circular-saw operator reached over the running saw to pick up a piece of scrap. His hand touched the blade, which was not covered, and his thumb was severely lacerated.

ANALYSIS

Question	Analysis Category	Answer	Code
a	Nature of injury	Laceration	170
b	Part of body	Thumb	340
c	Source of injury	Circular saw	3750
d	Accident type	Struck against	012
e	Hazardous condition	Unguarded	510
f	Agency of accident	Circular saw	3750
g	Agency of accident part	Blade	3099
h	Unsafe act	Cleaning, adjusting moving machine	052

2 A forklift truck went out of control when one wheel struck a piece of stock lumber which projected into the aisle. It ran out of the aisle and struck a machine operator, breaking his leg between the ankle and knee.

ANALYSIS

Question	Analysis Category	Answer	Code
a	Nature of injury	Fracture	210
b	Part of body	Lower leg	515
c	Source of injury	Forklift truck	5635
d	Accident type	Struck by	029
e	Hazardous condition	Improperly placed lumber	420
f	Agency of accident	Lumber	5700
g	Agency of accident part	None	9999
h	Unsafe act	Unsafe placement of material	657

3 A warehouse employee jumped from the loading platform to the ground instead of using the steps. As he landed, he sprained his ankle.

ANALYSIS

Question	Analysis Category	Answer	Code
a	Nature of injury	Sprain	310
b	Part of body	Ankle	520
c	Source of injury	Ground	5810
d	Accident type	Fall from elevation	031
e	Hazardous condition	None indicated	990
f	Agency of accident	None indicated	9800
g	Agency of accident part	None indicated	9800
h	Unsafe act	Jumping from elevation	503

4 A laborer working in a trench was suffocated under a mass of earth when the unshored wall of the trench caved in.

ANALYSIS

Question	Analysis Category	Answer	Code
a	Nature of injury	Asphyxia	110
b	Part of body	Respiratory system	850
c	Source of injury	Earth	4300
d	Accident type	Caught under	064
e	Hazardous condition	Lack of shoring	530
f	Agency of accident	Trench	1630
g	Agency of accident part	None	9999
h	Unsafe act	Not indicated	999

5 A salesman stopped his car at an intersection and was waiting for the traffic light to change from red to green. Another car struck his car in the rear. The whip-lash effect fractured a vertebra in the salesman's neck.

ANALYSIS

Question	Analysis Category	Answer	Code
a	Nature of injury	Fracture	210
b	Part of body	Neck	200
c	Source of injury	Auto	5620
d	Accident type	Collision	323
e	Hazardous condition	Traffic hazard	720
f	Agency of accident	Environment	5900
g	Agency of accident part	None	9999
h	Unsafe act	None	998

6 A longshoreman stowing boxed cargo pinched his finger between two boxes.

ANALYSIS

Question	Analysis Category	Answer	Code
a	Nature of injury	Contusion	160
b	Part of body	Finger	340
c	Source of injury	Boxes	0630
d	Accident type	Caught between	069
e	Hazardous condition	Not indicated	990
f	Agency of accident	Not indicated	9800
g	Agency of accident part	Not indicated	9800
h	Unsafe act	Not indicated	999

7 A deckhand on a river towboat fell overboard and drowned while on duty. There were no witnesses and no indications as to why he went overboard.

ANALYSIS

Question	Analysis Category	Answer	Code
a	Nature of injury	Drowning	110
b	Part of body	Respiratory system	850
c	Source of injury	Water	2910
d	Accident type	Inhalation	181
e	Hazardous condition	Not indicated	990
f	Agency of accident	Not indicated	9800
g	Agency of accident part	Not indicated	9800
h	Unsafe act	Not indicated	999

8 A paving worker collapsed with heat exhaustion while placing hot asphalt in full sun on a very hot day.

ANALYSIS

Question	Analysis Category	Answer	Code
a	Nature of injury	Heat exhaustion	240
b	Part of body	Circulatory system	801
c	Source of injury	Atmospheric heat	2400
d	Accident type	Contact with temperature	151
e	Hazardous condition	Environmental hazards	299
f	Agency of accident	Environment	5900
g	Agency of accident part	None	9999
h	Unsafe act	None indicated	999

As indicated before, there are some inherent weaknesses in the Z16.2 approach. Analyses based on the standard are based on getting a detailed description of the circumstances surrounding the incident, rather than information on why it happened. Thus, utilizing the Z16.2 approach, many computerized analysis programs end up giving up a great deal of easily accessible, but not too usable, information, and very little insight into why accidents are occurring.

An example of a printout utilizing the Z16.2 approach is shown in Fig. 6–1. As a result of our Z16.2 type of input, the computer has provided an analysis showing a frequency of nine accidents of the category called "caught in or between object handled and other object" with a cost of $2,023. It is our major category of accident. The causes of these nine accidents? We have no idea. Where did they occur? We do not know. What was handled? We don't know. What should we do about them? We have no idea. This analysis does not lead us far in our future control efforts. We simply cannot select remedies from the information available.

Thus, while computers are extremely valuable to us in assembling and analyzing data and information, the value obtained is entirely dependent on what we put in and how we put it in.

While most companies and insurance carriers are able to provide computer printouts of accident information, too often the output is similar to Fig. 6–1, and thus difficult to use. Worse yet, the carriers often provide loss runs as shown in Fig. 6–2, which is merely a chronological listing of claims and costs. This information is almost unusable in assessing what the problem might be.

The computer is capable of summarizing and analyzing in a variety of ways. Examples are shown in Fig. 6–3, where the accidents have

CAUSE CODE	CAUSE OF ACCIDENT	FREQUENCY NEW	FREQUENCY YTD	CLAIM COST NEW CLAIMS	CLAIM COST YR. TO DATE	PREMIUM	LOSS RATIO
	10 WORKMEN'S COMPENSATION						
10 01	EMPL STRUCK-INJURED BY HAND TOOL OR MACHINE IN USE	1	1		46		
10 02	EMPL STRUCK-INJURED BY FALLING OR FLYING OBJECT		11		2,871		
10 03	EMPL STRUCK BY TIPPING SLIDING OR ROLLING OBJECT		8		11,726		
10 09	EMPL STRUCK-INJURED BY MISCELLANEOUS-UNCLASSIFIED		1		459		
10 11	EMPL STRAIN-INJURY LIFTING		2		202		
10 13	EMPL STRAIN-INJURY PUSHING OR PULLING		3		467		
10 17	EMPL CUT-SCRAPE BY . HAND TOOL-UTENSIL-NOT POWERED	1	1	15	15		
10 19	EMPL CUT-SCRAPE BY OBJECT BEING LIFTED-HANDLED		2		32		
10 21	EMPL FELL OR SLIPPED ON SAME LEVEL		5		1,903		
10 22	EMPL FELL OR SLIPPED FROM DIFFERENT LEVEL	1	8	2,831	22,217		
10 23	EMPL FELL OR SLIPPED SLIPPED-DID NOT FALL		1		42		
10 31	STRIKING AGAINST OR STEP ON OBJECT BEING HANDLED		3		65		
10 32	STRIKING AGAINST OR STEP ON STEPPING ON SHARP OBJ				83		
10 42	CAUGHT IN OR BETWEEN MECHANICAL APPARATUS		3		513		
10 43	CAUGHT IN OR BETWEEN OBJECT HANDLED-OTHER OBJECT		9		2,023		
10 70	ELEC SHOCK OR BURN ELECTRICAL - ON POLE		1		5,081		
16 80	EMPL INJURY FOREIGN BODY IN EYE		8		602		
10 91	EMPL INJURY VEHICLE ACCIDENT - MISCELLANEOUS				42,121		
	TOTAL	3	17	2,846	90,518	152,291	59.4%
	OPEN		7		42,450		
	20 GENERAL LIABILITY						
20 37	GEN LIABILITY FIRE OR EXPLOSION		1		325		
20 50	GEN LIABILITY MISCELLANEOUS - UNCLASSIFIED	1	2	15,000	15,000		
	TOTAL	1		15,000	15,325	3,165	484.2%
	OPEN		1		15,000		
	21 EXCESS LIABILITY						
21 09	EXCESS LIABILITY FROM AUTO FLEET	1	2	49,671	99,342		
21 10	EXCESS LIABILITY FROM OUR OPERATIONS	1			160,000		
	TOTAL		3	49,671	259,342	65,481	396.1%
	OPEN		1		160,000		

00512 COMPANY SUMMARY   CAUSE ANALYSIS   PAGE 1

CONTROL DATE 01-01-   ANALYSIS DATE 07-10-

Figure 6-1. Computer printout.

Accident-Prevention Method

	WORKMEN'S COMPENSATION ACCIDENT/LOSS RECORD						
	ACCIDENT/LOSS DATE	CLAIM NUMBER	CLASS CODE	CLAIMANT	CAUSE	NATURE	RESERVE OR CLOSING COST
1	10/9/	68 01 1	3400	MULLIGAN L.	BLADE SLIPPED	LAC L ARM	695
2	10/10/	68 01 2	3400	EDWARDS J.	METAL	FB EYE	27
3	10/15/	68 01 3	3400	JONES M.	METAL CUTTINGS	LAC FGR	19
4	10/29/	68 01 4	3400	GOMEZ P.	GRINDER	LACR CORNEA	2,310 R
5	10/30/	68 01 5	3400	SILVA M.	FELL	STRAIN BACK	2,700 R
6	10/30/	68 01 6	3400	JOHNSON E.F.	PUNCH PRESS	LAC FINGER	179
7	11/12/	68 01 7	3400	HERMAN G.	SLIPPED	TWIST ANKLE	46
8	11/6/	68 01 8	3400	SMITH R.E.	LIFT TRUCK	CONT HAND	350 R
9	11/13/	68 01 9	8810	SPARKS J.A.	SHELF EDGE	INFECT ARM	19
10	11/14/	68 01 10	3400	WERNER C.D.	LIFT SACK	STR. ARM	11
11	11/20/	68 01 11	3400	MILLER E.	DRILL	CUT THUMB	28
12	11/26/	68 01 12	3400	PICKETT E.H.	SAW	LAC FGR	±
13	11/21/	68 01 13	3400	HANSEN H.	HAND DRILL	LAC THUMB	44
14	12/11/	68 01 14	3400	FRANKLIN B.	BENT DOWN	STR BACK	32
15	12/14/	68 01 15	3400	CHAVEZ P.	PLASTIC CARTON	ABR FGR	20
16	12/18/	68 01 16	3400	DUNNE T.F.	SAW BLADE	INF CUT HAND	109
17	11/23/	68 01 17	3400	HERMAN G.	UNPACKING	LAC FGR	36
18	12/15/	68 01 18	3400	CUNNINGHAM P.	CUTTING BLADE	LAC FGR	76
19	12/26/	68 01 19	8215	CASSIDY H.	SPIDER	BITE ARM	±
20	12/26/	68 01 20	3400	EMERY S.A.	GRINDER	FB EYE	±
21	1/8/	69 01 21	3400	MCNAMARA S.	SLIPPED	TWISTED KNEE	±
22	1/12/	69 01 22	3400	PETERS A. J.	LATHE	LAC THUMB	±
23	12/13/	69 01 23	3400	HAMILTON C.	LIFTING	HERNIA	1,500 R
24	1/12/	69 01 24	8810	SPARKS J.A.	LIFTING	MUSCLE STR BAC	11
25	1/22/	69 01 25	3400	SPICER G.C.	HANDLING STOCK	POSS DISC BACK	3,725 R
26	1/30/	69 01 26	3400	BROWN M.E.	SAW SLIPPED	LACR THUMB	13
27	10/5/	68 41 1	3400	GREEN D. H.	PUNCH PRESS	LAC HAND	±
28	10/9/	68 41 2	3400	ALVARADO J.	CAR FENDER	LAC ARM	±
29	10/4/	68 41 3	3400	DENNIS D.	TRIPPED	FRACT FOOT	1,775 R
30	10/15/	68 41 4	3400	PARSONS J.	THREAD MACH	LAC THUMB	±
31	10/18/	68 41 5	3400	STANLEY B.	TIN	LAC L FGR	35
32	10/31/	68 41 6	8742	MAXWELL S.	AUTO ACC	STRAIN NECK	1,500 R
33	11/26/	68 41 7	3400	COOPER F.	PUNCH PRESS	LAC L THIGH	15
34	12/21/	68 41 8	3400	BUTLER K.	METAL BEAM	LAC L EYEBROW	62
35	1/12/	69 41 9	3400	STANDERS A. C.	UNLOADING STEEL	INJ BACK	27
36	2/21/	69 41 10	3400	FOX L.B.	SAW	LAC L FGR	±
37							
38							
39					# OF	CLAIMS	36	15,364
40							

Figure 6-2. Computer loss run.

been analyzed by age group and by part of body. Any other type of analysis is also possible provided the program has been written.

Computer programs to assist the safety professional do not have to be based on the Z16.2 standard. There are programs that have been devised and programs that are in use with different formats.

Here is a case history of a successful computer-based safety management information system (SMIS) that was born in the U.S. Department of the Interior as a specialized means for identifying organizational errors and managerial weaknesses through accident analysis. Bill Pope, former chief, Division of Safety Management of Interior, describes the system.

Thanks to the computer, it has grown to maturity and Interior's safety

SUMMARY OF ANALYSIS BY EMPLOYEE AGE

	ACTUAL VALUES					PROPORTIONAL	COMPUTED VALUES BASED ON DISTRIBUTION OF UNKNOWN		
	NO. CLAIMS	%TOTAL CLAIMS	VALUE	%TOTAL VALUE	AVERAGE VALUE	NO. CLAIMS	%TOTAL CLAIMS	VALUE	%TOTAL VALUE	AVERAGE VALUE
UNKNOWN EMPLOYEE AGE	711	12.27	$137,111	9.53	$192	689	11.89	$65,805	4.57	$95
AGES 0-20	604	10.43	$59,535	4.14	$98	1,068	18.44	$120,659	8.38	$112
AGES 21-25	937	16.17	$109,163	7.59	$116	797	13.76	$117,174	8.14	$147
AGES 26-30	699	12.07	$106,019	7.37	$151	1,213	20.94	$397,723	27.64	$327
AGES 31-40	1,064	18.37	$359,829	25.00	$338	1,133	19.56	$268,858	18.68	$237
AGES 41-50	994	17.16	$243,242	16.90	$244	894	15.42	$468,823	32.58	$524
AGES OVER 50	784	13.53	$424,154	29.47	$541					
TOTALS	5,793	100.00	$1,439,053	100.00	$248	5,794	100.00	$1,439,052	100.00	$248

SUMMARY OF ANALYSIS BY INJURED PART OF BODY

	NO. CLAIMS	%TOTAL CLAIMS	VALUE	%TOTAL VALUE	AVERAGE VALUE	NO. CLAIMS	%TOTAL CLAIMS	VALUE	%TOTAL VALUE	AVERAGE VALUE
UNKNOWN INJURED PART OF BODY	399	6.89	$203,407	14.13	$509	344	5.94	$53,357	3.71	$155
HEAD INJURY	320	5.52	$45,815	3.18	$143	867	14.97	$36,674	2.55	$42
EYE INJURY	807	13.93	$31,490	2.19	$39	1,393	24.05	$778,494	54.10	$558
BODY/BACK INJURY	1,297	22.39	$668,865	46.45	$515	350	6.04	$59,238	4.12	$169
ARM INJURY	326	5.63	$50,865	3.53	$156	165	2.85	$65,994	4.59	$399
WRIST INJURY	154	2.66	$56,666	3.94	$367	1,684	29.07	$287,916	20.01	$170
HAND/FINGER INJURY	1,568	27.07	$247,220	17.18	$157	189	3.26	$53,369	3.71	$282
KNEE INJURY	176	3.04	$45,826	3.18	$260	347	5.99	$52,770	3.67	$152
LEG/ANKLE INJURY	323	5.58	$45,311	3.15	$140	454	7.84	$51,239	3.56	$112
FEET/TOES INJURY	423	7.30	$43,997	3.06	$104					
TOTALS	5,793	100.00	$1,439,053	100.00	$248	5,793	100.00	$1,439,051	100.00	$248

Figure 6-3. Computer analysis.

officials now are far more articulate on the cost/effectiveness of the work being done—and their advice is being heeded increasingly by fellow managers.

Interior's basic accident report (source document) was revised to expose human errors and condition defects causing operating loss being recorded as "accidental." Reasons for mishaps were expanded in this report to include breakdowns in the way management "manages"—adding, for the first time, a new dimension to accident cause analysis.

A new definition was given to "industrial accident." An "accident" is now perceived as a convenient, cover-up explanation for a cluster of operational errors. Furthermore, errors (or accidents) are merely symptoms of problems created by ineffective management.

Moving forward on this premise, operating mistakes simply reveal improvements needed in the framework that makes up the management system. This system includes personnel, property, finance, legal, sales, distribution, research, engineering, and related staff services.

A new philosophy about causes of accidents was conceived. Faultfinding with individuals—employees or managers—added little or nothing to the improvement of the **system** of management.

Interior's first problem was to consolidate all existing reporting documents—and there were many—into a single form specifically designed for computer use. This is good paper-work management and cuts cost, both in reproduction and in storage handling.

The accident report form that finally emerged, nurtured carefully by various associated specialists, has stood the test well.

Interior's accident report form (see Fig. 6-4) occupies both sides of a government standard 8″ × 10.5″ paper. It is easy to follow and easy to fill out. Though quite simple, it is comprehensive enough to cover the important details about any kind of accidental loss. In itself, this simplification of what appears to be a very complicated task, may pose a problem to many safety people. The key to this construction is in the separation of "need-to-know" facts from the "nice-to-know" details.

The next step . . . was to develop an accident analysis code dictionary for translating language into numbers. This is the "key" to the reporting form. It lists all causes, costs, and related data translating the written word into machine language. This is essential, of course, to reduce data storage requirements. Letters and numbers make programming easier. But—and this is important—the computer can reverse itself and translate these numbers and letters back into words and phrases. If this isn't done, the numbers presented to management would still be a mystery and probably as unacceptable as when there was no system at all.

Interior's printouts—language provided by the computer when a question is asked —can be read easily, for they are combinations of narration and numbers. The official studying a printout does not require a code dictionary to understand what has been placed before him for analysis. [The] example [in Fig. 6-5] is typical of what to expect.

In computer-based systems, retrieval of accident-cause information, in specialized format, is rapid and inexpensive. For about $15 worth of computer time (a matter of seconds) Interior's equipment (IBM 360/65 at this writing) can tell managers, for example, how many employees of any given organizational unit (field station) of a bureau were injured on any given day, in any state, at any specific time, and under what conditions. It will reveal, also, the employment status, the work environment, the condition of the tool or equipment used, the training problem, a man's fitness for duty, and hundreds of other aspects management might wish to review: Injury costs, repairs to damaged equipment, and any tort claim expenses that may have been involved.

All this for about $15 . . . available on short notice from a magnetic tape containing details of more than 30,000 accidents. Thousands of new reports are added each year, and as the file grows bigger it becomes even more valuable. Interior's accident experience over four years fills one-half reel of magnetic tape.

The speed of retrieving data is remarkable. The computer "reads" the 8¼ million characters of data in this file in a minute and a half, spends a few more seconds "processing" the data, then prints the results at 1,000 lines per minute.

Here are a few examples, briefly stated, to show the cost/effectiveness decisions made possible by a safety management information system:

Case 1

Reviewing a cluster matrix of several hundred sources of accidents, the computer printout revealed that 168 employees were hurt over a 26-month period while using a machete. The medical and compensation payments totaled $17,000. No property damage costs were involved. A closer examination showed that these were "emergency" employees injured between August and October each year while fighting forest fires. So, while the computer study period was for 26 months, these injuries occurred in a span of less than eight months. Other findings: The problem was concentrated in two organizational units; the fires generally occurred in Alaska; about 47% of the injuries involved the thigh, leg, and knee; 25% occurred to the toes, hands, and fingers.

While the cost was not prohibitive, loss of manpower demanded a solution to this problem. When one firefighter is injured, another has to help him off the fire line. This reduces fire-fighting effectiveness. Training "emergency" employees was found impractical. Protective clothing was

UNITED STATES DEPARTMENT OF INTERIOR
(FOR SAFETY MANAGEMENT USE ONLY)

SUPERVISOR'S REPORT OF ACCIDENT

Refer to DI-134A for Instructions, Definitions and Standard Coding Details

Field Report No.

DATE OF THIS REPORT

SECTION A. IDENTITY *(Supervisor to Complete)*

(1) ORGANIZATIONAL UNIT *(Area, Region, etc.)*

Last
No.
Here

(2) REPORTING STATION *(Name & Address)*

(3) STATE IN WHICH ACCIDENT OCCURRED

(4) DATE OF ACCIDENT
Mo. ____ Day ____ Year ____

(5) NEAREST HOUR OF ACCIDENT
____ a.m. ____ p.m.

(6) NAME *(Employee, visitor or other involved)*

Last _____ First _____

Soc.
Sec.
No.

Use separate form for each employee involved

(7) EMPLOYMENT STATUS *(Circle one)*

1 Permanent*	5 Contractor	9 Public *(other)*
2 Y.O.C.	6 Vista	0 Other *(specify)*
3 Temporary	7 Job Corpsman	
4 Emergency	8 Public *(visitor)*	

*Include Job Corps Staff

(8A) CSC OCCUPATIONAL CODE   (8B) WORK ENVIRONMENT

Last
No.
Here

(9) RESULT OF ACCIDENT *(Circle one)*

PI—Personal Injury
PD—Property Damage

01 PI only	07 PF *(fire)*
02 PI with PF *(fire)*	08 PM *(motor vehicle)*
03 PI with PM *(motor veh.)*	09 PB *(boat)*
04 PI with PB *(boat)*	10 PA *(aircraft)*
05 PI with PA *(aircraft)*	11 PO *(all other)*
06 PI with PO *(all other)*	

(10A) IDENTIFICATION OF PROPERTY DAMAGE *(if any)*. *(Give name, model, number, size, make, type, etc.)*

Employee operated:

"Other" operated:

(10B) ____ Make of Govt. Vehicle _____

(11) PROPERTY OWNERSHIP *(Circle one)*

0 No prop. involved	5 Employee-owned on O.B.
1 Interior *(owned or leased)*	6 Inter-agency (GSA) motor pool
2 Other Federal	7 Indian
3 Contractor	8 Privately owned
4 Concession	9 Other *(Explain in No. 18)*

(12A) AGE:
a. Of Employee *(Year of birth)*
b. Of Govt. Property *(Year of mfg.)* ____

(13) IS TORT CLAIM EXPECTED? Circle one: Yes or No
If "Yes", has Tort Claim Officer been contacted? Circle one: Yes or No

---

SECTION B. MEDICAL *(Supervisor to Complete)*

(14) NATURE OF INJURY

(15) PART OF BODY INJURED

(16A) SEVERITY OF INJURY *(Circle one)*

0 No injury involved
1 First aid attention only
2 Medical attention only
3 Disabling injury *(fatal)*
4 Disabling injury *(temporary)*
5 Disabling injury *(permanent)*

(16B) HAVE C.A. FORMS BEEN SENT TO B.E.C.?

Circle: Yes No

(17)
a. Leave date: Mo. ____ Day ____ Year ____
b. Return date: Mo. ____ Day ____ Year ____
c. Death date: Mo. ____ Day ____ Year ____
d. No. days lost ____ e. Scheduled charge ____

SECTION C. STORY OF ACCIDENT *(By Supervisor or Employee)*

(18) NARRATIVE: *(What led up to accident? How did it happen? Facts are important—fault finding not.)*

NOTE: Use separate sheet, if it is needed. But condense story here!

SECTION D. SUPERVISORY OPINION *(Supervisor to Complete)*

(19) I THINK ACCIDENT MIGHT HAVE BEEN PREVENTED IF:

Employee Fault Finding Adds Nothing to Management Improvement.

(20) I SUGGEST THE FOLLOWING POLICY OR PROCEDURE CHANGE BY MGMT. TO HELP PREVENT SIMILAR ACCIDENTS:

SIGNATURE AND TITLE OF REPORTING SUPERVISOR

Figure 6-4. Interior's source document.

FORM DI-134 *(back)* If the title you need is not in DI-134A, tell your safety officer so that it may be added. Information on this side represents a "team" effort by safety and other functional managers to assist the Line Officer to discover underlying causes of accidents and to plan for their ultimate correction. All items require an entry. Follow bureau instructions for analyzing the problems on this side.

SECTION E. LINE MANAGEMENT PROBLEM *(Operations, Design, Construction, Maintenance, Plant Mgmt., etc.)*

(21)		TYPE OF ACCIDENT *(Event)*	(22)			WHAT WAS USED, DONE, CONTACTED *(Source)*
(23A)		HUMAN ERROR *(First selection)*	(23B)			HUMAN ERROR *(Second selection)*
(24A)		CONDITION DEFECT *(First selection)*	(24B)			CONDITION DEFECT *(Second selection)*
(25)		REVIEW OF THE MGMT. PROBLEM CITED IN SECTIONS D AND E				

Sig. _____
REVIEWING MGMT. OFFICIAL

SECTIONS F, G, H, I, and J to be completed by responsible mgmt. analysis identifying problems in the system to be resolved that will reduce accident loss. Keep remarks brief. Use coded information. Leave no blanks. In each section, consultation with supervisor and appropriate mgmt. official is desired.

SECTION F. PERSONNEL PROBLEM *(Consultation with Supervisor and Personnel Official is Desired)*

(26)		SUPERVISORY CONTROL AND TRAINING	(27)		FITNESS-FOR-DUTY EVALUATION
(28)		OPINION OF THE MGMT. PROBLEM RELATED TO PERSONNEL SERVICES			

Sig. _____
REVIEWING PERSONNEL OFFICIAL

SECTION G. PROPERTY/EQUIPMENT/ENVIRONMENTAL PROBLEM *(Consultation with Engineering and Property Official Desired)*

(29)		MAINTENANCE AND ENVIRONMENTAL CONTROL	(30)		FITNESS-FOR-USE EVALUATION
(31)		OPINION OF THE MGMT. PROBLEM RELATED TO PROPERTY SERVICES			

To be repaired: Yes or No Est. Cost $ _____
To be replaced: Circle: Yes or No Tentative date: _____

Sig. _____
REVIEWING ENGINEERING OR PROPERTY OFFICIAL

SECTION H. FINANCE PROBLEM *(Consultation with Administrative and Finance Official Desired)*

(32) AMOUNT OF PROPERTY LOSS	To Govt. Prop. $ _____ To "Other" Prop. $ _____	(33) AMOUNT OF TORT CLAIM AWARD	To Government $ _____ To "Other" Party $ _____
(34)	OPINION OF MGMT. PROBLEM RELATED TO FINANCIAL SERVICES		

Sig. _____
REVIEWING FINANCE OFFICIAL

SECTION I. LEGAL PROBLEM *(Consultation with Tort Claim or Legal Official Desired)*

(35)		OPINION OF THE PUBLIC SAFETY PROBLEM	(36)		POSSIBILITY OF RECOVERY FROM A 3RD PARTY? Circle: Yes or No. If "No", explain why in (37)
(37)		OPINION OF THE CAUSE, NOT RELATED TO A GOVERNMENT EMPLOYEE OR OPERATION			

Sig. _____
REVIEWING TORT CLAIM OR LEGAL OFFICIAL

SECTION J. CORRECTIVE ACTION *(Taken to Make Less Probable the Recurrence of this Accident)*

(38) LOCAL CORRECTIVE ACTION TAKEN OR PLANNED: When: Now _____ Fiscal Year _____

Sig. _____   _____
MGMT. OFFICIAL TAKING ACTION   TITLE

(39) RECOMMENDED BUREAU OR DEPARTMENT ACTION TO ASSIST IN SOLVING IDENTIFIED PROBLEMS. *

A Bureau response is expected if request for action is made here.

SIGNATURE OF REVIEWING SAFETY OFFICER	DATE	SIGNATURE OF REVIEWING AUTHORITY	DATE	Initial of Bureau Safety Officer	DATE

```
BUREAU
DATE          06-29-70
NAME          JOHN SMITH      RESULT        PI WITH PD
STATE         S. DAKOTA       SEVERITY      DISABLED
HOUR          11 A.M.         NAT INJ       CONTUSION
EMP STAT      TEMPORARY       PART INJ      BACK
OCC CODE      LABORER         SOURCE        FORK LIFT
WORK ENV      WAREHOUSE       HUM ER A      INEXPERIENCE
AGE EMPL      25              HUM ER B      FAIL TO CHECK DEFEC
PROP OWN      INTERIOR        DEFECT A      MECH-MATERIAL DEFEC
AGE PROP      15 YEARS        DEFECT B      WORN, CRACKED, FRAYED
MGMT PROB              INSTRUCTIONS NEEDED        COMP COST $778
SUP, TRNG, CONT        NO WRITTEN INSTRUCT        PROP DMGE $350
FIT DUTY              NONE                        TORT PAID NONE
PER SERV             LABOR-MGMT PROBLEM
MAINT INV CONT       POOR REPAIR FACILITY
FIT FOR USE          OVERAGE, OBSOLETE
PROP SERVICE         EQPT. OPERATIONS             TOTAL $1,128.
```

Figure 6-5 Computer printout.

found costly and not too effective. A management cost/effectiveness decision was made to "outlaw" the use of machetes in favor of a Sandvik brush cutter, named after its inventor. Injuries from machetes have been reduced to a trickle. By monitoring with the computer, any violations of the no-machete directive are spotted quickly and are quickly corrected.

Case 2

Fifteen fatalities connected with tractor operations cost the Interior Department more than $840,000 over a span of 12 years. This disconcerting fact was determined **without** the aid of a computer. But, when we did go to the computer for data collected . . . some important facts came to light. Printouts revealed that employees assigned as tractor operators were not qualified to be doing this kind of work. Employees injured while on tractors were painters, laborers, engineering technicians, foresters, surveyors, and other non-qualified heavy equipment men. Some of the problems stemmed from unguarded equipment, but it was apparent that we had a very serious problem with a lack of seasoned heavy-duty equipment operators.

Another problem was disclosed through the computer. One organizational unit was found to have more than 1,400 tractors that were not a part of its authorized equipment allowance. Investigation showed that line managers were obtaining tractors as surplus equipment under the Federal Government's excess property disposal program.

While the tractors were "free," they often were in poor shape. Since tractors were not part of the normal complement, no money was available for maintenance or repairs. Furthermore, no qualified tractor operators had been assigned to the unit. Thus, the computer's identification of "obsolete," "beyond economical repair," and so forth as causes of tractor accidents. This disclosed a serious misunderstanding between the operating and planning managers.

This mixup ended with tractor-operator failures.

Management's decision was to make a feasibility study to determine whether tractors actually were necessary to the organization's program, and, if so, whether it would not be cheaper to contract for such services than to face the cost of accidents.

Case 3

Park rangers, a dedicated group of employees with a specialized talent for taking care of many tens of millions of visitors annually in National Parks, appeared as one of the Department's poorest groups of motor vehicle drivers. Their vehicle accident frequency rate was well over acceptable limits. Defensive driver-training seemed the only answer until computerized accident data told us differently. A printout of a three-year driver experience for park rangers gave up these facts:

- Low injury costs among operators (most vehicle accidents involved no injury at all);
- Minor property damage (only nicks to fenders; small damage under the vehicle; no serious collisions with other vehicles).

These facts finally brought out this story: Park rangers must cruise continuously in their vehicles to aid visitors. Rangers often drive their vehicles off surfaced roads to reach camp-sites. In doing so, the vehicles frequently strike trees, rocks, and other obstructions. The resulting damage was reported as "motor vehicle accidents." Computerized causes for park ranger accidents backed these findings.

The low cost of injury, property damage, and tort claims reduced the priority of this problem with safety management—even though the rate of such accidents remained high. Management cost/effectiveness decision-making went this way:

- We could change the kind of vehicle being used to one that could take off-road travel with less damage—but the cost of such special vehicles would be prohibitive;
- We could simply direct park rangers to leave their vehicles on the main roadside and walk the 100 to 200 yards into the wooded areas to reach a camper—but this would add to patrol time and would require the

hiring of many more park rangers. This solution would cost more than the accidents.
- We could do nothing, but watch the total cases in printouts. Then, if we found park ranger "repeaters," we would take individual action as necessary.

This was the solution management accepted. Mass defensive driver-training for park rangers was discontinued because it was the best cost/effectiveness answer to this problem.

Case 4

Lest the reader think the computer deals only with "hardware type" problems, we will examine one centering on the management function of "training." By building into the source document (accident report) the capability to isolate training problems, the Safety Management Information System can help career development officers determine where training is needed, who needs it, and what instruction should be given. Numerous studies have been prepared to support this phase of safety management. For example, Fig. 6-6 shows an itemized account of supervisory control and training problems extracted from more than 2,000 accident reports from one organizational unit over a three-year period.

The Safety Management Information System now is producing so much data about where improvements can be made that it is becoming difficult to set priorities. The important point here is that Interior is operating on a **total** accident reporting concept.

"This data processing approach to accident reporting," says IBM's Donald Priest, "has made it possible to identify all industrial injuries instead of disabling cases only. The probable key benefit of IBM's computerization of loss-control information is that data generated in this program provide a more meaningful breakdown of industrial accidents, plus more accurate details relative to type, cause, and severity of specific accidents."

The automated information system used by Interior is called QUERY. It was originally developed for the National Institute of Health to manage its grants program, but it is a generalized program and operates on safety data just as well.

QUERY makes no real decisions by itself. The safety manager must supply the answers, basing them on experience and the ability to analyze data retrieved by the computer.

QUERY is relatively uncomplicated and a safety manager or analyst can use the computerized data file without calling on the services of a professional programmer. The safety manager specifies what he wants done on a set of forms right at his desk. He then has the option of putting the information on these forms into punched cards himself (it takes but a

Supervisory/Training Problem Cited in Accident Reports	No. of Occurrences	Paid for Injuries	Paid for Damages	Total Cost
Needed better trained employee	55	$ 5,696	$ 7,013	$ 12,709
Emergency—no time to train	135	6,801	none	6,801
Work pressure—no time to train	34	14,666	2,478	17,144
Could not provide close supervision	70	8,480	574	9,054
Work skill needs upgrading	54	5,532	3,673	9,205
Trained, but did not follow directions	119	25,460	7,059	32,519
Solo assignment	23	64,972*	none	64,972
Work unauthorized—needed help	3	478	380	858
Written instructions needed	76	12,264	3,384	15,648
More craft training needed	15	1,128	929	2,057
Safety training for supervisor	18	2,324	3,215	5,639
Need orientation-to-the-job trng.	71	76,367*	2,503	78,870
More mgmt. training for supervisor	16	13,100	1,365	14,465
Technical training needed	3	150	none	150
Communication problem	4	150	759	909
Defensive driver training needed	46	2,724	33,332	36,056
Total	742	$240,292	$66,664	$306,956

*Fatality charges included

Figure 6-6 Training problems found.

few minutes when you know how) or having a card punch operator do the work. Either way, the cards are fed into the computer and in minutes the QUERY report is printed.

Interior's QUERY system, with its infallible memory for facts and its quick response to questions, pays immediate dividends when faced with management problems. Let us assume, for example, that an organization is preparing for the first time a program designed to fight forest fires.

QUERY will aggregate and display accident history in any sequence or pattern chosen by the safety manager. By analyzing this, the safety manager can get clues and insights as to geographic areas that pose special problems; the kinds of equipment most likely to fail; the response of employees in various occupational categories to stress; the effective life span of certain types of equipment; what hazards appear most often and associated with what tools and equipment; and how much accidents are likely to cost in terms of manpower and expense.

QUERY permits the user to tailor each computer printout to his specific demands. The technique avoids unwanted details and comes precisely to the point every time. While only one set of data may be requested, many details unintentionally overlooked can be requested again. The data are always there for recall.

QUERY can be used effectively by any safety manager, inspector, or engineer. It can be a tool for safety committees, any line supervisor, administrative official, or staff service manager. None needs any specialized knowledge of computers. All that is required is a minimum amount of nontechnical instruction.

There is still another plus: the safety manager in charge of this dynamic information service requires a minimum of training outside his own field of safety expertise. His basic training in computer use can be assimilated in about a month's time, provided he is already well-founded in the principles of **systems** safety management.

Examples of the virtues of a computer-based system are many. The U.S. Department of Labor recently sought from Interior's computer bank a report on all accidents involving employees 21 years or younger, with a breakdown of 26 kinds of work performed in the age span. Interior's computer provided all the necessary facts in readable format **in just under two minutes.**

Another entity asked Interior for facts about tractor accidents over a two-year period. QUERY was asked to list the kinds of tractors, age of equipment at time of accident, condition of the vehicle, and reason for the accident.

The information, collected from all parts of the country, was compiled in a computer printout **in less than 60 seconds.** The cost of all damage repairs was inserted and totaled along with the medical and compensation expenditures for each injury involved.

The points here are that the accident experience among 21-year-olds-and-under might have taken months to compile by hand manipulation; the tractor accident data, if hand sorted, would have been far too costly an exercise to respond to a noncritical request. Then, too, without the exactness of the computer memory, hundreds of vital facts might have been misplaced, miscounted, or just plain lost.

Those who are fortunate enough to be using QUERY, or some similar retrieval program, will find themselves no longer tied down to a cyclical timetable in collating and issuing a display of accident facts. The traditional monthly reports, quarterly reviews, and annual summaries are no longer required. Deadlines are out. The ad hoc capability of the computer system goes deeply into the past or will pick up in detail last week's accident report. Neither time span nor quantity of data need bother the analyst who is aided by a computer.

Here the outstanding qualities of a total-reporting, totally responsive information system are clearly visible. It is not likely that the harried safety manager would keep statistics so detailed, so neatly packaged, and so always ready for delivery. But the computer will.

By themselves, accident statistics are easy to come by. But they really only relate to after-the-fact events. The idea, of course, is to **avert** the events by removing the causes. The computer, aided by a well-constructed reporting system, can lead the way by using the past as a guide to future actions. Here are some examples:

- Identification of occupations that produce unacceptably high numbers of operating errors in accident situations . . .
- Highlighting repetive failure modes of certain kinds of tools and equipment with a history of unusually high repair cost . . .
- Tabular accounting of who needs training, for what particular reason, and how much the lack of know-how costs are related to the cost of "training out" the operational defects . . .
- Assessment of public relations problems in terms of suits for tort claims—who is involved, where, and for what correctible reasons . . .
- Critical management service area profiles showing the repetitive breakdown in staff services found to be causing human errors and condition defects found in accident situations . . .
- Computer tabulations of shortcomings related to specific organizational units—regions, districts, area offices, plants, laboratories, and the like . . .
- Utility data to support product liability claims, improve employment practices, install health units, make design changes, alter purchasing specifications, initiate research studies for cost reduction, and so on.

Interior's system is a workable computerized information system

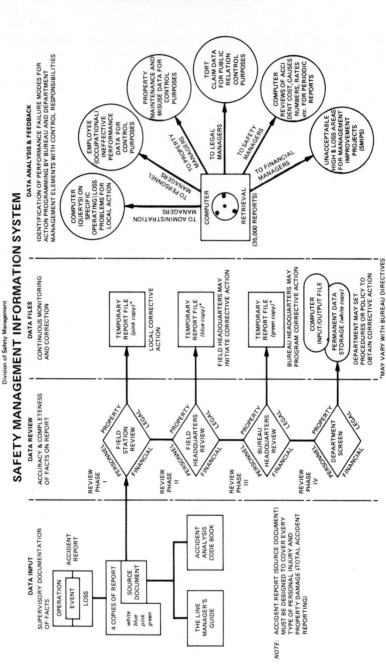

Figure 6-7. Flowchart of Interior's system.

which is not based on the Z16.2, but on the domino theory of accident causation. The system flow chart is shown in Fig. 6-7.

Accident analysis, discussed in this chapter, is the third method of collecting and analyzing data for decision making. In the next chapter we'll discuss selecting remedies.

REFERENCES

Method of Recording Basic Facts Relating to the Nature and Occurrence of Work Injuries, Z16.2, American National Standards Institute, New York, 1962.

Pope, W., and E. Nicolai, *In Case of Accident, Call the Computer*, U.S. Department of the Interior, Washington, 1969.

SEVEN

Selection of Remedy

In earlier years in safety this subject was somewhat simpler than it is today. That is because we had (earlier in our axioms) rather arbitrarily limited our remedy selection choice to four: engineering revision; instruction, persuasion, and appeal; personnel adjustment; and discipline. This made the selection of remedy somewhat simple; merely choose the most appropriate of the four as follows:

Reasons	Indicated Remedy
Improper attitude	Personnel adjustment
1 Willful disregard, reckless, lazy, disloyal, uncooperative, fearful, oversensitive, egotistical, jealous, impatient, absent-minded, excitable, obsessive, phobic, inconsiderate, intolerant, mentally unsuited in general.	Placement and medical attention, including psychology. Discipline in rare cases and as a last resort. Engineering revision always should be considered.

Lack of knowledge or skill. . . . Training, instruction, persuasion, and appeal

2 Insufficiently informed, misunderstands, not convinced of need, indecisive, etc.

Instruction and reinstruction, training and practice, persuasion and appeal based on motivating characteristics, psychology, and human engineering. Engineering revision always should be considered.

Physically unsuited. . . . . Personnel adjustment

3 Hearing, sight, age, sex, height, illness, allergy, slow reaction, crippled, intoxication, physical handicap in general.

Placement and medical attention, including psychology. Engineering revision always should be considered.

Improper mechanical or physical environment . . . . Engineering revision

4 Space, light, heat, arrangement, ventilation, materials, tools, equipment, procedures, company policy, routing, etc., make it awkward, difficult, inconvenient, embarrassing, or impossible to follow safe-practice rules.

And we provided these guidelines for the selection:

1 Assemble the facts as provided by cause analysis.
2 Consider first and always the application of "engineering revision."
3 Consider simple "instruction" coincidentally with "engineering revision" for first instances of personal causes, when no obstacle is encountered or suspected.
4 Consider the subcause in all cases where it is known without special inquiry, also in all cases where the problem does not yield to simple "instruction" or "engineering revision."
5 Consider the underlying cause where the problem does not yield to items 2 to 4.

In other words, in selecting remedies, always first try to engineer the hazard out, couple that if necessary with training, and only when these fail, look at the more difficult or drastic remedies of attitude change and/or discipline.

While this is obviously oversimplistic, it does contain some good thoughts. Whenever it is economically feasible, it is always, even today, a good approach to engineer the problem out. At times when the task is quite crucial, it is obviously the best approach even if it is economically costly. An example was the safety of the astronauts in the space program.

There are, however, many situations where the cost of engineering out the hazard just doesn't make sense in terms of benefit received. The notion of safety at all costs is simply not managerially sound. Similarly the notion that training is always the best device to use is also unsound. As we shall discuss in Chap. 12, training is an effective accident control technique only if the worker does not know the difference between the safe method to perform a job and the unsafe method. If the worker knows this difference and knows how to perform safely but chooses to perform unsafely, then we simply do not have a training problem. Training is a costly, time-consuming, and ineffective solution in this case.

Similarly in the areas of personnel adjustment through placement and medical attention and in discipline, we've thought through some of our earlier ideas. Placement is often simply not a possible control for us to use: there may be no other opening, or unions may not allow a job change, or the employee may not wish to be replaced. Discipline is extremely thorny as a viable accident control technique. Research indicates it may not even be very effective in accident control.

Selection of a remedy is not the simple process of picking one of four simple solutions in most accident situations. Rather it is the sifting and sorting of quite a number of possible, feasible solutions, and in some way making a decision that shall provide us with the best results of all the alternatives considered. It is this sifting and sorting process and the actual choosing that this chapter discusses.

In Chap. 1 we discussed several decision models that can be elaborated on in this chapter. One is the Kepner-Tregoe model shown in Fig. 1-3. A decision is always a choice among various ways of getting a task accomplished. There are always alternative methods of accomplishing a task. The final decision may be a compromise between what is really wanted and what can actually be done due to constraints of time and money. The right decision is the one that gets the most done at the least cost with the least adverse consequences.

Most decisions involve simple choices and require only small amounts of information to arrive at a good decision. These involve relatively simple mental gymnastics and can be done in your head. Major decisions will involve large amounts of information and possible uncertainties that must be processed in some way. A systematic process of arriving at such a decision is a necessity.

The process of decision making involves seven concepts, according to Charles H. Kepner and Benjamin B. Tregoe, authors of The Rational Manager. These concepts are:

Selection of Remedy

1 The objectives of a decision must be established first.
2 The objectives are classified as to importance.
3 Alternative actions are developed.
4 The alternatives are evaluated against the established objectives.
5 The choice of the alternatives best able to achieve all the objectives represents the tentative decision.
6 The tentative decision is explored for possible future adverse consequences.
7 The effects of the final decision are controlled by taking other actions to prevent possible adverse consequences from becoming problems, and by making sure the actions are carried out.

First each objective is classified as either a "want" item or a "must" item. Must items, those items that are absolutely required, are listed separately (see Fig. 7-1). Since these are "musts," they are easy to identify in a Go/No Go fashion. Figure 7-1 shows the process of deciding which house to buy. In this case, the only "Musts" are the down payment not over $10,000,

DECISION ANALYSIS WORKSHEET

Objectives	Wt.	A INFO	Sc	Wt·Sc	B INFO	Sc	Wt·Sc	C INFO	Sc	Wt·Sc
MUST		INFO	Go/No		INFO	Go/No		INFO	Go/No	
Down payment not over $10,000		$7,500			$9,500			$6,000		
Monthly payment not over $300		300			370	NO GO		280		
Minimum of four bedrooms		4						4		
Minimum of two bathrooms		2						2		
Occupancy within 60 days		45 days						45 days		
WANT	Wt.	INFO	Sc	Wt·Sc	INFO	Sc	Wt·Sc	INFO	Sc	Wt·Sc
Minimum down payment	.6	$7,500	.9	54				$6,000	1.0	60
Lowest monthly payment including taxes	1.0	300	.9	90				280	1.0	100
Location convenient to work	.7	good	1.0	70				O.K.	.8	56
Shelter for two cars	.4	carport	.7	28				garage	1.0	40
Public transportation nearby	.4	bus	.9	36				bus close	1.0	40
Location convenient to elementary and high schools	.8	1/2 mile	.7	56				1/4 mile	1.0	80
Location convenient to shopping center, stores and facilities	.7	1 mile	.7	49				1/2 mile	1.0	70
Workshop and storage space avail.	.2	large	1.0	20				poor	.3	6
Stable resale value	.7	good	1.0	70				good	1.0	70
Attractive; modern style and appearance	.5	good	.8	40				excellent	1.0	50
Good landscaping; trees, shrubs	.4	new	1.0	40				O.K.	.7	28
Large play area for kids	.5	unfenced	.7	35				fenced	1.0	50
Large, comfortable family room	.2	good	1.0	20				none	.0	0
Location on quiet street, in good neighborhood	.4	good	1.0	40				fair	.8	32
Minimum maintenance cost to house	.7	good	1.0	70				average	.8	56
Minimum risk-tax increase or special assessments	.4	high	.4	16				low	1.0	40
Weighted Score				7.44						7.78

Figure 7-1. Kepner-Tregoe decision analysis worksheet.

total monthly payment not over $300, minimum of four bedrooms, minimum of two bathrooms, and occupancy within 60 days.

Already perhaps 90 per cent of the houses that are for sale have been eliminated and need not be given further consideration.

Figure 7-1 illustrates concepts 3, 4, and 5.

The choices that appear to be reasonably acceptable are quickly narrowed to three houses—alternatives A, B, and C. B is immediately eliminated when it cannot fulfill one "Must"—the minimum monthly payment.

Next, each "Want" is weighed as to importance compared to all other wants and given a weight. The alternatives are then matched against how well they fill each "want" and given a score from one to 10.

To find the *Weighted Score*, the weight in the "want" column is multiplied by the score in the "info" column and the result entered in the "weighted score" column. The weighted scores are totaled and alternative C represents a tentative decision because it fulfills the "must" requirements and satisfies the weighted "wants" better.

However, there is one more step—evaluating the possible adverse consequences (Fig. 7-2).

Again, weighted scores are used. House A has fewer adverse consequences, awkward social situations, and few playmates for the children. Each circumstance, for each house, is weighted as to the *probability* of being an adverse consequence and the *seriousness* of the consequence.

As Fig. 7-2 shows, House C had greater adverse consequences than did House A.

POSSIBLE ADVERSE CONSEQUENCES

Alt. A	P	S	PxS	Alt. B	P	S	PxS	Alt. C	P	S	Pxc
Awkward social relationship - subordinate lives across the street	.8	.4	.32					Risk of basement flooding--major repairs	.9	.9	.81
Few playmates for children in neighborhood	.7	.7	.49					Heavier traffic due to addition to new shopping center	.5	.9	.45
								Inconvenience of country club across town	.9	.9	.81
								Poor land for wife's gardening	.9	1.0	.90
Weighted Score:			.81	Weighted Score:				Weighted Score			2.97
DECISION:											

Figure 7-2. Kepner-Tregoe adverse consequences analysis sheet.

In this case, House A is the best alternative of those available.

The same process is applicable to making decisions for the job of preventing accidents. And the probability of getting management action on the most advantageous action is enhanced as the process of arriving at such action was rational.

DECISION TABLES

Jack Gausch has developed several similar approaches to safety decision making. The first of these is the use of decision-making tables.

Most of the decisions made in safety today are made subjectively. Perhaps even politically. Since these are the "dirty words" of science, the safety professional seems to be in diametric opposition to the newest rage—scientific decision-making, a technique which uses mathematic models for arriving at desirable solutions to problems.

One of the popular names given to this scientific decision-making is "Operations Research." "OR," as it is called by the practitioners, got its start during World War II. It was born out of necessity—the necessity to select a winning military strategy from an overwhelming set of alternatives. As the computer became better developed and its use more and more widespread, acceptance to scientific decision-making increased across the broad spectrum of business and industry.

The acceptance of such techniques has been slow in the safety field. Development of methods requires two kinds of expertise, persons knowledgeable about the scientific method itself and others informed about the particular area where scientific approaches will be applied. Rarely do these occur in the same individual. Safety professionals generally are not attuned to high-objectivity methods.

One of the most easily digested tools is the *Decision Table*. It is oriented to simple words. Inspectors and the plant manager can immediately grasp it. Yet it is logic-oriented and will be appreciated by the computer specialist as well.

All information is in front of the user, and it assures a more uniform administration of actions based on conditions. It is better than seat-of-the-pants experience and not as greatly influenced by day-to-day emotions, personal prejudices, or similar human variables.

The Decision Table, when constructed properly, is a simple but marvelous tool for improving decision-making capability. It is extremely versatile and can be applied to management responsibility in any area, even to safety. Normally, we consider the decision table as being divided into two major sections: *conditions*, and *actions*.

Since all of the necessary rules for constructing tables cannot be shown

APPLICANT SELECTION (Minimum Requirements)								
Conditions	(*)							
1. Graduate engineer								All other
2. Certified safety professional								
3. Five years' experience								
4. Salary $15,000 plus								
5. Professionally active								
Actions								
A. Interview for director								
B. Interview for assistant director								
C. Interview for inspector								
D. Send "sorry" letter								

Figure 7-3. Decision table.

here, we will use very simple explanations that will provide a good general orientation. Figure 7-3 illustrates the two divisions of a table.

The condition is considered an "if" statement and the action the "then" statement. "If" the right *condition* or combination of conditions exist, "then" specified *action* or actions are taken. In the example, Fig. 7-3, one reads the table as follows: (Start at the (*) and read *DOWN*).

> "If" the applicant is a graduate engineer, has five years' experience in responsible charge, is presently making more than $x a year, and is a recognized contributor to safety efforts, *"then"* he will be interviewed for the job of Director or Assistant Director.

Obviously, this is a simplified example to show the use of tables. If one has 10 variables and each has a simple yes or no answer, then it is possible to have more than 1,000 different combinations of these variables. One could commit a small number of variables, such as the table illustrating applicant selection, to memory. One cannot, however, conceive a mental model of 1,000 variables. Consider, a simple two-dimensional table with 20 squares on each of two sides. There are thus 400 squares on the grid, and 400 categories of conditions can be shown on one page.

Now that we have seen a simple example, let's take a more realistic one.

Suppose you were recently put in charge of a large corporate safety department and decided to use 20 engineers to perform safety inspections of all facilities. In the interest of maintaining a program packed with action, you want the inspectors to find the hazards, appraise them, and specify the corrective action that the plant should take "on-the-spot."

Thus, you are preparing the inspector to be able to say, "If" this condition(s) exists, "then" this action(s) will be taken. As corporate safety director, you have certain concerns and wants:

1 *Uniformity*—The program should be administered uniformly, since you intended to measure management performance on the basis of the results.
2 *Administrative feasibility*—The engineer's job should be as easy as possible, as long as the task is performed accurately.
3 *Flexibility*—Rotation of inspectors from one group of plants to another without dramatically changing the emphasis on various conditions is preferred.
4 *Objectivity*—The subjective influence that is always present due to personal prejudices should be reduced.

It now becomes necessary to provide categories for hazards and to specify the actions that will be taken when hazards are discovered.

When a violation of standards is discovered, it will not always have the same significance. Consider for instance the violation of the requirement for guard rails on working platforms. This is very hazardous when the railing would prevent a person from falling into an acid-dip tank. The lack of guard railing is also very hazardous around a polar bear pit at the zoo. But, would you put railings around your swimming pool, or should we require railings at every loading dock? We require hand railings for every broad flight of steps in an industrial plant. Should we apply this to public buildings such as the Franklin Institute in Philadelphia? It obviously becomes necessary not only to find the violation of the standard but to decide just what category of hazard is involved.

In order to show how a decision table might be utilized, we must first decide on some arbitrary category of hazards; and since we are selecting categories purely for the purpose of demonstrating the use of a table, we will use some already advocated for system safety. First, we assume that the violation of a standard creates a hazard. We will classify this hazard according to how often we judge it will cause accidents, *frequency*, and how bad the losses will be, *severity*.

SEVERITY

For our severity levels we will use those four levels that are advocated by MIL-STD-882. This is a military standard for system safety programs. The four categories of severity selected from this standard are as follows:

1 NEGLIGIBLE (will not result in personal injury or system damage).
2 MARGINAL (can be counteracted or controlled without injury to personnel or major system damage).

3 CRITICAL (will cause personal injury or major systems damage or will require immediate corrective action for personnel or system survival).
4 CATASTROPHIC (will cause death or severe injury to personnel or system loss).

Now we have one parameter relative to classifying our hazard. We have identified four areas of severity.

FREQUENCY

We must know how often we can expect that a loss will occur. For frequency categories we will design some of our own using terminology that is somewhat relative to system safety but simplified by being quantified with days instead of hours.

a EXTREMELY REMOTE (one failure in more than 3,000 days of operation).
b REMOTE (one failure in 301 to 3,000 days of operation).
c REASONABLY PROBABLE (one failure in 31 to 300 days of operation).
d PROBABLE (one failure in 30 days or less of operation).

CONDITIONS

Now we are able to construct the first part of our decision table, the *conditions*. Since we have four categories of frequency and four categories of severity, there are 16 possible levels of hazards. These are illustrated in Fig. 7-4.

ACTION

Now we must decide what actions we would like our inspector to take, since we desire to make this man a decision-making representative of our office. Basic actions that we have selected are:

1 Submit to design engineering for long-range study.
2 Complete within one year.
3 Establish plan for completion within three months.
4 Initiate repair order and complete correction within 30 days (immediate shutdown not required).
5 Shutdown of operations, immediate repair.

Once again, these are just categories selected for the purpose of illustrating the technique of decision-table construction. A company should *not* adopt

HAZARD ACTION TABLE												
	Severity											
Frequency	Negligible			Marginal			Critical			Catastrophic		
Extremely remote												
Remote												
Reasonably probable												(*)
Probable												
Actions												
Forget it												
Long-range study												
Correct (1 year)												
Correct (90 days)												
Correct (30 days)												
Shutdown												

Figure 7-4. Hazard action table.

these categories for its own use. Categories of action should be selected by group participation. Since the decision table imposes burdens upon operating people, it is prudent that they be involved in the designation of appropriate actions.

Now let us give an example of the method of reading this table. Taking the example marked (*), one would say to himself, "if" the hazard created by the violation standard would cause an accident estimated to occur in less than 30 days and will cause severe injury or death, "then" the Action specified is to shut down the operation. After we classify the level of hazard, we read down following the arrow to the indicated decision below. If one can understand this example, then the rest of the chart becomes immediately obvious.

VALUE ENGINEERING

Jack Gausch's second systematic aid to decision making in safety is value engineering.

A decision to select a particular method of controlling hazards should consider **many** things. Decision tables are useable for such problems, especially for priority rating hazards and assigning expediency for corrective action. They handle **two** factors, such as frequency and severity, very readily. It is possible to enter **three** factors on one table.

The simplest way to expand the considerations which can be evaluated on one table is to apply **numbers**. This forms a payoff matrix. Value engineering can meet this need. It offers some exciting new sensitivities to safety decisions and will handle many factors.

To provide a quick review of value engineering, a step-by-step description is helpful. You have a defined problem and a choice of solutions. Each **choice**, however, has different values in respect to specific desired benefits, and thus an aggregate value can be determined:

1 *Define important benefits.* Consider that you are trying to make a decision regarding your future occupation. After giving the problem careful thought, you select three benefits you must have: **salary, challenge,** and **security.**
2 *List possible choices.* As of this moment you are not aware of the great benefits a profession in safety offers. You have selected as occupations you could handle: **richman, poorman, beggarman, thief,** or **doctor.**
3 *Construct a table and apply numbers.* In the following illustration (Fig. 7-5), a table is shown with values of 1 to 10 applied to each benefit and choice. In this example, it would be better to pick an occupation of richman.

In this example, a possible value of **one** to **ten** (1 to 10) is used. One is a **low** value and ten is **high.** A value of 70 to 100 has been suggested in some writing and is easily understood because it is related to school grading systems. Most of us, however, will not need more than 10 relative

WEIGHT BENEFITS	SALARY	CHALLENGE	SECURITY	
CHOICE ⟶	10	2	6	
Richman	10 / 100	2 / 4	10 / 60	164
Poorman	4 / 40	4 / 8	5 / 30	78
Beggarman	2 / 20	6 / 12	4 / 24	56
Thief	6 / 60	10 / 20	2 / 12	92
Doctor	9 / 90	8 / 16	8 / 48	154

Figure 7-5. Value table.

numbers for typical safety work. This illustration demonstrates the technique.

Benefits rarely have equal relative weights. Such an approach is not realistic for different corporations, points-in-time, or specific problems. When selecting an automobile, some people might be highly interested in "comfort" and place a high value on it. Others might feel "cost" is the most important item. You **might** believe neither is significant and "miles per gallon" should be added to a list of benefits. These differences create a need for **weighting** benefits. This permits adjustment to such variables as: time, corporate differences, or specific problems. If an energy crisis occurs, you **will** feel "miles per gallon" should be added to a chart for selecting automobiles. You will probably assign a high relative weight to this variable.

There are complex ways to arrive at a weighting factor. A simple method is to use the same 1 to 10 approach. One (1) is **low** in importance, ten (10) is **high** in importance. As demonstrated in the example, the assigned value is multiplied by the appropriate weight, and the new values are totalled for the final score. In the case of thief, a value of ten (10) for highest possible **challenge** was multiplied by an assigned weight of two (2) to obtain the relative value of twenty (20).

Now that we have reviewed the method, further consideration can be given to a fundamental model which is applicable to safety. We will deal here with the use of value engineering as a technique of appraising control choices. This can be achieved by following the same general procedure illustrated in the previous example. Each use of a value engineering technique will provide values for control choices for specific problems. For this reason, **choices** cannot be identified in advance of the recognition of the problem. **Benefits** tend to be somewhat constant in definition although their weighting factors will change.

This is where the technique offers dynamic contributions to the safety profession. It prompts consideration for a wide range of factors not normally addressed in informal selection of controls. It imposes a thoroughness complementary to the concept of total safety and respectful of the complex interrelationship of factors which influence the quality of human life.

On the following chart (Fig. 7-6) some benefits are shown. The main benefits include: cost, safety, morale and social. These factors demonstrate how a wide variety of considerations can be applied.

COST

Cost is shown here in three separate categories. When evaluating alternate choices, it must be remembered that a **low** cost will result in a **high** value.

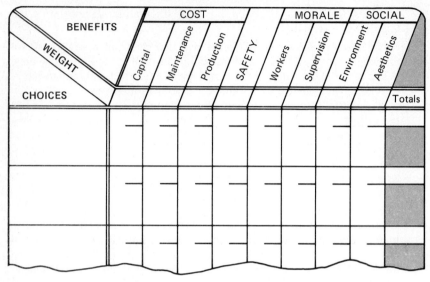

Figure 7-6. Benefit table.

Likelihood	Value
Might well be expected	10
Quite possible	6
Unusual but possible	3
Only remotely possible	1
Conceivable but very unlikely	0.5
Practically impossible	0.1
Virtually impossible	0.1

Figure 7-6*a*

Likelihood	Value
Continuous	10
Frequent (daily)	6
Occasional (weekly)	3
Unusual (monthly)	2
Rare (a few per year)	1
Very rare (yearly)	0.5

Figure 7-6*b*

Possible Consequence	Value
Catastrophe (many fatalities, or 10^7 damage)	100
Disaster (few fatalities, or 10^6 damage)	40
Very serious (fatality, or 10^5 damage)	15
Serious (serious injury, or 10^4 damage)	7
Important (disability, or 10^3 damage)	3
Noticeable (minor first-aid accident, or $100 damage)	1

Figure 7-6c

Risk Score	Risk Situation
400	Very high risk; consider discontinuing operation
200 to 400	High risk; immediate correction required
70 to 200	Substantial risk; correction needed
20 to 70	Possible risk; attention indicated
20	Risk; perhaps acceptable

Figure 7-6d

Capital—This includes construction, fabrication, establishing procedures, and the initial capital cost of equipment.

Maintenance—Basically a service expense. Program maintenance, training, and promotion would be included.

Production—Many choices of controls have an effect on production. Sometimes a safety procedure, machine guard, or protective equipment will curtail production. Sometimes it will increase it. These benefits are measurable.

SAFETY

A relative value can be assigned as an appraisal of the amount of safety obtained. Abating noise at the source would perhaps have a value of **10** as compared with compulsory use of ear protection, which would be less in value. More protection is obtained by abatement. Similarly, requiring continuous use of respiratory protection might provide a safety value of **6**, increasing ventilation **8**, substitution of non-toxic solvent **10**.

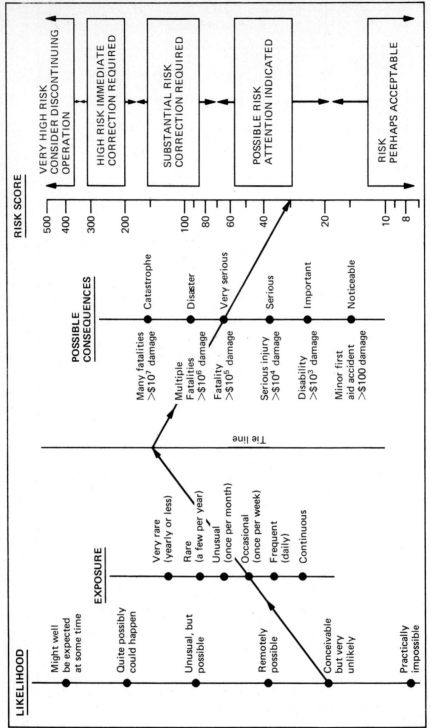

Figure 7-6e. Risk analysis.

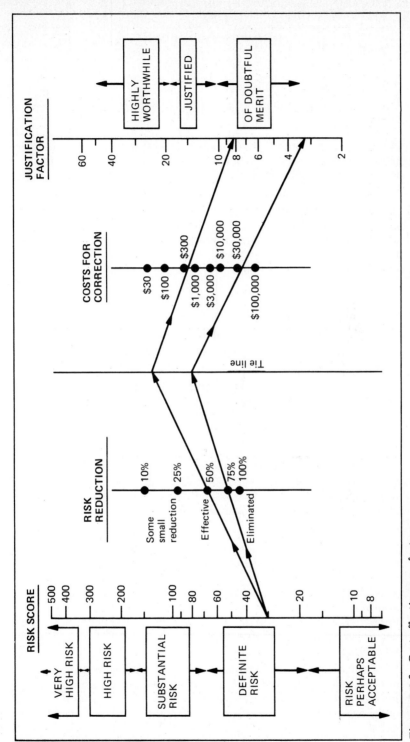

Figure 7-6f. Cost-effectiveness factors.

MORALE

Control measures will have various relative effects on the morale of employees.

Workers—Safety controls have an effect on worker morale. Requiring use of safety belts or harnesses in a tank could be a **4** in value as compared to having a bottom entry manhole which could rate **9**. Automatic cleaning could be **10**.

Supervision—Having Supervision enforce a rule that employees must shut off the machine when cleaning jams will cause more morale problems for supervisors than providing controls which require two-hand activation. The supervision of pull-backs on punch presses has less of a morale value than automatic feeding!

SOCIAL

A comprehensive assessment of control choices can address problems of social impact. Environment, community relations, and even aesthetics have value.

Environment—Think about the noisy air compressor! We can enclose it and reduce in-plant noise to 80 dBA, perhaps a value of **9**. We could also move it outside, eliminating worker exposure but polluting the environment for plant neighbors for a relative value of a lower order of magnitude, perhaps **5**.

Aesthetics—A problem exists related to discharging toxic fumes in the community environment. A tall discharge stack can be installed. This would have a low value from the standpoint of aesthetics when compared to collecting the material in charcoal scrubbers or even substituting safer solvents.

Practicing safety professionals will see many ways that this technique can be applied. They will recognize benefits which are not included in these brief examples. Typical charts could be developed for appraising consumer products, selecting security controls and fire protection installations, or even measuring aspects of corporate safety programs. The use of such techniques will strengthen the service of safety professionals. They offer documentation to support decisions for management, the worker, and the safety specialist in his defense of choices.

Professional integrity must be maintained to assure that negative benefits, or those which are counter-opposed to responsibility, are not applied in the development of a matrix for safety decisions. Such a factor is visibility of the hazard. Some practitioners advocate priority rating hazards on the basis of their visibility to compliance officers or others enforcing codes. This indicates a less than professional approach.

RISK ANALYSIS

Hugh M. Douglas of Imperial Oil Ltd., of Canada, and author of several texts on loss control, has provided an interesting method of decision making in risk analysis. The method is based upon three maxims:

- All risks (uncertainties of loss) can never be completely eliminated.
- Wise use of available knowledge can reduce potential loss.
- Efforts to reduce loss should be geared to achieve maximum cost effectiveness.

These maxims have led to a system of quantitatively identifying risks and evaluating proposed risk reduction procedures. The system uses numerical values for comparison purposes. It is based on the contents of two papers: "Mathematical Evaluations for Controlling Hazards" by William Fine and "Practical Risk Analysis for Safety Management" by G. F. Kinney.

According to Fine and Kinney, risk or uncertainty of loss imposed by a particular hazard will increase (1) with the likelihood that the hazardous event will actually occur, (2) with exposure to that event, and (3) with possible consequences of that event. To calculate the risk, numerical values are assigned to each of these factors. From these, an overall risk score is computed as the product of these three separate factors. The numerical values are arbitrarily chosen, but provide a realistic but relative score for the overall risk.

The likelihood of occurrence of a hazardous event is related to the mathematical probability that it might actually occur. Likelihoods range from the completely unexpected up to an event that might be expected (see Fig. 7-6a).

The greater the exposure to a potentially dangerous situation, the greater the associated risk. The value of unity is assigned to the situation of a rather rare exposure, and the value of 10 is assigned to continuous exposure (see Fig. 7-6b).

Injury or asset loss from a hazardous event can range all the way from minor injury or loss that is barely noticeable up to the catastrophic (see Fig. 7-6c).

The risk score for some potentially hazardous situation is given numerically as the product of three factors: one numerical value for likelihood, for exposure, and for possible consequences. (See Fig. 7-6d for suggested controls attached to the risk score.)

Risk scores can be calculated graphically as shown in Fig. 7-6e. The likelihoods are listed on the first or left line of the nomograph. Exposure factors are listed on the second line, and factors for possible consequences appear on the fourth line. To calculate a risk score using this nomograph, locations corresponding to each factor involved are established. Then a line is drawn from the point for the likelihood factor through that for the exposure factor and extended to the tie line at the center. A second line is drawn from this point on the tie line through that for the consequence factor and extended to the scale for the risk score.

The larger the risk score for a situation, the more effective a proposed corrective action, and the less that action costs, the greater is the justification. A quantitative index for the justification can be derived from numerical values assigned to each of its three component factors.

The effectiveness value assigned to a proposed risk reduction action is taken as unity for complete elimination of risk and as zero for an action that has no effect.

Cost and justification bear an inverse relation. A cost factor is expressed as a divisor whose numerical value increases with cost so

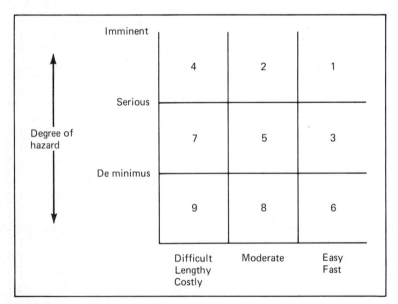

Figure 7-7. OSHA matrix.

that increased cost gives lesser justification. The justification factor provides an index for cost-effectiveness, and like the risk score can be calculated graphically as shown in Fig. 7-6f. Entry to this nomograph is by three factors: one numerical value each for risk score as calculated previously, for degree of risk reduction that the proposed measure provides, and for its cost divisor. Lines through these points give both a numerical value and a descriptive term for the justification factor.

				Law violations			
#	Dept.	Violation	Priority		Fix Schedule		
			#	6/16	6/23	6/30	7/7
306	Paint rm	Repaint lines on floor for 5,000 gal.					
307	Paint rm	Provide for bonding while pumping					
308	Paint rm	Too many barrels stored					
309	Paint rm	Open barrels in paint room					
310	QC rm	Ungrounded electric plate					
311	Pl #3	Bag sprinkler heads in booth					
312	Pl# 3	Holes in booth roof					
313	Pl #3	Nonstandard handrails over PL					
314	Pl #3	Unguarded Ch & Sp 3rd level					
315	Pl #3	Handrail missing top PL					
316	Pl #3	Guard conveyor (unused) nip point					
317	Pl #3	Guard back of B & P on conveyor					
318	Pl #3	Ground lights on deionizing tanks					
320	Pl #3	Water on floor is excessive					
321	Pl #3	Hard hats in area					
322	Pl #3	Unguarded coupling in oven area					
323	Pl #3	Reversed polarity ext cord to PH tester					
324	Pl #3	Uninspected ladders					
325	Pl #3	Construct bar for swinging hooks					
326	Pl #3	Side shields for all hangars					
327	Pl #3	Hooks on floor					
328	Metals	Bottom B & P of fixture unguarded					
329	Metals	Oil on floor					
330	Metals	Guard back of B & P on B-5					
331	Metals	Guard back of B & P on P-11					
332	Metals	Guard back of B & P on P-12					
333	Metals	Guard back of B & P on edge trimmers					
334	Metals	Guard back of B & P on S-4					
335	Metals	Guard Ch & Sp of P-12 edge trimmer					
336	Metals	P-14 totally unguarded					
337	Metals	P-14 has improper controls					
338	Mntn.	Guard belt sander					
339	Mntn.	Guard back of B & P recip. saw					
340	Mntn.	Need dead man switches on drills					
341	Mntn.	No underslung guard radial saw					
342	Mntn.	Guard B & P, Ch & Sp on lawnmower					
343	Mntn.	Guard B & P on motorized cart					
344	Mntn.	Ungrounded fan					
345	Mntn.	Unguarded fan					
346	Mntn.	Provide approved hand rails on rolling scaffold					
347	Mntn.	Provide platform that covers top on scaffold					
348	Mntn.	Provide safe ladder to get up on scaffold					
349	Mntn.	O$_2$ cap off					
350	Mntn.	Guard the back of B & P portable hacksaw					
351	T & D	Guard the B & P on file					
352	T & D	Ungrounded light on vertical mill					
353	Receiving	Steel pallets stored too high					
354	Receiving	Wood pallets stored too high					
355	Receiving	RR chain down					
356	Receiving	F-7 leaking oil					
357	Receiving	Int. rails platform by truck dock					

Figure 7-8. OSHA fix list.

DECISION MATRICES

Many types of decision matrices are used in industry and in safety. One type used by some companies for OSHA compliance is shown in Fig. 7-7. This matrix was used in the following way: (1) by determining, as a compliance officer might, the degree of hazard (imminent, serious, or de minimus); and (2) by estimating the costliness of the change in terms of time, manpower, and money. The number in each box—which indicates priority—was arbitrarily assigned at the outset. It could be shifted depending on management's decision, money available, manpower, etc.

As supervisors or inspectors report to safety any physical violations they cannot correct themselves, the violations are given a priority number ranging from 1 to 9 and are put on the maintenance OSHA task group's list and scheduled for correction (Fig. 7-8).

OTHER DECISION TOOLS

We have only touched on a few representative systematic decision-making tools. There are others available, all kinds. Perhaps, however, these will give the reader an indication of some of the various ways available to help in selecting remedies.

REFERENCES

Fine, W., "Mathematical Evaluations for Controlling Hazards," in *Selected Readings in Safety*, Academy Press, Macon, Ga., 1973.

Gausch, J., "Safety and Decision-Making Tables," *Journal of the ASSE*, November 1972.

Gausch, J., "Value Engineering & Decision Making," *Journal of the ASSE*, May 1974.

Kepner, C., and B. Tregoe, *The Rational Manager*, McGraw-Hill, New York, 1965.

Lahey, J., "Problem Solving & Decision Making," *National Safety News* reprint, National Safety Council, Chicago.

Petersen, D., *The OSHA Compliance Manual*, McGraw-Hill, New York, 1975.

EIGHT

Application of Remedy—Corrective Action

C orrective action, or application of remedy, is the active step in accident prevention and must be taken invariably on the basis of *the pertinent facts of accident causation.*

WHO SHOULD APPLY CORRECTIVE ACTION?

First and foremost when it comes to *action* in applying corrective measures are the supervisors. Inasmuch as they are management's direct representatives and the immediate contacts for the workers, it is their orders and instructions that are disregarded when safe-practice rules are not followed. Also, when unsafe mechanical conditions exist, it is their responsibility except in cases where, because of monetary expenditure, etc., they have no authority. Therefore it is an inherent part of their work to apply corrective action.

The work of the safety engineer is primarily that of staff, not of line. Safety engineers are fact finders, planners, analysts, advisors, organizers, salespeople, and enthusiasm builders. They themselves are seldom authorized to control the performance of workers directly, or personally to correct mechanical conditions.

Production managers or superintendents apply corrective action through their assistants and especially through their supervisors.

Individual workers as a result of proper supervision, safety education, and personal initiative, correct their own unsafe practices. They also can maintain safely a good part of their physical environment.

Basically the application of corrective action is the *job of management* including its entire executive, supervisory, and specialist staff. The *initiative* and *follow-through* are almost wholly the responsibility of management.

FOLLOW-THROUGH

In sports, the expression "follow-through" has vital significance. It refers to that particular bodily action that comes about through a combination of coordinated muscular and mental effort, stance, poise, grip, and balanced determination, and that produces the much-to-be-desired accuracy and power. Power alone is not enough. There must be accuracy as well, and accuracy depends upon the complete knowledge, ability, and practice that permit follow-through to accomplish its objective. The term is here used, however, chiefly as it relates to the continuity of applied and effective force in dealing with persons in the prevention of accidents.

It is not enough to issue orders and to demand results; it is not enough to express interest and to promise support. The orders must be enforced by subsequent procedure, the interest must be sustained, and the promised support must be forthcoming. There is no need to explain or defend follow-through in accident control, but there may be value in a description of some of its more interesting variations, taken from records of successful application.

Failure to achieve results in accident prevention by executives has not been due to lack of interest or absence of will so much as to the absence of facts concrete enough to necessitate or to permit customary follow-through efforts to be applied. When prevention work was confined largely to the establishment of a safety organization, the appointment of a safety engineer, the guarding of physical hazards, or the unassisted service of an outside agency, there was apparently little for an executive to do, other than to arrange for the inauguration of service and to see that it functioned as planned. If accident frequency increased, little if any responsibility was placed upon those in charge of safety, provided they were active. In other words, if meetings were held, inspections were made, physical hazards removed where practicable, and reports in order, accident frequency and accident cost were considered fortuitous.

When, on the other hand, real causes of accidents are known, and, furthermore, when these conditions and circumstances represent worker failure that is subject to supervisory control, the executives who have established a suitable corrective procedure are in a position to secure satisfactory results.

What executive, for example, could fail to develop a definite course of personal action if he realized the existence of an abnormal and unnecessary operating expense, representing a substantial part of the entire overhead cost and relating directly to efficiency and to volume and quality of product, caused by disregard of orders that he and his associates had issued?

If, in addition, the executive knew what particular instructions were disregarded, if he knew who was at fault and where and why the disregard existed, and if he knew also that the orders were of a practicable nature, it would be hard to conceive that anything short of death or paralysis could prevent executive action and follow-through.

Yet such action is not too general a practice, notwithstanding that the information about cost and cause is available. This situation is due chiefly to lack of appreciation of ready-to-hand facts.

A better understanding of the situation is rapidly spreading, however, and, wherever it is applied, results in consistent achievement.

In applying a remedy there is a choice of an almost endless variety of methods, as far as detail is concerned, but the principles remain fixed.

How best may executive support be utilized, assuming that predominant causes of accidents are known and that specific unsafe conditions or practices have likewise been located and specified? Contact with many industrial establishments where these principles of accident prevention have already been successfully applied provides the answer. A few examples may be of interest.

EXECUTIVE INITIATIVE AND FOLLOW-THROUGH

The cases cited here are typical examples of executive action in the successful enforcement of accident-prevention procedures.

Case 1
This is a story of a business executive who managed a string of hotels and who saved hundreds of employees from injury and thousands of dollars in accident cost. One feature is the executive's use of his very

efficient secretary. Another is the fact that the amount of executive time necessary to accomplish satisfactory results was not more than 10 *minutes a month.*

A good secretary is often irritating and, from one point of view at least, the more irritating the more efficient. Busy people often dislike to be reminded of unfinished tasks, of forgotten promises, and of procrastination. But at the same time they know that they must be so reminded or action may never be taken.

The executive's action followed overwhelming proof that accident occurrence was responsible not only for injuries but also for excessive insurance cost, waste, spoilage, delay, lowered morale, and other similar evils. His initial act took less than 5 minutes. It was expressed in the following instruction to his secretary: "Place before me each month *the essential facts* of accident occurrence." He further instructed the secretary—"and see that after the first report has been prepared I am reminded of it everyday until I take action." Thereafter the expenditure of approximately 10 minutes of his time was required each month for a period of 6 months, when it became feasible to refer the matter to an assistant.

The secretary was of the "sometimes irritatingly efficient" type. It was said of him that he had once had three house detectives forcibly eject his employer from a business conference and put him on board a train, notwithstanding that the trip had been postponed, all because he had been instructed "under no circumstances let me miss that train."

This secretary, after consulting the safety engineer, agreed that the essential facts of accident occurrence in this case undoubtedly were cost, frequency, kind, and cause.

He received reports on these matters covering the several branches and departments, had digests made in brief detail suitable for executive consideration, and presented his first monthly statement. It came back to him the next day marked "noted" but with no comment or other indication of action. Five times he sent the statement back to the executive, each time attaching a slip marked: "You have not yet taken action."

Perhaps such action was overefficient; perhaps it was a source of irritation, but bear in mind that the executive had decided that the matter should receive consideration. The executive knew that it was worthwhile, and furthermore he knew himself pretty well and also knew his secretary.

Finally, for the first time, 10 minutes was set aside for executive action. Heads of departments having the poorer accident records were called into conference. The meeting was brief and to the point. With the facts before him, the executive said to the head of one department: "Your workers persist in storing glasses on the ledges over the steam cookers, in violation of instructions. The glasses fall and break, get into the food, and cause injuries. I want you to see that this practice is stopped at once." Similar pointed remarks were made to each of the other department heads, and the meeting was over.

Subsequently, monthly interviews were held, all of them being of short duration. The department heads were impressed; they recognized their responsibility; they sensed the determination of the chief executive to demand results; and they got them.

The details of procedure are relatively unimportant except that they followed identically the procedure already established for the control of other matters of business routine. The main point is that accident prevention was thus so merged with management and supervision that it received, regularly and effectively, the share of executive consideration that it deserved.

The case here described clearly reveals the value of initial executive action and follow-through. It should serve as a stimulating example to the vast majority of businesspeople who do not want employees to suffer needless injury, who are concerned because of high accident costs, and who await only the suggestion of a simple and practicable step before proceeding to do everything within reason to prevent accidents. It is worthy of note that simple and direct corrective action by the executive was made possible only because *causes of accident occurrence* were found and made available to him.

Case 2

This case illustrates one of the most difficult problems that is encountered in safety work—the enforcement of the rule requiring the use of guards for operations where there is commonly some doubt concerning the practical value of the guards. Practicability and common sense must prevail in safety as in other things, and it must be granted that it is sometimes hard to provide an effective guard for the point of operation on a circular saw used for a variety of operations.

Many excellent circular-saw guards have been devised and are available for purchase. Some of them are adaptable to a wide variety of operations, but much resetting and adjusting is necessary to provide constant protection.

The necessity for readjustment of guards, the inherent difficulties in the work, and a certain amount of prejudice on the part of skilled workers—all these things combined to create a real problem in this cabinet shop with the result that cuts and amputation of fingers occurred to employees from unguarded saws, jointers, and shapers on which guards were misadjusted.

The prejudice of old-time employees against protective appliances and their lack of sympathy with orders to keep guards properly adjusted were disturbing factors. But when the executives realized that worker failure was the root of the trouble and that they had previously encountered and solved many production problems arising from the same cause, they were no longer in doubt about the solution of the difficulty.

The situation was frankly explained by the mill superintendent to the supervisors at a group meeting. The attitude of the company with regard to safety was explained. So, too, were the specific causes of accidents (unguarded point of operation and misadjusting guards). The supervisors were asked to cooperate and specifically were requested *not to sympathize* with unwarranted complaints about the delay and inconvenience involved in readjusting guards for short jobs. They were instructed to make it a part of their daily routine to see that orders concerning the use of guards were effectively enforced. In short, the "impossible" was attempted and, as it so often turns out, was found to be not only possible but an accomplished fact.

At first the workers complained, but gradually by constant insistence the new program was put in force and became a fixed part of routine. The accident frequency dropped appreciably in the first 6 months. During the next 6 months a still lower frequency prevailed, and there were no lost-time accidents. The record for the full year, for both frequency and severity, was more than 50 percent better than that for the preceding year.

The particular variation of follow-through used here consisted in personal inspections of the mill room by the mill superintendent. These visits were for the express purpose of observing misadjusted guards and enforcing corrective action on the spot. They were made semi-weekly for the first month, and once a month for the next 5 months, when they were discontinued because they were no longer considered necessary. The employees were impressed, and the supervisors were keyed up and made alert by the manifest sincerity and interest of their superiors.

Case 3

The type of accident, in this instance, was not unusual in trap-rock quarry operations. After blasting, it was customary to scale down the working face to remove loose rock that might otherwise fall unexpectedly upon the people working below. This was often carelessly done.

At regular intervals, as the working face was cut back, the topsoil or overburden was removed. In this stripping operation, loose rock and soil were often left close to the edge of the face and sometimes fell down upon the employees. Cars were loaded, at times in a haphazard manner, rock fell off them, and workers were injured. Employees stood between cars while coupling them.

If these unsafe practices and conditions had received proper consideration, the accident-prevention problem would have been comparatively simple, but for a long time no suitable corrective action was taken because of an improper conception of the situation. The injured employees were considered to be at fault, whereas, in reality, the supervisor of the scaling crew and those in charge of the stripping and loading operations were chiefly responsible. This responsibility was indicated by the fact that the supervisors not only permitted the unsafe practices to go unchecked but had so ordered and arranged the work that it was often difficult for the employees to follow safe procedures.

When the actual facts had been determined by analysis, orders were issued that safe practices must be enforced, and the executives in following through adopted a method that is a bit unusual in quarry operation.

It was customary for the general manager and all superintendents to meet weekly, for 1 hour, to discuss problems concerning cost, production, new orders, personnel, and other related matters. The timekeeper acted as secretary at these conferences. He kept the minutes of the meetings and presented them for approval and, in addition, prepared an agenda for discussion. He was instructed to include a new item (accident prevention) in the agenda for all future meetings and was authorized to provide data relating to it.

At the next meeting he read the following rather startling note:

Accident prevention: Major trouble is "failure to enforce instruction" with regard to safe practices in scaling quarry face, removing overburden, and loading cars. Progress has not yet been made. It has not been necessary to remove overburden since our last meeting, but neither has the crew been instructed as to proper practices. Workers still load cars in such manner that rock is jolted off; and a fall of rock caused by improper scaling occurred Wednesday noon, after scaling had been reported as completed. Fortunately no one was below at the time. Injuries have occurred, however, from rock falling off cars. The general situation remains as unsatisfactory as at our last meeting.

As can well be imagined, a spirited discussion followed, and a plan for more effective action was devised and put into operation at once. In a short time accident frequency was brought under control, and subsequently a splendid record was made in the reduction of lost-time accidents and in reduction of expense. This type of follow-through may be described as "making a place for accident prevention" on the official records of the company.

An even more important point is that the agenda for the safety meeting *was properly planned*. It was not confined merely to a discussion of such accidents as had occurred since the last meeting. Instead, the main topic was what might be called the principal uncorrected hazards of the quarry.

Case 4

Cause analysis in connection with *a tunnel-construction contract* had indicated that poor health, ignorance of safe practice, and lack of conviction that certain activities were really unsafe resulted in the following unsafe practices: Sand hogs entered air locks while in temporarily poor physical condition; barrows of rock were dumped from runways without making sure that all was clear below; workers rode up and down shafts in material buckets, exposed themselves at the bottoms of vertical shafts, and rode the bumpers of muck cars.

Time was at a premium on this particular contract, and the workers were engaged in work that was physically exhausting. It was therefore impracticable to hold meetings at which executive participation in and support of accident prevention might be expressed in talks. Consequently, the expedient of circulating letters, bulletins, and warning notices was chosen as the method to be followed in the new safety program.

The general superintendent, who was also part owner of the company, wrote a personal letter to each superintendent and supervisor. Each unsafe practice that had been selected as typifying one of the chief causes of accidents was made the subject of a bulletin which was posted at strategic locations. Each supervisor received weekly a list showing the number, causes, and types of the injuries that had been sustained by his workers. When orders for work were given out, they were accompanied by an admonition to avoid the practices that had already caused injuries.

Supervisors were instructed to caution workers more frequently and to be more insistent that they follow safe practices—the particular practices requiring special emphasis being named in each case. The general superintendent and her assistants, when making inspections, invariably included in their comments to supervisors remarks concerning specific unsafe conditions as well as the amount and quality of the work performed. In short, the particular conditions that were responsible for the existing accident frequency were emphasized repeatedly by the higher executives, until the desired impression was created. In this way official support was extended without introducing red tape or causing extra clerical work or unnecessary delay.

Inasmuch as the underlying causes of accidents are of a managerial or supervisory nature and also include "outside-of-the-plant," home, social, and environmental circumstances, safety directors or servicing safety engineers are somewhat handicapped in their efforts *directly* to control frequency and severity. What they *can* do is to present the facts as tactfully and impressively as possible to responsible and authorized executives.

Their presentation is most effective when it employs language with which the executives are already familiar. For example, accident cost may be expressed in dollars per ton or other unit of production, or it may be shown as a percent of payroll, cost of raw material, maintenance, or overhead. In one case the safety engineer won executive support by proving that several carloads of lettuce would have to be received, handled, and sold to individual customers in grocery stores before the net profit on the lot equaled the cost of a single serious accident.

Industry as a whole (also large establishments that occupy dominant positions in a community) has an opportunity practicably to improve public attitudes and social and environmental conditions which in turn have a direct bearing on the accident record of employment. As shown in the following examples, management executives have both the opportunity and ability to take corrective action based on causes.

Case 5

In a fertilizer-manufacturing plant, the symptom behind most accidents was known to be violation of rules to use lifelines when entering storage bins.

Corrective action consisted in reprimands and warnings of penalty, and also repeated orders to supervisors that they must strictly enforce the rule.

Further violations occurred, and additional injuries were reported.

The cause was eventually traced to an improper attitude on the part of two supervisors. When one of these employees was asked why he thought an injured employee violated the rule in question he said, "How the———do I know? I wasn't even in the department at the time. I can't stand over every worker and wet-nurse him continually. If the workers haven't sense enough to take care of themselves, they ought to get hurt." It then developed that the unsafe acts of the workers merely reflected improper supervision.

The management, having found that this expressed attitude was more deep-seated than superficial and that in some degree it existed in one other supervisor, made certain supervisory adjustments that effected an immediate improvement in the violation of the safe-practice rule under discussion.

Case 6

A high frequency and severity of employee accidents existed in a group of chemical-manufacturing plants.

These plants were but one group of several subsidiaries of a large concern. When the problem became so acute that it necessitated consideration by the home-office executives, facts were developed as follows:

The local executives were "too busy" to initiate and direct safety work. Production was placed so far above safety in importance that the latter was practically ignored. A wrong conception existed of the relation of safety and production and also of the effect of accident cost on net profit. The local plant executives were reluctant to listen to the advice of the safety engineer. In short, at these plants no attempt had ever been made to establish safety work on an orderly and systematic basis, nor was there the probability that this would ever occur unless because of a "jolt" from some respected source in the industrial organization itself.

Such a jolt was finally forthcoming. An executive order was received direct from the president of the parent company which required:

1　The chief executive to establish immediately a *central* safety committee.
2　The chief executive herself to act as chairperson of the committee.
3　The appointment of a safety inspector and also of a safety committee *in each plant.*
4　That each plant superintendent spend one full day of each week *wholly on accident-prevention work.*
5　That necessary forms be provided, accidents be properly investigated, job analyses be made, responsibility be placed, safeguards be installed,

and in general that safety work be carried on in the future in an orderly, systematic, and effective manner.

Finally, provision was made whereby the chief executive of the subsidiary group of plants was to make a personal report of progress to the home office of the parent company.

Needless to say the subsidiary executives were not "too busy" to attend to safety work thereafter. As a matter of fact, their attitude was not *inherently* faulty, and when they were finally aroused to the importance with which safety was regarded by their superiors, they entered into the program with zeal and managed to accomplish amazingly good accident-prevention results while at the same time improving the quality and volume of production.

It should be concluded from these examples that it is either necessary or advisable to search for underlying managerial and supervisory faults in *all* accident problems.

The cases described here embody the general principles of accident prevention discussed in preceding chapters. In each one of them, however, the chief aim has been to indicate the way in which some of the simpler methods of executive follow-through are employed under different conditions. Further exposition of executive methods of obtaining results would be beyond the scope of this book. It is enough, perhaps, to repeat that executive machinery, as it exists today for the control of quality and volume of product in any well-organized industrial establishment, is adequate to secure and maintain reasonable freedom from accidents—it merely needs to be extended to include accident prevention, without change in method. The chief essential is to *provide facts* of real value in accident prevention, to determine the unsafe practices and conditions causing the accidents and the reasons why such practices and conditions exist.

SAFETY ADMINISTRATION

It must be granted that the inauguration and subsequent functioning of any industrial procedure, be it cost accounting, selection and training of personnel, manufacturing process, bookkeeping, or sales, should have executive direction. The most nearly perfect remedy for accident occurrence, likewise, would suffer in application and successful continuance if executive direction were lacking.

Second, just as it is true with regard to production and sales that responsibility must be placed upon capable and trustworthy persons and that these individuals must be charged with, and made to account for, their records of achievement, just so is it true that persons with similar characteristics should be held responsible for accident prevention.

In the larger industrial operations there are one or more safety engineers who devote all or a part of their time to safety work. There may also be safety committees of various kinds, perhaps an employment, insurance, or personnel manager who supervises accident prevention, or a manager of a department of health and safety; and there are executives who are in a position to spend at least a small part of their time in the administration of safety activities. In short, there is an organization already in existence and available for the administration of an effective accident-prevention program. In the smaller plants a single executive may be obliged to assume the duties (or, at least, the responsibilities) that under other circumstances would be taken over by several employees. Foremen or supervisors, as such, are not always included in the personnel makeup of all branches of industry.

In all cases, however, the principles governing the successful application of a remedy in accident prevention remain the same.

THE PHYSICALLY IMPAIRED WORKER

Inasmuch as accident prevention is basically a humane work, it must take into account the partially disabled or physically impaired workers as well as those who are physically normal. Impaired workers have been proved to be safe workers, but this is contingent on *their proper placement.* Obviously, it would be unwise and unsafe for persons having serious back lesions, lower-extremity disability, artificial limbs, or heart disease to be placed in work requiring extensive walking, standing, or climbing. It is equally clear that persons with chronic bronchial conditions should not be assigned to work where they are exposed to dusts and fumes, that persons having chronic ear impairments would be best suited to noisefree areas, and that individuals with hernias should not be given work requiring heavy lifting.

Placement of the impaired requires:

1 Job analyses—to show the personal and physical requirements
2 Determination of the physical capacities of the disabled person
3 Matching the requirements of the job with the capacity of the person
4 Special instruction and training of the person
5 Revision, as necessary, of the tools and procedures of the job

THE SMALL PLANT

Executive support is available even in a typical one-person plant where the employer himself is production manager, salesman, superintendent, and supervisor. He must personally select the remedy for accident occurrence or delegate the task to one of his employees. He must also issue and enforce orders with regard to the application of the remedy, and he alone must accept the responsibility for failure. If the limited amount of time required for accident-prevention work cannot be found in a plant so small as to fit this description, then there is little hope of accomplishment. This situation, however, does not reflect on the methods herein proposed but rather on the management of the plant, which would fail as regards sales or any other procedure if proper executive supervision were omitted because of cost or lack of time.

EMPLOYEES ARE EXPECTED TO WORK SAFELY

The preceding examples illustrate a few of the many ways in which executives may lend support to a plan for accident prevention in a practicable and businesslike way. One point of considerable value may be expressed as the necessity for asking for what we want if we actually hope to get it. It may be pleasanter, although less direct, to allow employees to *infer* that safe practice is wanted, with the hope that eventually they will themselves decide to do something about it. In business generally, however, these unnecessarily diplomatic preliminaries are often omitted. Producers are expected to produce, and sales people to sell, and definite requests are made by authorized executives along these lines. Good management and diplomacy, of course, require the use of tact, common sense, and variation in the method of making these requests. Granting that supervisors are paid to run their jobs and supervise their workers in all respects, employers clearly have a right to expect of them that proper measures be taken to control worker performance. In turn, the supervisors have a perfect right to expect the workers to follow safe-practice rules.

RESPONSIBILITY MUST BE PLACED

When cause and remedy have been determined and executive interest and support have been extended to the accident-prevention plan, the next step is one of placing responsibility upon the foremen or supervisors.

By and large, employees, in return for wages, are expected to carry on their work according to practicable and fair instructions, and if the executives know exactly what they want and express their desires clearly, they have every right to expect compliance. The whole matter of getting results in accident prevention through the supervisors, therefore, once the causes and conditions have been specified, may be summed up as that of utilizing the four basic forms of executive procedure, namely, *training and instruction, issuing orders, checking for compliance,* and *prevention and correction.*

Without further discussion of these methods at this time, it may be of interest to show how certain individual executives have succeeded in convincing the supervisors that responsibility was fairly placed on them and that action was expected of them.

Case 1

A case in point is that of a hardware plant where the frequency of accidents from burrs and sharp edges while handling sheared material was abnormally high. All ordinary methods had failed to reduce this frequency. Unusual care had been taken to keep the dies and shears sharp and properly adjusted. This had proved helpful, but the conditions had not been improved sufficiently. Gloves with wire insertion in the palms were supplied, and automatic handling equipment was provided for certain operations, with little effect upon the occurrence of injuries.

Finally, at a safety committee meeting, the plant superintendent emphatically stated that something was decidedly wrong, but she didn't know just what it was, and that she was going to put it squarely up to the supervisors to produce results. She said that she did not propose to tell them how—that was their business and was one of the things they were paid for. She demanded improvement and left the meeting in high dudgeon. Within the week one of the supervisors advanced the suggestion that certain objects, then being manufactured out of sheet metal, be cast in molds. Although at first thought this seemed to be utterly impracticable, experiments were made and it was found that the manufacture of malleable iron castings was cheaper, the product was better in every way, and the process was safer. Nowhere is it more true than with regard to accident prevention that necessity mothers both invention and accomplishment.

Case 2

In another case, the management of a large paper mill had become thoroughly convinced of the humanitarian and economic significance of accidents. The executives were informed about causes, specific unsafe conditions and practices, and remedy, but they found it hard to convince the supervisors that it was fair to charge them with responsibility. The supervisor of the machine-repair department was the most obstinate of the group, and it was believed that his conversion would have a beneficial effect upon many of the others who looked upon him as their model in many ways.

This supervisor was highly intelligent. He was a clear thinker and possessed no mean ability to argue. He was converted by drawing an analogy between his position and that of a police official in charge of traffic—he himself having drawn a similar analogy, but from the opposite point of view. Discussion had taken place concerning one of his workers, a lathe hand, who persisted in the practice of brushing off chip accumulations with the heel of his palm, where they curled up, hot and jagged from the carbide cutting tool, while he was dressing down rolls. A hand brush was available for this purpose, and he had also been told that on roughing cuts it was permissible to stop the lathe, but he openly disregarded the instruction and had already twice been injured.

The supervisor said: "What am I going to do except fire the man? None of us wants to do that." Then he added: "All people break rules occasionally. Look at the apparently intelligent pedestrians who cross · streets against the traffic lights. I do it myself at times. Everybody does it. You just can't get people to do what they should do."

The secretary and treasurer, with whom he was arguing, then related the following incident:

He said: "While visiting a nearby city not long ago, I was guilty of jaywalking, just as you have indicated. I wanted to cut across a street to get to a cigar store directly opposite, in the middle of the block. People, hundreds of them, were lined up on the sidewalks but the street was bare. I looked both ways. No traffic was in sight, and as in my judgment it was entirely safe to cross, I stepped off the curb. I was within 15 feet of the other curb when, with a tremendous clatter, a mounted policeman drove his horse between me and the sidewalk, reared the prancing and spirited animal up on its hind legs, and then caused him to sidestep so close to my toes that I was forced to retreat. I moved slowly and with dignity at first, but as the horse kept crowding me I was obliged to turn tail and make a dash for the place I had come from.

"The officer pursued me over the sidewalk and to the store fronts and then dismounted, while the appreciative crowd, that had apparently gathered in the hope of witnessing just such an event, roared with laughter. The officer said: 'Do you happen to be from out of town, sir?'

'Yes, I am a stranger here,' I replied. 'I thought so. Well, let me make it clear that we are campaigning against jaywalking in this city. We have traffic rules and they are made to be obeyed and not to be ignored. We mean business. If you want to cross this street, go up to the next corner, wait for the green light, and cross in safety. Good-by, sir, and good luck to you.'

"I maintain, therefore," continued the secretary, "that it is entirely possible to enforce even traffic regulations if the will and executive support exist, and if the enforcement officers are capable. They had good enforcement in that town—good supervisors, if you please—and I may add that I follow the police regulations to the letter—at least whenever I am in that particular place."

The supervisor agreed that although handicapped a bit by lack of a spirited mount, he nevertheless could probably find some way to get his workers to obey instructions.

In these illustrative cases there was executive interest, and knowledge of specific unsafe practices, and in all cases responsibility was placed on the supervisors.

The supervisor upon whose shoulders the burden now rests must begin his work. In trying to control injury frequency, however, he is often puzzled because the injuries are not frequent. He may, in fact, wait for one to occur so that he may then make use of his opportunity to improve conditions. He may have few such opportunities in a limited period, however, and, meanwhile, interest is likely to wane, the impression created by his superiors may wear off, the injuries (when they do occur) may be widely different as to cause and type, and he may feel that for the moment there isn't much to do. This is a grave mistake. There are imperfections in every industrial operation, and successful accident prevention requires unceasing thought and effort. The trouble generally lies in the failure to appreciate the marked distinction between an accident, its cause, and the injury. If this distinction is fully understood, much can be done to stop accidents before they result in injuries.

MATERIAL-HANDLING ACCIDENTS

It used to be that unguarded machines and unprotected mechanical exposures were the chief factors in causing industrial accidents. Conditions have changed, however, and although machines still cause accidents, machine-guard manufacturers and employers who purchase machines and put them to use, working cooperatively, have materially reduced the hazards and incidentally the accidents resulting from these hazards.

Of all accidents, not more than about 10 percent are fairly attributable to machines, mechanical processes, dangerous liquids, gases, and materials, and to unsafe building conditions, improper dress, lack of safety wearing apparel, lighting, ventilation, and other conditions in the physical or mechanical group of accident causes. On the other hand, about 88 percent of all accidents are attributed chiefly to unsafe practices that may be controlled by good management. These practices include the unsafe handling of material, and it is therefore particularly appropriate at this point to consider ways and means of applying the principles of accident prevention to such operations.

Application of Remedy in Material Handling

Two highly successful methods may be employed in preventing material-handling accidents. One lies in the study and improvement of methods and procedure to the end that some safer and better way may be found to conduct any particular operation than that in use. In addition, it is of value to utilize that proper degree of supervision— from the chief executives down to the supervisor—that will result in the observance and correction of unsafe material-handling practices on the part of the workers, *before* injuries occur.

Process and Procedure Revision Engineering revision has already substituted mechanical power for human power—automation, the feedback system, mechanical feeding and handling equipment in general, for the slower, more costly, and more dangerous manual handling. Power shovels have been devised of such capacity that one of them will, in a given time, move as much rock, slag, ore, or other similar material as could be handled by more than 1,000 laborers working with hand shovels, picks, and barrows. Machines have been built that convert logs to toothpicks and matches, cotton to cloth, sand to bottles, and paper and ink to newspapers, with no more hand labor than is necessary to adjust, repair, and maintain the equipment. Belt and bucket conveyors, cranes and hoists, concrete mixers, and forklift trucks are available to handle material mechanically.

More encouraging still is the fact that the workers thus released from hand labor have found profitable employment in the constantly increasing operations that have resulted from the public demand for manufactured products.

Supervisory Methods

If engineering revision fails to provide an answer to the problem of preventing material-handling accidents, dependence can safely be placed upon supervisory methods—provided these methods are extended in the direction of safety with the same thoroughness with which they are applied to production.

Careful workers can handle material safely. Even rough and splintered lumber, sharp-edged objects of all kinds, and heavy and cumbersome pieces can be handled without accidents if the workers are safety-conscious, that is, if they are careful and attentive, get a good firm grip on the material, and do not let it slip out of their hands. If employees were to ruin expensive products or jam and break tools, dies, and other costly machinery and thus cause unnecessary delay and loss because of their fumbling and bungling—if they were doing these things as frequently as they do the unsafe things that result in material-handling accidents—industrial executives would not be at a loss to provide a remedy.

The remedy that they would apply is the one here advocated for safety, namely, supervisory methods based on facts. The supervisor would be held responsible for issuing and enforcing orders for careful procedure, and all executives would share in this responsibility.

All that is required is executive recognition and interest, followed by supervisory activity directed at the causes of the chief sources of trouble, be they encountered in storing or taking down material in racks, in piling lumber, in feeding machines, or in any one of the other countless operations associated with the handling of material.

The experience of accident-prevention engineers proves conclusively that remarkable results can be attained in this way and that they are being attained wherever the facts are presented to interested executives who are desirous of cooperating and who have the will to achieve.

Suggested Executive Procedure

Executives should give due consideration to the following suggestions in their effort to reduce the frequency (and, incidentally, the cost) of material-handling accidents. They should:

1 Determine from an analysis of past injuries the kind of operation conducted, the type of accidents, and the specific causes of the accidents.
2 Analyze existing operations to determine if they are conducted in an unsafe way.
3 Determine the frequency as well as the number of injuries, by operation, location, and supervisor.
4 Study the entire situation carefully and select the place or places and the operations most worthy of attack.
5 Having determined the types of accidents and the specific unsafe practices and conditions at fault, proceed first with the application of engineering revision—asking the question: Can this operation be done in another, safer, and better manner?
6 Proceed next (or coincidentally) with improvement of supervision directed to the elimination, or at least the mitigation, of the predominating unsafe practices or conditions in handling of material.
   a. Charge supervisors with responsibility for the prevention of accidents.
   b. Enforce instructions for safe-handling practices.
   c. Keep in mind that an unsafe act occurs several hundred times before a serious injury results and that there is, therefore, an excellent opportunity to detect and correct unsafe practices before injury occurs.
   d. Select, train, and instruct employees.
   e. Place accident prevention on a par with production, with respect to supervisory methods of improvement and executive interest, participation, and follow-through.

APPLICATION OF REMEDY FOR SLIPS AND FALLS ON FLOORS

A concern specializing in nonslip and electrically conductive floor maintenance reached the following conclusions as a result of many years' experience in office buildings, hospitals, warehouses, schools, and industry in general:

1 Floors receive less attention with regard to safe and efficient maintenance than any other industrial item of equal cost.
2 In the great majority of cases decisions about maintenance materials and methods are left wholly to uninformed personnel.
3 Maintenance programs and systematic methods are conspicuous for their absence.
4 Floors *should be* and *can be* nonslip, electrically conductive where necessary, attractive in appearance, and inexpensive to maintain.
5 Floor maintenance is worthy of consideration by persons trained in floor-maintenance engineering.

The following typical example serves to illustrate these points:

A large company maintained an extensive engineering department in which plans, specifications, and blueprints were prepared and materials selected for all the departments of the company with the exception of floor maintenance. This department was given no attention or guidance as to how the floors were to be maintained or materials to be used. The problem of maintaining 250,000 square feet of asphalt-tile floor with all its complications was left to uninformed personnel to do as best they could.

As a result no planned method or program was followed. Mistake after mistake in the selection of materials was made. Buying was haphazard. The department was completely lacking in organization or leadership. The selection of materials depended entirely upon low first cost.

A straight mopping method was followed for a period of 2 years at great cost in labor and material, because of deterioration of the asphalt tile, before they discovered a system of nonslip floor maintenance. During this mopping period surface deterioration was such as to lead to consideration of replacing 150,000 square feet with new asphalt tile. The frequency of slip and fall accidents was high.

Safety engineers who were specialists in floor maintenance were then called in. They found a method of reclaiming the floors making it unnecessary to purchase new flooring, and for the next 7 years this system of nonslip floor maintenance was diligently followed with economy formerly unheard of, with an accident rate always at a minimum, and with a most desirable floor appearance.

After 25 years of experience the following conclusions seem to be justified:

1 A planned system of floor maintenance is just as important as any other functional operation.
2 In the selection of materials the following rules apply:
   a. Materials should not be selected wholly on a low first-cost basis.
   b. Only safe cleaners should be selected to protect the flooring and the personnel handling it.
   c. Only nonslip finishes should be selected, and this should be on a performance basis.
3 On-the-job training classes for porters should be conducted by specialists in floor maintenance—to train them to use and apply materials and equipment effectively and safely.
4 Floor finishes should accomplish the following:
   a. Be nonslip at all times.
   b. Meet NFPA standards for electrical conductivity.
   c. Be economical to maintain.

    d. Form a tough durable film to protect the floor from abrasion of traffic.

    e. Eliminate the necessity for frequent strippings and other expensive operations.

    f. Accentuate the beauty of the floor.

5 Specialists in floor maintenance should be called in. Building, school, hospital, and industrial floors differ, and the initial problems presented should have the attention of a trained specialist so that a safe underfoot surface will be ensured.

CORRECTIVE ACTION FOR OSHA VIOLATIONS

Finding OSHA violations is considerably easier than correcting them. Regardless of how a company has gone about finding them, it no doubt ends up with quite a list of things wrong that must be corrected. Obviously each company handles this somewhat differently, but a few general principles seem to apply:

1 First some priorities have to be set. In the last chapter we discussed this, utilizing a decision matrix; the system is shown in Fig. 7–8.

2 Then someone has to be assigned to get things done. Often this takes a special task force within maintenance, in order not to overload them.

3 There must be continuing and constant follow-up by the safety professional.

4 There will have to be documentation of everything done.

Should you rewire the presses or the exit lights first? Should you build collapsible handrails for your loading dock or handrails around your roof first? These are the kinds of decisions that someone will have to make. You will have a list (Fig. 7–8, for instance), and for each item on it there should be a start and a finish time.

    Figure 7–7 is the matrix used by one organization. As supervisors or inspectors reported to safety personnel any physical violations they could not correct themselves, the violations were given a priority ranging from 1 to 9 and put on maintenance (task force) lists (see Fig. 7–8). This assured early completion dates for imminent and serious hazards (particularly those of moderate or small cost).

    But first you will need to assess the capabilities of your maintenance people. Can they do your work as well as their regular tasks within the time you want? Many organizations have found they need to set up a separate OSHA maintenance force. This force can report to staff safety or to maintenance as long as their efforts are directed *solely* to OSHA items.

    In our previous example, once priorities were established, the maintenance task force scheduled each (Fig. 7–8), setting a start and finish date. This schedule became the work assignment schedule for the task force, the report to management, and the needed documentation for OSHA. Updated weekly, it showed the progress in compliance.

Some companies have expressed concern over showing compliance officers a chart of this nature, fearing that it would show them violations they might not otherwise have found. If you have this fear, you might feel a little better if you build the chart with job numbers rather than descriptions. The officers may not ask to look further than that.

Obviously, follow-up is always needed. Schedules have to be adjusted, new items are added, completed items removed. The chart, however, makes follow-up easy. Progress (or lack of it) can be seen at a glance, and any problems in completing can be discussed.

The first task is to get in compliance. This is the easier task. The second one, to keep in compliance, is considerably more difficult.

For instance, the law requires that the tool rests of abrasive wheels be adjusted to within ⅛ inch of the wheel. We can make that adjustment on Monday at 8 A.M. and the adjustment will no doubt be changed by 8:05, putting us again in violation of the law. The trick is to devise plans and systems that keep the tool rest at ⅛ inch.

Any person who has seen both the standards and an industrial organization knows the difficulty of keeping in compliance. First of all, this task is really no different from any other managerial task. To achieve it, management must state what it wants (set policy), state who is to do it (assign responsibility), and check to see that it is being done (fixing accountability through inspections, audits, and so forth).

Keeping in compliance with OSHA requires that we regard the supervisor as the key person and that middle and top management do whatever is necessary to ensure that the supervisor will be the key person. Therefore, in this section we looked generally at what makes supervisors perform, and specifically at what makes them perform in the area of keeping their departments in compliance with the law. We have always felt that the supervisors were the key people in safety work, and modern thinking does not change this, although it places additional emphasis on the "key chain-holders."

One important aspect of the OSHA compliance task is the extreme need for documentation. Much, almost all, of what is done must be documented. First, the standards themselves require a number of specific types of documentation. At last count we found some 208 specific places in the standards that require some type of inspection and documentation thereof; 82 places where they require some type of record to be kept of training that has been given; etc. In addition the law requires that plans and procedures and programs be documented in various categories. And finally, good documentation has been found to be evidence of "good faith," and thus will reduce the amount of fine levied. Documentation is a must.

The final step in our accident process is monitoring, discussed in the next chapter.

REFERENCES

Grimaldi, J., "Industrial Rehabilitation," *Safety Training Digest*, New York University, New York, 1945.

Petersen, D., *The OSHA Compliance Manual*, McGraw-Hill, New York, 1975.

NINE

Monitoring

Monitoring is the final step in the process; it is also the first step, for with good measures, we spot problems and symptoms for detailed analyses and additional remedies. Monitoring as we'll discuss it in this chapter refers to three distinct activities:

1 First it is a method of gauging our performance. It tells us roughly how we are doing.
2 It also provides us with red flags—tells us when and where something in our system is going wrong.
3 And finally it is a measure of performance for the individuals in our organization that are charged with doing the job—the line organization.

APPRAISING THE EFFECTIVENESS OF ACCIDENT PREVENTION

Application of remedy, or corrective action, is of value only as it results in control of the conditions and circumstances it seeks to correct. A yardstick of some kind must be provided to check effectiveness and to justify the time and effort expended.

Disabling-Accident Frequency Formula

The lost-time or disabling-accident frequency formula is commonly used as a measure of effectiveness, and where its limitations are appreciated it is quite satisfactory. This formula is expressed as the number of disabling injuries times 1,000,000, divided by the total number of work-hours worked. It provides what is called the "frequency rate of disabling injury," or the number of disabling injuries per 1,000,000 work-hours of work exposure.

Example
Five hundred people work 8 hours each day for 30 days and during this time 2 disabling injuries occur. According to the formula the number of disabling injuries (2) times 1,000,000 equals 2,000,000, and this divided by the number of work-hours worked (500 × 8 × 30, or 120,000) results in a disabling-injury frequency rate of 16.6 plus.

Limitations of Disabling-Injury Frequency Rate

A small establishment will find this frequency rate of little value except for a year or more of exposure. For week-to-week or even month-to-month periods so few disabling injuries occur that frequency rates are highly erratic and noninformative. Small establishments would do better to use *all* accidental injuries instead of disabling cases only and to adapt the formula so as to produce a readable end figure. This can be done by decreasing the factor of 1,000,000 in the numerator to 30,000, this representing approximately the comparative frequency of minor to major injuries.

Example
One hundred people work 8 hours each day for 20 days. During this period there are 6 minor injuries, none of a disabling nature. According to this *revised* formula, the number of minor injuries (6) times 30,000 equals 180,000, and this divided by the work-hours worked (100 × 8 × 20, or 16,000) results in a minor injury frequency of 11.2 plus. Obviously it would not be proper to compare this frequency with one obtained in another plant where disabling injuries only were consid-

ered and where the factor of 1,000,000 was used. However, it would serve to permit comparisons of past to present experience *in the same plant.*

Accident Severity Formula

The severity or extent of accidental injury provides an additional method of appraisal. It is expressed as the days lost per 1,000,000 hours worked or

$$\frac{\text{No. of days lost} \times 1,000,000}{\text{No. of hours worked}}$$

If, as in the preceding example on accident frequency, 2 disabling injuries occur in a plant where 500 people work 8 hours a day for 30 days and the two injuries cause a total of 6 days lost time, the severity rate would be calculated as follows:

$$\text{No. of days lost} = 6$$
$$\text{No. of hours worked} = 500 \times 8 \times 30 = 120,000$$
$$\frac{6 \times 1,000,000}{120,000} = \text{severity rate of 50}$$

Severity is largely fortuitous, as shown by instances where one person may lose an eye when struck by a flying object while another in identical circumstances may sustain but a glancing blow on the forehead. The severity formula must therefore be used with discretion. In combination with the frequency formula, it is of value, over long periods of exposure, as a means of evaluating the injury hazards of varying occupations.

There is much disenchantment with the frequency and severity rate measures of safety performance. As noted before, they should not be used as indicators of performance in safety for smaller companies. The data base is simply too small for them to be valid measures. In larger companies they are somewhat better, but still have weaknesses. A number of other measures have recently been discussed.

Injury frequency and severity ratios measure only the *end results* of accident-prevention work and as such do not fully qualify as appraisal methods. It takes an appreciable amount of time for even the most effective safety programs to "take hold." Especially is this statement true of safety educational activities. In certain cases, executives who were but partially convinced and who somewhat reluctantly initiated a series of safety meetings have been discouraged to the point of discontinuing their efforts when serious accidents occurred following close on the heels of the very first meeting. Such action is as indefensible as to discontinue sales advertising because a customer is lost the very day the first advertisement is printed.

A *complete* appraisal attempts to evaluate *activities* as well as injury frequency and considers organization, executive attitudes, committees, accident investigation, safety inspections, safety posters, and other safety educational work, first aid and hospital, analysis, records, etc.

Some of the other measures available are:

1 *Frequency severity indicator.* A combined frequency and severity rate. FSI equals the square root of the frequency rate times the severity rate divided by 1,000.

$$FSI = \sqrt{\frac{F \times S}{1,000}}$$

For example,

Frequency Rate	Severity Rate	FSI
2	125	0.5
4	250	1.0
8	500	2.0

2 *Total cost of first-aid cases.* In this appraisal the pro-rata cost per case handled in-plant would be considered. For example, say that a first-aid unit in a plant costs $10,000 to operate. Forty percent of the unit's time is used to treat first-aid cases. An average of 1,000 cases are treated; therefore, the cost is $4 per case. This example combines industrial and nonindustrial first-aid cases. If average time per case is substantially different for each of these categories, a separate average cost per case may be better.

3 *Cost incurred.* Includes the actual compensation and medical costs paid for cases which occurred in a specified period plus an estimate of what is still to be paid for those cases.

4 *Estimated cost incurred.* An estimate of item 3 above, based on averages.

5 The *cost factor.* Equals the total compensation and medical cost incurred (see item 3 above) per 1,000 work-hours of exposure.

$$\text{Cost factor} = \frac{\text{cost incurred} \times 1,000}{\text{total work-hours}}$$

6 *Insurance loss ratios.* Equal to the incurred injury cost divided by the insurance premium.

$$\text{Loss ratio} = \frac{\text{incurred costs}}{\text{insurance premium}}$$

Some of the above measures are dollar-oriented. These are perhaps more meaningful than some of safety's more traditional measures. To line managers the dollar is well understood. When we talk dollars, we are talking management's language.

Management is first of all interested in the relationship of the safety professional's ideas to the profits of the organization. That is, what will management get in return for the money it is being asked to spend? Thus we ought to be dollar-oriented when we talk to management. Even if management understands the language of frequency and severity rates, we ought to talk in terms of dollars. What dollar indicators can be used with management? Here are some possibilities:

1 Dollar losses (claim costs) from the insurance company
2 Total dollar losses (insurance direct costs) plus first-aid costs not paid by insurance
3 So-called hidden costs
4 Estimated costs
5 Insurance loss ratio
6 Insurance premium
7 Insurance experience modification
8 Insurance retrospective premium
9 Cost factor (see Chap. 2)

Losses

Dollar losses (items 1 and 2) are good indicators. (It seems more realistic to include first-aid cost than to ignore it.) Many companies use these figures or some measurements based on them, such as cost per work-hour.

Actually, of course, the direct cost of accidents (claim costs) is not money paid out directly by the company. It is money that the insurance carrier pays to the injured employee. Hence these figures are in some sense "unreal" to management. In addition, they are not easily or accurately obtained. It may be years before actual costs of serious losses are known. Furthermore, using such figures for multiplant companies involves unfair comparison, since compensation benefits vary widely in different states.

Hidden costs are real, but they are difficult to demonstrate.

Estimated Costs

The actual direct cost of a work injury—compensation and medical benefits—often is not established until long after the injury occurs, especially for the more severe cases. Thus, the loss statements provided by insurance carriers do not serve the purpose of effective, *current* cost evaluation. The only information that such statements can provide on relatively recent injuries consists of the amounts of medical and compensation benefits paid to date, plus the outstanding reserves. These reserves are primarily established to assure that sufficient funds are set aside for the eventual cost of the claims. It is not until the healing period has ended, however, and the degree of disability has been positively determined, that any accurate estimation of the cost can be made. And the more complicated the injury, the longer it takes to establish the final cost.

In recent years there has been an increasing demand for some method by which employers can determine promptly the approximate cost of compensable occupational injuries. Several such estimation systems are now in use in various companies in the United States. Estimated costs have proved themselves in industry to be an effective method of obtaining cost information for line measurement.

Insurance Costs

Management pays an insurance premium for workers' compensation coverage. This is a real cost to it. How much it pays is directly dependent on the company's past and present accident record. This premium is directly out of management's pocket. Why not then measure the results in terms of this real out-of-pocket cost to the management? The insurance premium is based on industry averages, adjusted by each company's record in the past and in some cases again adjusted by the currect accident record. These adjustment factors, known as the experience modifiers and the retrospective adjustment, are excellent indicators of past and present performance. Safety people ought to know, understand, and use these adjustment modifiers, since they represent the true costs of safety to management. Chapter 17 describes these rating systems.

Audits

As indicated before, using frequency and severity rates, or even using dollar measures, is to utilize measurements of only end results. This is just not enough. Complete appraisals are also necessary to evaluate activities. Some checklists and systems that can be used to audit safety performance were discussed in Chap. 4.

One excellent activity measuring device was presented by Roman Diekemper and Donald Spartz, in an article in the December 1970 *Journal of the ASSE,* quoted below:

> Before any measurement device can be used with reasonable accuracy to evaluate safety activities, there must be a set of rules or standards developed; that is, the measurement device must be standardized and capable of objective and continued application. We must use the same "yard stick," and use it the same way, each time an evaluation or measurement is made.
>
> Secondly, the thing to be measured must possess certain characteristics. It must be structured so as to be measurable.
>
> This is vital if an evaluation of safety activity is to be made at various locations of a multi-location corporation. Each plant or location must be structured with the same ground rules; i.e., a uniform "safety program."
>
> This "program" must clearly define the structure and nature of expected activity. The responsibilities of the various management-levels must be defined, and essential, basic activities (standards) which are to be implemented must be outlined; for example, accident and/or incident investigation, analysis of data for causal factors, internal self-inspection programs, etc.
>
> Third, the measurement technique must be designed so that the line-managers can personally relate their activities to the standard.
>
> The weighted values incorporated into the measurement device must realistically reflect the degree of importance placed by management on each of the various safety activities to be measured. This is demonstrated in the measurement technique which follows later, and is determined by in-depth analysis of past occurrences.
>
> The criteria and the weighted values should be reviewed and changed to reflect progress. (As the application of a sound safety program becomes more sophisticated, the need to change the criteria and the weighted values increases). Again, it must be stressed that for valid comparative

purposes the same set of criteria and values must be used throughout the organization.

The following measurement technique could be used for a single-plant operation or with a multi-plant operation, where the level of safety activity is comparable and a standard corporate safety "program" exists. It also lends itself to the evaluation of safety activities where the degree of sophistication is not far advanced. This would apply, unfortunately, to a wide area of American industry.

The mechanics are simple, and the process is easily understood by plant management. The format consists of three parts: (1) the activity standards, (2) a rating form, and (3) the summary sheet to compute the final score. As outlined below, these are self-explanatory.

The benefits of a valued measurement of safety activity are many. It provides a vehicle to convey to management in understandable, concrete terms the status of the safety effort. A measurement "score" is easily-understood.

A valued measurement will bring into proper perspective the various loss-control measures. On a continuing basis, this enables the safety professional to promote through plant management, in an understandable way, those loss-control measures which are considered to be the most effective.

The evaluation process incorporated in the technique requires a degree of guided self-evaluation. The local plant management must examine their own operation under the guidance of an evaluator to determine the status of safety activities. Weaknesses and areas needing additional emphasis are readily recognized.

By participating in the evaluation process, the local management group becomes familiar with the expectations of the company, and they are motivated to strengthen those activities in need of strengthening.

A measurement which provides a basis for comparative evaluation is achieved in the above technique. A comparison of the scores from one year to the next is valid to determine progress. Within a multi-plant operation a comparison of the various plants can be made.

The criteria are known in advance and the weighted values are provided. Thus a plant knows ahead of time what is going to be measured, how it is to be measured, and that the results will be compared with previous results. The local management is also quite aware of the comparisons with other plants which might be made at the corporate level.

While the technique outlined tends to maintain objectivity, the evaluator must also be objective in his assignment. His ability to be objective in guiding local management through the self-evaluation portions of the measurement process will enhance the whole effort and make it a useful tool for future control of accidents.

ACTIVITY STANDARDS

A. ORGANIZATION & ADMINISTRATION

	Activity	Poor	Fair	Good	Excellent
1	Statement of policy, responsibilities assigned.	No statement of Loss Control policy. Responsibility and accountability not assigned.	A general understanding of Loss Control, responsibilities and accountability, but not written.	Loss Control Policy and responsibilities written and distributed to supervisors.	In addition to "Good" Loss Control policy is reviewed annually and is posted. Responsibility and accountability are emphasized in supervisory performance evaluations.
2	Safe operating procedures (SOP's)	No written SOP's.	Written SOP's for some, but not all, hazardous operations.	Written SOP's for all hazardous operations.	All hazardous operations covered by a procedure, posted at the job location, with an annual documented review to determine adequacy.
3	Employee selection and placement.	Only pre-employment physical examination given.	In addition, an aptitude test is administered to new employees	In addition to "Fair" new employees' past safety record is considered in their employment.	In addition to "Good" when employees are considered for promotion, their safety attitude and record are considered.
4	Emergency and disaster control plans.	No plan or procedures.	Verbal understanding on emergency procedures.	Written plan outlining the minimum requirements.	All types of emergencies covered with written procedures. Responsibilities are defined with backup personnel provisions

	Item				
5	Direct management involvement.	No measurable activity.	Follow-up on accident problems.	In addition to "Fair," management reviews all injury and property damage reports and holds supervision accountable for verifying firm corrective measures.	In addition to "Good" reviews all investigation reports. Loss Control problems are treated as other operational problems in staff meeting.
6	Plant safety rules.	No written rules.	Plant safety rules have been developed and posted.	Plant safety rules are incorporated in the plant work rules.	In addition, plant work rules are firmly enforced and updated at least annually.

B. INDUSTRIAL HAZARD CONTROL

	Item				
1	Housekeeping—storage of materials, etc.	Housekeeping is generally poor. Raw materials, items being processed and finished materials are poorly stored.	Housekeeping is fair. Some attempts to adequately store materials are being made.	Housekeeping and storage of materials are orderly. Heavy and bulky objects well stored out of aisles, etc.	Housekeeping and storage of materials are ideally controlled.
2	Machine guarding.	Little attempt is made to control hazardous points on machinery.	Partial, but inadequate or ineffective, attempts at control are in evidence.	There is evidence of control which meets applicable Federal and State requirements, but improvement may still be made.	Machine hazards are effectively controlled to the extent that injury is unlikely. Safety of operator is given prime consideration at time of process design.
3	General area guarding.	Little attempt is made to control such hazards as: unprotected floor openings; slippery or defective floors; stairway surfaces; inadequate illumination, etc.	Partial but inadequate attempts to control these hazards are evidenced.	There is evidence of control which meets applicable Federal and State requirements—but further improvement may still be made.	These hazards are effectively controlled to the extent that injury is unlikely.

Activity	Poor	Fair	Good	Excellent
4 Maintenance of equipment, guards, handtools, etc.	No systematic program of maintaining guards, handtools, controls and other safety features of equipment, etc.	Partial, but inadequate or ineffective maintenance.	Maintenance program for equipment and safety features is adequate. Electrical handtools are tested and inspected before issuance, and on a routine basis.	In addition to "Good" a preventative maintenance system is programmed for hazardous equipment and devices. Safety reports filed and safety department consulted when abnormal conditions are found.
5 Material handling—hand and merchanized.	Little attempt is made to minimize possibility of injury from the handling of materials.	Partial but inadequate or ineffective attempts at control are in evidence.	Loads are limited as to size and shape for handling by hand, and mechanization is provided for heavy or bulky loads.	In addition to controls for both hand and mechanized handling, adequate measures prevail to prevent conflict between other workers and material being moved.
6 Personal protective equipment—adequacy and use.	Proper equipment not provided or is not adequate for specific hazards.	Partial but inadequate or ineffective provision, distribution and use of personal protective equipment.	Proper equipment is provided. Equipment identified for special hazards, distribution of equipment is controlled by supervisor. Employee is required to use protective equipment.	Equipment provided complies with standards. Close control maintained by supervision. Use of safety equipment recognized as an employment requirement. Injury record bears this out.

C. FIRE CONTROL AND INDUSTRIAL HYGIENE

1 Chemical hazard control references.	No knowledge or use of reference data.	Data available and used by foremen when needed.	In addition to "Fair" additional standards have been requested when necessary.	Data posted and followed where needed. Additional standards have been promulgated, reviewed with employees involved and posted.
2 Flammable and explosive materials control.	Storage facilities do not meet fire regulations. Containers do not carry name of contents. Approved dispensing equipment not used. Excessive quantities permitted in manufacturing areas.	Some storage facilities meet minimum fire regulations. Most containers carry name of contents. Some approved dispensing equipment in use.	Storage facilities meet minimum fire regulations. Most containers carry name of contents. Approved equipment generally is used. Supply at work area is limited to one day requirement. Containers are kept in approved storage cabinets.	In addition to "Good" storage facilities exceed the minimum fire regulations and containers are always labeled. A strong policy is in evidence relative to the control of the handling, storage and use of flammable materials.
3 Ventilation— fumes, smoke and dust control.	Ventilation rates are below industrial hygiene standards in areas where there is an industrial hygiene exposure.	Ventilation rates in exposure areas meet minimum standards.	In addition to "Fair" ventilation rates are periodically measured, recorded and maintained at approved levels.	In addition to "Good" equipment is properly selected and maintained close to maximum efficiency.

	Activity	Poor	Fair	Good	Excellent
4	Skin contamination control.	Little attempt at control or elimination of skin irritation exposures.	Partial, but incomplete program for protecting workers. First-aid reports on skin problems are followed up on an individual basis for determination of cause.	The majority of workmen instructed concerning skin-irritating materials. Workmen provided with approved personal protective equipment or devices. Use of this equipment is enforced.	All workmen informed about skin-irritating materials. Workmen in all cases provided with approved personal protective equipment or devices. Use of proper equipment enforced and facilities available for maintenance. Workers are encouraged to wash skin frequently. Injury record indicates good control.
5	Fire control measures.	Do not meet minimum insurance or municipal requirements.	Meet minimum requirements.	In addition to "Fair" additional fire hoses and/or extinguishers are provided. Welding permits issued. Extinguishers on all welding carts.	In addition to "Good" a fire crew is organized and trained in emergency procedures and in the use of fire fighting equipment.
6	Waste—trash collection and disposal, air/water pollution.	Control measures are inadequate.	Some controls exist for disposal of harmful wastes or trash. Controls exist but are ineffective in methods or procedures of collection and disposal. Further study is necessary.	Most waste disposal problems have been identified and control programs instituted. There is room for further improvement.	Waste disposal hazards are effectively controlled. Air/water pollution potential is minimal.

D. SUPERVISORY PARTICIPATION, MOTIVATION AND TRAINING

1	Line supervisor safety training.	All supervisors have not received basic safety training.	All shop supervisors have received some safety training.	All supervisors participate in division safety training session a minimum of twice a year.	In addition, specialized sessions conducted on specific problems.
2	Indoctrination of new employees.	No program covering the health and safety job requirements.	Verbal only.	A written handout to assist in indoctrination.	A formal indoctrination program to orientate new employees is in effect.
3	Job hazard analysis.	No written program.	Job hazard analysis program being implemented on some jobs.	JHA conducted on majority of operations.	In addition, job hazard analyses performed on a regular basis and safety procedures written and posted for all operations.
4	Training for specialized operations (Fork trucks, grinding, press brakes, punch presses, solvent handling, etc.)	Inadequate training given for specialized operations.	An occasional training program given for specialized operations.	Safety training is given for all specialized operations on a regular basis and retraining given periodically to review correct procedures.	In addition to "Good" an evaluation is performed annually to determine training needs.
5	Internal self-inspection.	No written program to identify and evaluate hazardous practices and/or conditions.	Plant relies on outside sources, i.e., Insurance Safety Engineer and assumes each supervisor inspects his area.	A written program outlining inspection guidelines, responsibilities, frequency and follow up is in effect.	Inspection program is measured by results. i.e., reduction in accidents and costs. Inspection results are followed up by top management.

215

Activity	Poor	Fair	Good	Excellent
6 Safety promotion and publicity.	Bulletin boards and posters are considered the primary means for safety promotion.	Additional safety plays, demonstrations, films, are used infrequently.	Safety displays and demonstrations are used on a regular basis.	Special display cabinets, windows, etc. are provided. Displays are used regularly and are keyed to special themes.
7 Employee/supervisor safety contact and communication.	Little or no attempt made by supervisor to discuss safety with employees.	Infrequent safety discussions between supervisor and employees.	Supervisors regularly cover safety when reviewing work practices with individual employees.	In addition to items covered under "Good" supervisors make good use of the shop safety plan and regularly review job safety requirements with each worker. They contact at least one employee daily to discuss safe job performance.

E. ACCIDENT INVESTIGATION, STATISTICS AND REPORTING PROCEDURES

Activity	Poor	Fair	Good	Excellent
1 Accident investigation by line personnel.	No accident investigation made by line supervision.	Line supervision makes investigations of only medical injuries.	Line supervision trained and makes complete and effective investigations of all accidents; the cause is determined; corrective measures initiated immediately with a completion date firmly established.	In addition to items covered under "Good" investigation is made of every accident within 24 hours of occurrence. Reports are reviewed by the department manager and plant manager.

	Poor	Fair	Good	Excellent	Comments
2 Accident cause and injury location analysis and statistics.	No analysis of disabling and medical cases to identify prevalent causes of accidents and location where they occur.	Effective analysis by both cause and location maintained on medical and first-aid cases.	In addition to effective accident analysis, results are used to pinpoint accident causes so accident prevention objectives can be established.	Accident causes and injuries are graphically illustrated to develop the trends and evaluate performance. Management is kept informed on status.	
3 Investigation of property damage.	No program.	Verbal requirement or general practice to inquire about property damage accidents.	Written requirement that all property damage accidents of $50 and more will be investigated.	In addition, management requires a vigorous investigation effort on all property damage accidents.	
4 Proper reporting of accidents and contact with carrier.	Accident reporting procedures are inadequate.	Accidents are correctly reported on a timely basis.	In addition to "Fair" accident records are maintained for analysis purposes.	In addition to "Good" there is a close liaison with the insurance carrier.	

RATING FORM

A. ORGANIZATION & ADMINISTRATION

		Poor	Fair	Good	Excellent
1	Statement of policy, responsibilities assigned.	0	5	15	20
2	Safe operating procedures (SOP's.).	0	2	15	17
3	Employee selection and placement.	0	2	10	12
4	Emergency and disaster control planning.	0	5	15	18
5	Direct management involvement.	0	10	20	25
6	Plant safety rules.	0	2	5	8
	Total value of circled numbers	(—)	+—	+—	+—) (.20) Rating

B. INDUSTRIAL HAZARD CONTROL

	Poor	Fair	Good	Excellent	Comments
1 Housekeeping—storage of materials, etc.	0	4	8	10	
2 Machine guarding.	0	5	16	20	
3 General area guarding.	0	5	16	20	
4 Maintenance of equipment guards, hand tools, etc.	0	5	16	20	
5 Material handling—hand and mechanized.	0	3	8	10	
6 Personal protective equipment—adequacy and use.	0	4	16	20	
Total value of circled numbers	(__	+__	+__	+__)	(.20) Rating

C. FIRE CONTROL & INDUSTRIAL HYGIENE

	Poor	Fair	Good	Excellent	Comments
1 Chemical hazard control references.	0	6	17	20	
2 Flammable and explosive materials control.	0	6	17	20	
3 Ventilation—fumes, smoke and dust control.	0	2	8	10	
4 Skin contamination control.	0	3	10	15	
5 Fire control measures.	0	2	8	10	
6 Waste—trash collection and disposal, air/water pollution.	0	7	20	25	
Total value of circled numbers	(__	+__	+__	+__)	(.20) Rating

D. SUPERVISORY PARTICIPATION, MOTIVATION & TRAINING

	Poor	Fair	Good	Excellent
1 Line supervisor safety training.	0	10	22	25
2 Indoctrination of new employees.	0	1	5	10
3 Job hazard analysis.	0	2	8	10
4 Training for specialized operations.	0	2	7	10

5 Internal self-inspection.	0	5	14	15
6 Safety promotion and publicity.	0	1	4	5
7 Employee/supervisor contact and communication.	0	5	20	25
Total value of circled numbers	(___	+___	+___	+___) (.20) Rating

E. ACCIDENT INVESTIGATION, STATISTICS & REPORTING PROCEDURES

1 Accident investigation by line supervisor.	0	10	32	40
2 Accident cause and injury location analysis and statistics.	0	3	8	10
3 Investigation of property damage.	0	10	32	40
4 Proper reporting of accidents and contact with carrier.	0	3	8	10
Total value of circled numbers	(___	+___	+___	+___) (.20) Rating

SUMMARY

The numerical values below are the weighted ratings calculated on rating sheets. The total becomes the overall score for the location.

A Organization & Administration _____

B Industrial Hazard Control _____

C Fire Control & Industrial Hygiene _____

D Supervisory Participation, Motivation & Training _____

E Accident Investigation, Statistics & Reporting Procedures _____

TOTAL RATING _____

219

By measuring on a valued basis against activity standards, the results provide a basis for setting future objectives. When a weakness can be pin-pointed, a results-oriented objective can be established, along with a target date for reaching the objective.

The very process of evaluation required in making a valued measurement of safety activity provides a clear understanding of what the standards mean, and how they apply at the local level.

RED FLAGGING

A second thing that monitoring will do for safety professionals is to red-flag them—to inform them quickly when there is a problem in their company or in some department. For this we might use a statistical control chart.

The control chart is based upon the *bell-shaped curve*, which is applicable to many things, such as height and weight of people, frequency of accidents (Fig. 9–1), etc.

We want to put limits on our chart—limits that we decide we should stay within in the future—that is, an *upper control limit* and a *lower control limit* (Fig. 9–2). We choose those limits such that if any month's record falls outside them, we know that it is not strictly because of chance—something is wrong.

Suppose we set our limits so that there is only a 1 percent possibility that a month's record will lie outside those limits by chance;

Figure 9-1. Normal distribution curve.

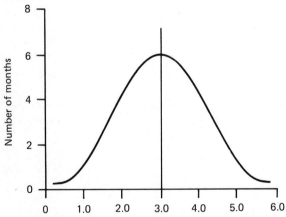

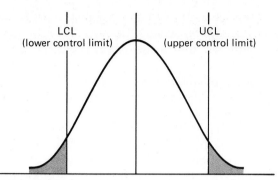

Figure 9-2. Control limits.

we arrive at the shaded area in Fig. 9–2. We then turn our curve on its side, plot our months along the bottom, and we have a control chart (Fig. 9–3). Each period (biweekly) is plotted. Any time our record goes outside the limits, we know that it is not because of chance—something is wrong—and a red flag goes up.

The control chart is the working tool of statistical control. On it the observed accident rates are plotted against time, with the overall accident rate or mean for the entire period. Finally, upper and lower control limits are computed such that the probability of an accident rate exceeding the limits by chance alone is very small.

From a study of statistical control chart data we can tell whether the system is a relatively constant one. If it is, we have a stable situation. Conversely, we can tell whether the system has changed. If

Figure 9-3. Control chart.

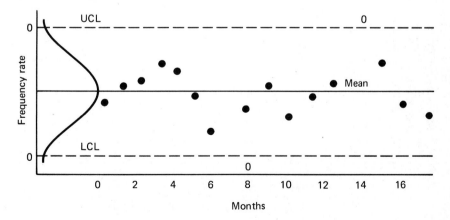

an accident rate exceeds the upper limit, it signals a change for which there is an assignable cause. Similarly, when a point falls below the lower limit, we infer that there has been a significant change for the better.

Statistical control techniques offer a means for making the work of accident reduction more effective and efficient. They cannot assign cause, but they can point out where and when to look for causes.

It takes little more than a glance at year 1 of Fig. 9–4 to see that the accident rate was very stable throughout the year. No rate falls outside the limits. At this moment in time the safety director would have had every assurance of a well-controlled accident situation, but she would have had no knowledge of what the future might hold.

Let us suppose that she continued to use this chart and plotted data points as the months passed. By the middle of year 2 the chart would have taken on quite a different appearance (Fig. 9–4).

The rate of April of year 2 can be seen to signal a change. This was the time to examine the situation closely. In this particular instance the records disclosed that a plant expansion program had been instituted for which no provision had been made in the safety program.

Once a rate has fallen outside the limits, signaling a change to the chart user, there is no way of telling at that time when stability will be reestablished and at what level. If subsequent points fall within the limits, it would indicate one of two possibilities: either (1) the cause of the change was determined and corrected, or (2) the cause was transitory and self-correcting. Should subsequent points continue to fall outside the limits, one might infer that the reestablishment of stability has taken place at another level, but this could not be

Figure 9-4.   Control chart for accident frequency.

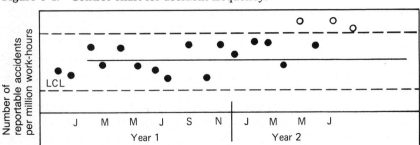

Order of occurrence (each point represents frequency for a month)

concluded until several points have accumulated. For this reason, practice calls for delaying the computation of a new mean and new control limits until at least 10 such points have accumulated.

To distinguish between instability and a new level of stability, one can apply control limits. If all the points falling outside the original limits are contained by these new limits, one can assume a new level of stability; but if they are not so contained, there is continuing instability.

MEASURING INDIVIDUAL PERFORMANCE (ACCOUNTABILITY)

The third function of monitoring is to hold the line organization individually accountable for performance in safety. This is no doubt the most important of the three functions of monitoring. At the lower management level (supervisory level) it is perhaps the primary, at times the only, reason for monitoring.

When people are held accountable (are measured) by their bosses for something, they will accept the responsibility for it. If they are not held accountable, they will not accept the responsibility. They will put their efforts in the direction where the bosses are measuring them.

In safety work, there are three ways of fixing responsibility. We can fix accountability based on the *results* of the activities the line is doing, we can fix accountability based on the *activities* themselves, or we can use both methods.

Accountability for Results

Figure 9-5 shows a partial listing of the things that we might consider measuring in fixing accountability for results. One of the simplest means of doing this is to charge accidents to the department in which they occurred. If supervisor A has a worker who suffers a disabling injury, it shows up on supervisor A's record. An adaptation of this is to charge the claim costs back to the line. Here we are measuring the line supervisor in terms of *dollars*. He then will pay for his accidents; the cost will come out of his working budget.

Some companies choose to prorate their insurance premiums. When an accident occurs, the insurance company pays the direct costs of that accident. However, the insurance premium that the company pays is influenced by what the costs of the accidents are. If a

---

ACCOUNTABILITY FOR RESULTS

1 Charge accidents to departments
  a Charging claim costs to the line
  b Including accident costs in the profit and loss statements
2 Prorate insurance premiums
3 Put safety into the supervisor's appraisal
4 Have safety affect the supervisor's income

Figure 9-5

department or a location is charged a specified amount of the insurance premium based on its percentage of the total accident costs, this will be a more accurate dollar measuring stick.

Let us assume, for example, that the XYZ Manufacturing Company has three plants: plant A has 2,000 employees; plant B has 500 employees; and plant C has 5,000 employees. Prorating the insurance premium based on how they used it suggests that regardless of size, they be charged the proportion of that premium on the same rate as their losses. The XYZ Manufacturing Company insurance premium is $100,000. Thus:

	Accident Loss Record	Premium Charged
Plant A	$15,000 or 30%	$30,000
Plant B	$20,000 or 40%	$40,000
Plant C	$15,000 or 30%	$30,000

It will be difficult for plant B above to show a profit this year because of its large accident losses.

When line supervisors are appraised on safety as well as on their production records, they generally become far more interested in accident prevention and begin to do something about it. This factor is too often overlooked today in the appraisal of line managers.

When line supervisors' paychecks are in some way influenced by their accident records as well as by their production records, they become far more interested in safety as a primary thing to worry about and consider.

This has been very effectively revealed in some construction companies. When job superintendents must subtract accident costs from their profit figure for a job, and when their personal bonuses for the job are reduced or wiped out because of this, they rapidly become more interested in accident prevention in the future.

All the above are keyed to the idea that the accident record is the line supervisor's—not the safety department's or the insurance manager's.

Accountability for Activities

Figure 9-6 lists some of the items that management might measure the line organization against to determine what it is *doing* to prevent accidents from occurring. This measurement is *accountability for activities*. This is perhaps more important than the measurement of results because it measures line competence in controlling losses *before* the accidents have happened.

Management can measure line supervisors to see whether or not they are utilizing such techniques of accident control as toolbox meetings, job hazard analyses, inspections, accident investigations, incident reports, safety committees, and safety meetings. Management may require line supervisors to submit activity reports. All these are activities that are known to be effective in safety. When management measures these activities, it is setting up a system of accountability for activities.

PAST PERFORMANCE AS A STANDARD

The past performance of any group is the best standard to use as a guide for present performance. The people in the group understand it and will accept it, whereas they are often reluctant to accept an arbitrary standard set by someone outside the group. Often they believe that competing against a different group is unfair, since no two groups face the same challenges, hazards, or situations. Management is principally interested only in its own company and whether or not it is improving.

MONITORING AND MOTIVATING

In the next chapter we'll discuss creating and maintaining interest in safety, and later we'll look at motivation under the chapter entitled "A

ACCOUNTABILITY FOR ACTIVITIES
Management measures what supervisors are doing
1   Safety meetings that they hold
2   Toolbox meetings
3   Their activity reports on safety
4   Their inspection results
5   Their accident investigation
6   Their incident reports
7   Job hazard analysis

Figure 9-6

Motivation Model." Before we leave this subject of monitoring, however, we might point out the close relationship between the two subjects. Many believe that at certain levels of the organization the primary motivator is measurement. Research in the behavioral sciences tends to bear this out, for particularly at the first line supervisor and at the middle management level a person will react most to how and where he or she is being measured. While less true at the worker and executive level, it seems to be an organizational fact of life that managers are first and foremost motivated by their bosses' measures of their performance.

REFERENCES

DieKempe, R., and D. Spartz, "A Quantitative and Qualitative Measurement of Industrial Safety Activities," *Journal of the ASSE,* September 1970.

Dunlap, J., *Manual for the Application of Statistical Techniques for Use in Accident Control,* Dunlap & Associates, Stamford, Conn., 1958.

Method of Recording and Measuring Work Injury Experience, American National Standards Institute, New York, 1967.

Petersen, D., *Techniques of Safety Management,* McGraw-Hill, New York, 1971.

Tarrants, W., "Applying Measurement Concepts to the Appraisal of Safety Performance," *Journal of the ASSE,* May 1965.

TEN

Creating and Maintaining Interest

The practice of creating and maintaining interest, as a first principle of accident prevention, applies to employers as well as employees. In this chapter, however, emphasis is placed chiefly on methods whereby management, supervisors, instructors, and safety engineers may create, stimulate, and maintain the safety interest of *employees*.

APPEAL TO PERSONAL CHARACTERISTICS

It is characteristic of human behavior that the interest of people may be aroused by appealing to one or more of their stronger senses or desires, such as self-preservation, loyalty, pride, or responsibility. The amount of interest thus created in individuals varies according to their reaction to the kind of appeal that is made.

For example, a machine operator who subconsciously fears personal injury may be induced to keep a guard in its proper place on her machine if an appeal is made to her fear of being injured. She may have little interest in protecting the property of her employer or in the safety of her fellow employees or in the safety record of her department. If, however, she is deeply concerned with regard to her *own* safety, she can be persuaded to keep a guard in place on the machine, if convinced that its unguarded condition is dangerous.

Conversely, another machine operator may scorn fear of personal injury and in a spirit of bravado may deliberately expose himself to danger if self-preservation is featured too strongly or too obviously. However, this employee may take pride in the safety record of his department. If so, he too can be made to keep his machine guarded, if he is approached in a way that appeals to this side of his nature.

Another individual may have a strong tendency to "follow the leader" and to conform to the practice of the majority. In his case interest in safety may be aroused by pointing out that the maintenance of machine guards is the customary and general practice followed by other employees. The spirit of "team play" is strong in many individuals.

Of course, keeping guards in place is but one of the many phases of industrial accident prevention in which the work of creating and maintaining interest is a vital factor. Even more important is the part that an employee plays as a member of a safety committee, or a leader in safety meetings, in accident investigation, in finding and reporting unsafe practices and conditions, and in many other activities. The necessity for individual interest, however, and the methods of creating and maintaining it are fundamental with respect to all phases of industrial safety.

Naturally a person who is keenly interested in accident prevention will play a more constructive part in a safety program and will be less likely to be injured or to cause injury to others than one who is indifferent. The interest of individuals may be aroused and maintained by making use of known incentives. Those of chief significance are here listed, and commented upon in subsequent paragraphs. They are not the conclusions of trained professional psychologists but are the summed-up beliefs of experienced accident preventionists.

1. *Self-preservation* (fear of personal injury)
2. *Personal and material gain* (desire for reward)
3. *Loyalty* (desire to cooperate)
4. *Responsibility* (recognition of obligations)
5. *Pride* (self-satisfaction and the desire for praise)
6. *Conformity* (fear of being thought different from others)
7. *Rivalry* (desire to compete)

8. *Leadership* (desire to be outstanding)
9. *Logic* (special ability to reason)
10. *Humanity* (desire to serve others)

It is probably true that all the foregoing motivating qualities or characteristics are present to some extent in almost everyone. It is also true that some of them will be more pronounced in certain individuals than in others. Therefore, the task of the safety director who is confronted with a problem of creating and maintaining interest is, first, to determine which particular characteristic predominates in the individual or groups of individuals whom she desires to interest and, second, to plan her program so as to make use of the knowledge thus obtained. Although such procedure is a form of psychology, it requires an understanding of human nature no greater than that which is usually possessed by industrial executives and supervisors. An observant supervisor will note the reactions of employees when encouraged, instructed, admonished, criticized, or praised. The good supervisor prides himself on his ability to know his workers and to get the best results from them. Instinctively, perhaps, he often selects methods that appeal to an individual through that individual's predominant motivating characteristic.

It is efficient procedure to recognize and identify predominating personal characteristics as far as possible and consciously to adopt specific methods of approach, rather than to rely upon arbitrary or "hit-or-miss" methods. The safety director or supervisor who knows which particular qualities are strongest in the individuals under his control is in a good position to reach his objective in accident prevention with less effort and greater speed than when he proceeds without regard for personal receptiveness.

Some of the factors to be considered when making a choice of the foregoing characteristics are home conditions, financial status, health, age, sex, race, tastes, hobbies, habits, likes and dislikes, disposition, character, prevailing moods, temperament, and reaction to various happenings, as well as the degree of attention given to oral and visual methods of educational approach.

Selection of the means of approach to a worker's interest may be likened to a salesperson's plan for arousing the interest of a potential customer. For example, the person who must support a family will undoubtedly have a far stronger sense of financial responsibility than one who has no dependents. The young ne'er-do-well, who previously was an accident problem, will be found to become interested in the "economics" of accident prevention when he starts planning definitely *for his own home.* The alert safety director will be alive to these changes in attitude and take advantage of them in carrying on her work of creating and maintaining interest.

Interest in self-preservation often increases following a serious injury to oneself or a fellow worker or to some well-known person. The desire for financial gain is stronger in the person who is buying a new home or an automobile on monthly installments. The spirit of conformity is an extremely strong motivator for almost all workers in a group. This motive, which we'll call "peer pressure," is perhaps the single greatest determinant of worker behavior on the job.

Pride, the desire for leadership, logic, humanity, and rivalry, or their absence, are readily observed in most persons. And so the supervisor or the safety director who depends on her powers of observation, common sense, and good judgment will come to know her workers; she will recognize the opportunities for special appeals and will select methods of appeal best adapted to secure favorable results.

BEHAVIORAL THEORY

Several times in this chapter we will refer to modern behavioral theory. Perhaps before moving on to further discussion of the 10 incentives we might mention some behavioral theory currently in use in management and indicate how it might relate to accident prevention.

First of all, there are five categories of behavioral theorists that seem to be particularly relevant to the field of safety management:

1 Motivation theorists make up the first group, and in this category we might look at some of the thoughts and theories of Abraham Maslow, Chris Argyris, Frederick Herzberg, and Rensis Likert. There are, of course, many other motivation theorists in the field of psychology, but the works and thoughts of the above four seem to be particularly applicable to the industrial situation and particularly usable to us in safety management.
2 The second category of theorists in the behavioral sciences that seem to be particularly usable to safety comprises the learning theorists: the first of these would be B. F. Skinner, and the second might be Robert Mager. While totally different in concepts, these two are saying some things that tie together very closely and are very usable to us in our efforts to communicate with and change the behavior of workers on the job.
3 The interpersonal skills theorists are the third group. There has been a lot written in recent years on the whole area of interpersonal skills, but two particular theorists and one particular approach stand out as being very applicable to the field of safety management and the industrial situation. The theorists are Eric Berne and Tom Harris. These names are probably

quite familiar to you as they are the people who have experimented with and have talked about the uses of transaction analysis.

4   The fourth category encompasses the ideas of the communication theorists; in particular, we might look at the works of Janis and Hovland. Since much of what we do depends on our communication skills, the research and theories from communications are highly important to us.

5   The final category of behavioral science people that we might look at is made up of the management theorists. Some of these people are in the behavioral sciences, and some of them are in the field of management. Especially useful for our purposes are the theories of Douglas McGregor, the works of Blake and Mouton, and the models of supervisory behavior that come to us from Lyman Porter and E. E. Lawler.

Let's then take a look at what some of these people have said and how it might relate to us. First of all are the motivation theorists. All tend to agree on a definition of motivation as a starting point as being fulfilling of a person's needs at his or her current level. This is in contrast to our traditional definition in the safety field of motivation, which is the utilization of numbers of gimmicks and other kinds of things that we have viewed as motivation.

The first of the motivation theorists is Abraham Maslow. His particular contribution to the field is his hierarchy of needs. The lowest need level, according to Maslow, is the physiological level, the needs for water and drink and other basic needs. Once these are satisfied, a person is no longer motivated by them, but moves up to the next level, the security level. At this level, a person makes sure that his or her basic physiological needs will continue to be met.

The next level is the belonging level, where a person is motivated by the need to be a part of a group. Once the need for belonging is satisfied and the person is a usual part of a group, he or she moves on to the next level of needs, which is that of self-esteem; here, of course, one is primarily interested in satisfying oneself. There is a final level in Maslow's hierarchy which is beyond that of self-esteem: the level of self-actualization—a person wanting to do those things that he or she is capable of. In Maslow's hierarchy of needs, the key point is that if you are going to motivate a person, you have to hit her with things that will be satisfying to her current need level. For a man who is at the self-esteem level, it is ridiculous to try to motivate him with things that have to do with belonging, security, or physiological needs that are quite a bit below where he is at. What does all this mean to safety and the management of safety programs? Probably this, we ought to examine our current safety programming and see what need level we are aiming at. Most people in most industrial organizations today are not at the physiological need state or at the security need state. They are somewhere above that—the belonging level as a minimum or probably at the self-esteem level. People in today's industrial organizations are looking for jobs with meaning; they are looking to be a part of a peer group, and they can be motivated by things that satisfy those needs. If we are appealing to them in our safety program at a lower level, we are going to be totally unsuccessful.

A second motivation theorist who gives us considerable insight is Chris Argyris. Argyris looks at industrial organizations and how people fit into those organizations. First of all, he looks at human beings as they mature and finds that they go through a change from certain sets of activities and characteristics as children to other sets of activities and characteristics as adults. For instance, as children people tend to be somewhat passive; as adults they are very active. As children they are very dependent upon other people; as adults they are very independent and believe they can stand on their own. As children they exhibit relatively few behaviors; as adults many. As children they have a relatively shallow and short-term interest in things, and as adults they are capable of deep and long-term interest. As children they find themselves in every interpersonal situation as a subordinate; as adults they like to think of themselves as at least an equal in every interpersonal relationship. So as people mature, they go through this process of change. These changes are what maturation is all about. We find that adults have these kinds of characteristics: they are active and independent, exhibit many behaviors, are capable of deep and long-term interest, and perceive themselves as at least an equal in all situations. They move into an industrial organization that has certain characteristics that run counter to those characteristics. All industrial organizations have a chain of command, which is a series of subordinate-superior relationships. This creates dependency, passivity, and inequality. All organizations are structured around a principal of unity of command, which means that one person should have only one boss, which creates dependency, subordinacy, etc. And finally a principal of all organizations is the principal of specialization, which means that the total work that must be done within an organization is broken down into small subunits and one small subunit is given to one individual. This of course runs counter to the adult's characteristic of having deep and long-term interests.

Therefore, most individuals, if they are in fact mature and adult, perceive themselves in an industrial organization, particularly as they start at the bottom of that organization, as being in a situation that is incongruent with their own personal needs. This results in all kinds of problems. Some of the problems that happen are: (1) people quit—there's a turnover problem; (2) people quit mentally—they become apathetic; (3) people daydream on the job; (4) people lose their motivation—they are disinterested in the corporate goals as they don't seem to be at all in harmony with their own personal goals; (5) people tend to climb the organizational ladder; they want to get out of the situation they're in; (6) they gather together in informal groupings, and they set their own group standards and group norms; and (7) they evolve a psychological set against the company in the belief that the company is wrong in most things that it does. These kinds of behaviors lead on the part of management to a feeling that "if they are going to act like babies, we will treat them like babies," and that results in more management pressure, more managerial control, more specialization. This of course makes people exhibit the above behaviors even more, and we have a circular process. The answer, according to Argyris, is less specialization, less control, less superior-subordinate relationships, and a leveling process between various levels of the organization. In other words, more treating of individuals on the job as human beings and less treating of them as a part of the machine. How does Argyris's theory relate to safety? It seems that if our safety programs are structured with the normal characteristics of organizations—chain of command, unity of command, and specialization—chances are that our safety programs are coming across counter to the needs and characteristics of mature human beings also. And most safety programs are structured this way.

The third motivation theorist that holds some insight for us is Frederick Herzberg. Herzberg came up with a concept that disputes some of our previous theories of motivation of people. It seems that we used to view everyone as being somewhere on a scale that might stretch between being dissatisfied on the job and being motivated. For many years we manipulated all kinds of variables to try to move people up the scale and be less dissatisfied and more motivated on the job. Herzberg's key insight was the fact there is no such thing as a single continuum that stretches from dissatisfaction to motivation; rather there are two: (1) There is a continuum which reaches from not being satisfied on the job at its worst up to being satisfied on the job. (2) There is a separate and distinct continuum which stretches from unmotivated at its worst to motivated at its best. Every individual lies somewhere on both lines. Individuals can be satisfied on the job and not motivated; conversely, they can be dissatisfied and motivated. The key in Herzberg's theory is that there are different variables that regulate where we are on each of the two continuums. The things that determine whether or not we are satisfied on the job are things having to do with company policies, supervision, working conditions, interpersonal relations between the worker and the boss, money, status, security, and things of this nature. "Satisfiers," as we can call them, tend to determine whether or not we are going to stay on the job and be satisfied in the situation we exist in. They are not going to determine whether or not we are going to work. Different things determine our level of motivation, and those things have to do primarily with work itself: Does it offer a chance for achievement? Is there any recognition connected with it? Is the work itself any fun? Does it give a sense of responsibility? Does it give personal and professional growth? These kinds of things determine our level of motivation.

What does this mean to safety management? It means that if we have a safety program that is built primarily on things that relate to where we stand on the satisfaction scale, we are not going to turn anybody on with the safety program. At best we can have people that are not totally dissatisfied. At worst they will be totally dissatisfied with the safety program. If we are going to turn people on with the safety program, we have to build motivators into it.

A final motivation theorist that we have to look at is Rensis Likert. Likert has major research in behavioral science in two areas that relate very importantly to safety management. First of all, he researched the area of supervisory style and how that relates to productivity. He studied, for instance, the relationship between closeness of supervision and the amount of productivity received, and he found that the closer a supervisor supervises, the less people work, and the more loosely the supervisor supervises, the more people work. He studied the relationship between the supervisor's reaction to a poor job and the amount of productivity, and he found that the more punitive that a supervisor was and the more critical he was toward the job done, the less work that got accomplished. He studied the relationship between freedom that employees feel to set their own work pace and the amount of work they put out, and found that the more freedom employees had, the more work they did. He studied the relationship between the pressure employees feel for production and the amount of production they give, and he found that the more pressure that was applied by the boss, the less work that got done. What he studied in total might be summed up by his description of two kinds of supervisors, an employee-centered supervisor and a job-centered supervisor. The employee-centered supervisor puts emphasis on the good of his or her people and gets the highest production. The job-centered supervisor puts emphasis on production and gets relatively low production.

How does this relate to safety management? This gives some real insights into what gets people to want to do what we want them to do. If we are truly interested in them, we get results. If we are truly interested in the task and not in them, we tend not to get results.

The second major group of theorists from the behavioral sciences that give us some insights on possible future uses in safety management are the learning theorists B. F. Skinner and Robert Mager.

Most of you are familiar with the work of B. F. Skinner, which has been translated in the industrial situation into concepts of behavior modification. These are being used in different kinds of industrial situations today even in safety situations. In behavior modification, if you reinforce positive behavior (that is, behavior that you want), it is more likely that you will get that behavior again. If, on the other hand, you ignore the kind of behavior you want and pay attention, even punitively, to adverse behavior or the kind of behavior you do not want, you are going to get negative results. This means that if we pay attention to unsafe acts, we are not going to get very good results. Rather, if we reinforce and reward people for what we want, that is, safe acts, we are going to build safe behavior.

Robert Mager also has done some work in learning theory that is useful to safety managers. He talks about the kinds of situations that determine whether or not a person is going to build a good attitude toward safety by looking at the kinds of situations that have to do with the learning situation for that subject. His work applies directly to the field of industrial training and is very usable to us in that field.

A third group of behavioral scientists whose work is very useful to us are the interpersonal skills specialists, notably Eric Berne and Tom Harris and their work in transaction analysis. In transaction analysis, the focus is on the relationship between two people, a worker and a boss. And it makes a particular point of the fact that you can only accomplish results or solve problems between two people when each of these people is in the adult stage. If one assumes the role of a parent and the other a child, as is typical in a supervisor-worker situation, you cannot accomplish results. Only when people are talking person to person can you really accomplish anything in safety or anything else.

Communications theorists compose a fourth major category of behavioral scientists whose theories are usable for us in the field of safety management. Communications theorists, Janis and Hovland talk about and have developed communication models and then try to identify what makes for good, solid communications, communications that result in behavior change. They first developed a model which says that if a communication is to take place, there must be a sender, in our case management, a medium (a way to send it), a message (what we want to say), and a receiver (who is the worker). And they examine the variables that determine what makes for effective communication in each of those four categories. One insight that comes out of their research is that the success of a communication depends, first of all, on management's credibility. Do the employees believe that we as managers know what we are talking about, and do they believe that we have the right to be saying these things to them? If credibility is lacking on the part of management or on the part of the safety director, there is no sense communicating because the workers are not listening. As for the message itself, there are some variables that have been identified from the research which are very important, the most important being whether or not the message is meaningful—whether the material can be used by the workers on their jobs. If these two aspects are there, the workers will listen. As far as variables in the receiver we find that if we are talking to workers in opposition to what their peer group believes, they are not listening. If they have had any participation in what that message will be, they will be listening. So we have aspects of peer pressure and participation in the decision-making process that determine whether or not the listeners will be listening. These kinds of things are very important, as we do a lot of work in safety management with communications. We are talking as managers to our employees all the time. About 85 percent of the time they are not listening and we are wasting our time. Communications research has some answers for us which safety managers should be listening to.

A fifth major group of behavioral scientists that we shall discuss are the management theorists. Many managers are familar with Douglas McGregor's X and Y assumptions about people. Although often times we talk about the X manager and the Y manager, McGregor never talked about an X or Y manager; rather he talked about a manager who has certain assumptions about what people are like. The X assumptions are that people basically do not like to work and that the managerial task is to force or coerce them in some way to do the task. Conversely, the manager who has Y assumptions about people believes that people do in fact like to work—that work is not foreign to their nature, that they perhaps are turned off by the work situation because of poor managerial practices of the past. The Y assumption leads the manager to believe that her role as manager is not to force her people to work but rather to clear away the barriers that lie in her way to effective job performance, a quite different role.

Blake and Mouton have developed a similar kind of thinking. Looking at a grid approach to managerial styles, they say a manager can be interested in both his people and in production, and in fact the best manager (which is McGregor's Y manager) is the person that has high performance goals and also is highly interested in the good of his people. He is the best manager.

A third useful concept from management theory is the model developed by Lawler and Porter on what motivates supervisors. Their model is very simple. It says that performance or accomplishment of any task by supervisors is dependent upon three things: (1) their abilities and traits (are they able to do what you're asking them to do, and have you trained them and taught them what it is that you want?), (2) their role perception (do they perceive that what you are asking them to do is truly in fact a part of their legal job as supervisor?), and (3) their effort (do they try?). Lawler and Porter go on and look at the effort aspect of performance and say that whether or not supervisors try to accomplish the task you want done is a function of two things: (1) value of the reward (is it really important enough that management has attached some rewards to it, either positive or negative?), and (2) the effort-reward probability (is there any connection between whether or not they do in fact try or whether or not they do in fact get that reward?). This model is very applicable to safety management because normally we have a situation where we have a lack of supervisory performance in the safety field not because supervisors don't believe it's part of their job, as management policy usually specifies this, but because they don't try. Performance or lack of performance, then, is a function primarily of their lack of effort. This model tells us that if this is true, we should look at the reward structure and at the probability of a connection between that effort and reward.

What do we get out of all this? We get a picture like this: (1) Maslow says that we can only motivate people if we recognize their need level (if we care enough to find out what their need level is); (2) Argyris says that the interpersonal characteristics of organizations turned people off in the organizational setting—lack of caring as expressed by the top of the organization to the bottom; (3) Herzberg says that only through caring about a worker's needs and creating a feeling of worth can we truly motivate people; (4) Likert says that only employee-centered managers get results—those managers that exhibit a caring attitude; (5) both Skinner and Mager identify the aversives and positives in every kind of a situation including a safety situation and show that only in a positive environment will people learn; (6) Eric Berne and Tom Harris are clear in saying that problem-solving progress between two people takes place only when they are in an adult ego state (when they are treating each other as equals); (7) communications researchers say that employees won't even be listening to us if we are not credible (unless we have demonstrated prior to the communications that we do in fact care); (8) Blake and Mouton and McGregor all indicate that the employee-centered supervisors of Likert get better production when they exhibit a caring attitude; (9) finally, Lawler and Porter in their model have clearly demonstrated that supervisors perform when (a) they know what to do, (b) they believe it's part of their job to do it, (c) there is a reward contingent upon their performance of the tasks.

All these people have messages which are not only very relevant in management generally, but notably relevant in safety management. All these people demonstrate that when trying to turn on employees we have got to (1) treat them as mature human beings and (2) give them jobs with meanings.

If we want to get performance from supervisors we have to (1) tell them very specifically what we want from them, (2) demonstrate through policy statement and by fixing of accountabilities that safety is a part of their job, and (3) have some kind of reward structure that has some contingency upon their actual performance. When these things are present, supervisors will do what we want them to do in safety and employees will not be turned off by our safety program.

All this means that (1) the key to employee safety performance is a management that cares and does something to demonstrate that caring and (2) the key to supervisory performance is doing something to make it matter to the supervisor whether or not he or she performs in safety.

THE TEN INCENTIVES

With the above background from the behavioral theorists, we'll take a look at the ten incentives mentioned earlier in the chapter.

1 *Self-Preservation.* One of the most common of all qualities or characteristics is that of self-preservation. Some of the more common methods of creating and maintaining interest through knowledge of this characteristic are:

   a *Featuring the injury.* Posters may portray serious bodily injuries. They may show that the loss of arms, legs, hands, fingers, eyes, the results of infection, hardships from the loss of income, and many other unfortunate circumstances result from accidents. Failure to report minor injuries can be associated with serious health impairment. Contrast can be drawn vividly between persons having full strength and vigor and those physically impaired whose ability to work, earn wages, and enjoy life consequently is reduced.

b *Accident incidentals.* Crutches, glass eyes, broken safety goggles, the beggar's tin cup, the hospital cot, the operating room and its paraphernalia, and many other things associated with accident occurrence can be played up with advantage.

Methods of appeal that feature the injury and the accident incidentals, such as described above, need not be confined to posters and notices but may include oral discussion at meetings and lectures. Sound slides and moving-picture films also may be used.

It should be understood that emphasis is not being placed here on *the desirability* of featuring bodily injury and accident incidentals. As has been pointed out, the appeal to the instinct of self-preservation is only one of the many methods that may be utilized in creating and maintaining interest. Further, it is suggested that this particular method be used only when it fits a specific case. There is much to be said in opposition to gruesome posters although use of this approach is defensible under certain conditions.

One of the main, and perhaps most interesting, areas of research in safety is the communications research on the use of fear in safety messages. Some studies have shown that minimal fear arousal in connection with the message rather than high fear arousal is more effective in producing attitude change. Other studies have only occasionally supported these findings. Some studies have found a positive relationship between the use of fear and attitude change. There have been a few studies showing a negative relationship. It seems that if a fear appeal is to be most effective, it should be directed at an audience already predisposed toward the recommended practices, and the recommended action should be of such a nature that the audience can engage in it immediately.

Example

In a hardware-manufacturing plant, accident frequency was quickly controlled when it was discovered that the majority of employees were susceptible to an appeal that was directed to "fear of injury." In this case, previous efforts of a general nature featuring competition, awards, etc., had been unsuccessful.

Infections from relatively minor injuries on screw-cutting machines constituted the chief problem.

The method of approach was to minimize the emphasis on numbers and cost of accidents, on poor records, and on lack of skill, etc. In lieu thereof, advantage was taken of all opportunities to feature bodily injury, up to but not including gruesomeness.

Posted notices described the results of accident occurrence as "serious infection," "loss of use," "painful lacerations," "mutilation of fingers," "crippling of worker," etc. Posters and films were shown depicting infected areas on arms and legs, resulting from neglect of minor injuries.

The chairperson of the safety committee used similar terms and illustrations in his discussions. Accident records were arranged so as to emphasize the *injury* rather than the *accident*, and in many other ways the entire oral and visual tone of safety work was altered so as to appeal to the selected predominating characteristic of self-preservation.

2 *Personal and material gain.* Many persons are more than ordinarily desirous of financial or other forms of material and personal gain. It is often true, therefore, that interest in safety may be created by associating a reward of some kind with recognition of outstanding safety performance. Here are a few ways in which this may be done:

a *Bonuses.* Monetary rewards may be given for good safety records.

b *Salary increases.* The possibility of a raise in salary is a powerful incentive. Certain types of employment plans—such as civil service—may map out a lifetime service career showing promotions and other benefits that result from a satisfactory record. It can be pointed out that accidents interfere with continuity of service and with the rewards that such service provides. More directly, quality and length of service may be coupled with an accident-free record, when the employee is considered for promotion or salary increase.

c *Vacations with pay.* These may be given to individuals or to whole departments on the basis of outstanding accident-free performance.

d *Days off.* Rewards can be given in the form of occasional days off.

e *Trips.* Trips to safety conferences may be offered. These help to maintain interest in the job and also provide opportunity for obtaining safety instruction.

f *Personal gifts.* Interest in a planned safety program or contest can be stimulated by awards of small personal articles such as billfolds, belt buckles, knives, and pencils.

g *More desirable types of work.* Some plans include assignment to jobs providing steadier or more desirable work, and these assignments are made as rewards. Many desirable jobs *necessitate* the employment of safe workers.

h *Banquets and picnics.* These form a convenient method of rewarding groups of employees for good records and have the added advantage of improving morale.

i *Appointments to safety activities.* A place on the safety committee may be made to appear highly desirable and to appeal to certain employees. Other appointments may be made as rewards for safety performance.

It is generally agreed that best results are secured when employees practice safety for safety's sake, and not primarily for gain. Thus, when awards are given, it has been found desirable to impress upon the employees the point of view that prizes are given not only to reward past safety performance but also to encourage and stimulate future active interest in the elimination of mechanical hazards and unsafe personal practices.

Some employers are of the opinion that prizes and bonuses for good no-accident records are unwarranted—that by their very acceptance of a job, employees automatically are obligated to work carefully and safely. Theoretically, this is true. But as long as human nature remains what it is, the judicious granting of awards for safety will often produce results that perhaps could not be attained by any other means. Especially is this true when the employees who are approached are more than ordinarily susceptible to appeals that are directed to their desire for personal gain.

In looking at this area, so common to safety programs traditionally, a few points need to be made. First is to keep in mind the newer behavioral theories which relate to this. One of these is the theory introduced and documented by Frederick Herzberg, which will be discussed in more detail later. In effect, Herzberg has shown that certain things that we in management might provide in the way of rewards are in fact truly motivators. On the other hand, much of what we traditionally use for rewards are actually not motivational at all. Rather they are items which fall on his "hygiene" scale. They serve only to keep employees from being dissatisfied on the job. Many of the items mentioned above fall on this "dissatisfaction" scale: bonuses, salary increases, vacations, all fringe benefits, banquets, picnics, etc.

In using the reward approach, when we appeal to personal gain, probably we are much better off using items on Herzberg's "motivation" scale, things that have to do with being able to achieve, with responsibility, with promotability, with making the work itself more fun, etc.

Secondly, according to Herzberg's theory, when we use the "hygiene" items as rewards, we find that they are only temporary in nature—we must continually add more and more goodies to achieve the same interest. It is in fact somewhat dangerous to build our reward structure on these hygiene items—and costly.

3 *Loyalty.* A well-developed sense of loyalty, plus an understanding and approval of safety rules, usually ensures observance of safe-practice rules. Some of the more common ways of creating and maintaining interest in safety, when the employee's sense of loyalty is well developed, are as follows:

a *Effect of accident occurrence on supervisor's record.* Employees can be shown that their failure to perform their job safely reflects on their supervisor's record.

b *Effect of accident occurrence on employer's overhead cost.* It can be explained that injury to the individual employee increases the plant overhead cost.

c *Effect of accident occurrence on quality of employer's product.* Employees can be shown that their unsafe actions not only may reduce the volume of production but also may result in output of inferior quality.

d *Effect of accident occurrence on fellow employees.* Employees who are loyal to their fellow workers are averse to the continuance of unsafe practices that cause accidents and thus spoil the safety record of all other workers in a given group or department.

e *Supporting the "boss."* Accidents occur largely from violation of commonsense safe-practice rules issued by the "boss." Loyal workers prefer to support the managerial and supervisory staffs and if appealed to on this basis will themselves obey the rules and can be induced to set an example that less loyal employees may follow.

It might be mentioned here that this particular appeal seems more limited today than ever before. It seems, and is somewhat documented by sociologists, that people in the work situation, and notably younger people, are less loyal to the company they work for than they used to be. Workers today have different ethics, different values, different motivations than they did, and this is important to us. We know that there is going to be a large shift in our workforce and that the people in their early twenties today are going to constitute a major portion of the workforce of tomorrow. And we know that those people tend to be different in attitude, approaches, and values than older people. It does not take a professional psychologist or sociologist to realize that people do not necessarily think and act as people used to. The question is not whether or not things are different; the question is how do we understand and manage people that are different.

Sociologists tell us why they are different, which helps at least to understand them. People's values and attitudes are formulated in their early years by two major areas: (1) the institutions that have been in charge of them; (2) what life experiences they have had. And as you look at these two categories, you find that there are quite different things that have molded the values and attitudes of the young versus the old. You find, for instance, family life has changed in terms of divorce rates, in terms of starting school at much younger ages, in terms of women working, in terms of men traveling, etc. We find that family influence is probably at a relatively low figure compared with previous years. You find the institutions of the churches have changed and the churches and their moral values are relatively less influential in the development of people. And we find that schools have changed considerably; today people are being taught to think things out for themselves rather than to accept the rules and regulations and rote learning approaches of past years. Institutions are different and have a different influence on people. In addition, life experiences are different. The older generations were influenced primarily by the experiences of depressions and major world wars, which developed the values in people of loyalty to an organization that stood by them during depression years and patriotism to a country that was developed during world wars. Younger people do not perceive the depressions and wars such as World War II as being major influences that shaped their lives. Rather they look at things like the ecology movement, the poverty movement, Viet Nam, television, and these kinds of things as major shaping influences. In short, the things that developed the values and attitudes of young people are different from the things that developed the values and attitudes of older people. So the young have different life-styles based upon different values and different attitudes. They believe in things like (1) instant everything—it used to be that a person in debt was viewed as a social outcast; today that is a way of life; (2) pleasure—there is a trend toward living in the present rather than in the future, having your fun now, doing your own thing; (3) a different work ethic—the puritan ethic stated that because work was good, it should be maximized and fun minimized. The younger ethic is different from that. People use to live to work, and now they work to live. Younger people often reject the old work ethic; (4) simplification—most people are interested in products and services that take the work out of life, that make things easier and quicker; (5) safety and health—younger people are considerably more interested in safety and health; (6) naturalism—the young tend to reject artificial behavior; they don't dig phonies; (7) personal creativity—people like to throw themselves into whatever kind of situation they are in, including their work situation; (8) relying on others—problems seem to be of such magnitude (pollution, energy, traffic, etc.) that people feel that individually they can do very little, that they must rely upon others (big business and big government) to get them done; yet at the same time (9) not trusting big institutions—there is almost total loss of confidence in big government, big business, big management, and so on; and (10) consumerism. In the light of these changes the appeal to loyalty is less effective today than it has been in the past. This does not mean it cannot be used; it can, but only for certain people in certain situations.

Example

In a plant department having a poor accident experience, several attempts to enlist the interest of the workers in accident prevention had failed. Scare posters, showing the results of accidents from both a personal-injury and a personal-economic standpoint, had been tried, films and inspirational speakers had been used, and even a costly "safety slogan" contest had been carried out, all with indifferent success.

A majority of the employees in this department were of the same nationality and of a very clannish nature. The supervisor too was of the same nationality and held important community and fraternal positions among his countrymen.

It was explained to the employees of this department, especially to those who were most prone to commit unsafe practices, that unless the accident experience of the department improved it would reflect seriously on the supervisor's record. In fact, the rumor got around that unless improvement was made shortly, a new supervisor, unknown to all of the workers in the department, might replace their present boss.

The response was immediate. Once his loyalty had been appealed to, each friend of the supervisor rallied around, and the accident experience of this department not only improved but became the best in the entire plant.

4 *Responsibility.* The sense of responsibility, both to self and to others, is an attribute that is found to a degree in the great majority of persons. It also is one that is capable of being easily utilized to promote interest.

"Responsible" persons will be inclined to work safely if they are shown that by persisting in unsafe practices they are likely, sooner or later, to cause an accident that will bring misery and economic loss to their own dependents, or to fellow workers or members of their families. Remorse is an emotion that no one, least of all the employees who fully realize their responsibilities, is desirous of experiencing, and the skillful dramatization of responsibility in connection with the aftermath of a serious accident is a powerful deterrent to unsafe practice.

Appeals may be made as follows:

a *Assignments.* Interest may be developed when additional duties in certain safety work are definitely assigned to persons in whom the sense of responsibility is well developed. These employees may be:

Charged with the care, maintenance, and fitting of goggles, gloves, masks, safety shoes, and other safety equipment
Given charge of housekeeping in specified areas
Given the duty of keeping bulletin boards neat and putting up new posters
Appointed as chairpersons of committees or leaders in other safety activities
Assigned duties in accident investigation and the keeping of records

b *Analogy.* Responsibility for safe use of tools and equipment and for the observance of safe-practice rules can be compared to responsibility for the care of a family or a motor vehicle, and for the observance of rules for the family budget, for health, old-age security, etc.

Responsibility is an appeal that can be used to a much greater degree than it usually is in safety programs. It fits with modern behavioral theory in that responsibility is one of the key motivators on Herzberg's motivation scale. It also fits with modern management theory relating to employee participation in management.

Example

A meat and grocery chain-store organization, finding that it was difficult to maintain the safety interest of its store managers, who were to a large extent its only steady personnel, decided to uncover the primary reason why these managers were so unconcerned. Certain interesting and justifiable conclusions were drawn. It was found, first of all, that because of the general nature of their positions the store managers were characterized by a strong sense of responsibility. In chain stores the managers of the stores must act largely on their own responsibility. Although they are visited more or less frequently by traveling supervisors and superintendents, they are to all intents and purposes the "monarchs of all they survey." Upon further investigation it was also found that the majority of these store managers did not sense their responsibility for the prevention of accidents either to employees or to the public and as a consequence did not take continuous and specific action of any great value.

The remark of one store manager was typical. When asked why he did not keep litter off the floors he said that, in accordance with his understanding of his responsibilities, it was up to him to develop a satisfactory volume of profitable business and that as far as house-keeping was concerned he felt that he had discharged his obligations when he had lived up to posted rules to the effect that the floors were to be swept four times each day. Inasmuch as the reaction of this store manager was typical, the management obtained results in safety by issuing specific instructions placing directly on the store manager full responsibility not only for sales volume and other phases of business management but also for the occurrence of accidents and the elimination of conditions that caused them. Responsibility was featured for several months in the daily bulletins issued from the main office, in correspondence, and in periodical accident progress reports. The drive thus made was in this case successful almost entirely because of its appeal to the predominating characteristic (responsibility) of the store managers.

Many employees have erroneously believed that accident prevention was the responsibility of someone else, as of the boss or the insurance inspector, and the correction of this improper thinking will solve numerous accident problems.

5 *Pride.* Pride in their work is one of the strongest incentives individuals can have for producing the best results within their ability. In many instances it is even stronger than reward or responsibility toward oneself and others.

Pride increases as achievement progresses and is often proportionate to the relative quality of the achievement, to the individuals' personal recognition of the worth of their work, and to the degree to which the work has improved. The ability to experience pride, therefore, cannot exist unless individuals are willing to compare and evaluate. This characteristic may at times be evidenced even more strikingly in persons of lesser intelligence than in highly skilled individuals whose standards of achievement are difficult to surpass. People who are capable of well-founded pride in their work are not necessarily also interested in the safety of their peers or the safety record of their plant. Lucifer was proud, but Lucifer was profoundly antisocial. When dealing with people who are proud of their own strictly material achievements, it may be necessary to demonstrate to them that the safety of the group with which they are associated is an important factor in their own accomplishments, and that safety, in the long run, is an aid to efficiency. Pride in group achievement often can be "sold" readily to individuals who already have pride in their personal achievements. Interest in safety may be stimulated through appeals to pride in the following ways:

a *Praise.* Approval of good work is a stimulus upon which pride thrives. Congratulations on safe performance therefore are a successful medium of approach.

b *Exhibits.* Display of craftsmanship is a form of showing the results of one's best abilities. Accident charts and statistics are exhibits of safety endeavor.

c *Awards and Insignia.* Pride in safety achievement is maintained by awarding cups, plaques, banners, and trophies for groups, and badges, buttons, and other similar insignia for individuals.

d *Differentiation.* In some industries it has been found practicable to provide distinctive uniforms or hats or jackets for employees whose safety record is outstanding or who are members of a group or department that leads in safety performance.

Employees who have a well-developed sense of pride will frequently excel in performance when they are given a part in plant administration. An electric manufacturing company found this to be true in experiments carried on in one of its plants. A small "control" group of employees, all women, was selected for research so that it might be determined which physical conditions were most conducive to highest production. When improved lighting was provided, production increased. When illumination was lowered, production increased again. When work-hours were shortened, production increased as it did when rest periods were instituted, when piecework was introduced, hot lunch was served, and other experiments were tried. The answer to this somewhat paradoxical situation was in the *attitude* of these women. They had been told that they were helping management to perform an experiment; they were no longer cogs in a machine but were helping to solve a problem. *They had pride.* And so they continually improved the speed and quality of their work, regardless of environmental circumstances.

Example

In an effort to improve an unsatisfactory accident record, the director of safety in an iron foundry resolved to appeal to the employees' pride. His choice of this method was not "hit or miss" but was made because of his knowledge of two predominating employee characteristics. The personnel of the plant was of such a nature that the majority of employees were inclined to follow the leadership of certain others who, for lack of a better term, may be described as "ring leaders." There were two people in this group who had shown evidence of pride in their ability as skilled workers, also in their ability to lead and influence their associates. Knowing this to be true, the employer called these two people into conference, one at a time. Inasmuch as they were valued employees, he could quite honestly compliment them upon their mechanical skill and ability. This he did and then proceeded immediately to compliment them also with regard to their qualities of leadership.

He then indicated that he wished to have them share with him one of the important plant problems, this being the prevention of accidents. In each case he was careful to point out the specific unsafe practices of the workers in the department of the individual with whom he was talking, but he did not make the mistake of belittling the difficulty of converting the violators of safe-practice rules to more common sense and safe-practice procedures. As a matter of fact, he exaggerated the difficulties and found, much to his satisfaction, that this was accepted as a challenge. In each case the employee whose assistance was enlisted by appealing to the pride of accomplishment and whose vanity presumably was flattered, obtained a far more comprehensive viewpoint of accident occurrence and accident prevention. These two employees also studied their individual departmental situations with great care and set good examples by their own performances, thus influencing their associates to do likewise.

An additional noteworthy feature in this case is that the individuals appealed to in the manner here described were so characterized by pride in achievement that they voluntarily reported progress to the director of safety. This provided further opportunities for deserved praise and kept the wheels of accident prevention turning in the right direction. This might be cited also as an example of appeal to an individual's sense of leadership (No. 8). Many of these motivating qualities will be found to overlap.

6 *Conformity.* Through the years, the experience of the majority has proved that certain patterns of living are productive of the greatest degree of happiness, material well-being, and social harmony. Conformity with these patterns is necessary to the maintenance of any well-ordered civilization. The social and material values of morality, hard work, consideration for others, and observance of the golden rule are universally recognized, and the person who habitually flaunts established conventions in any of these respects is likely to be declared a social outcast. Fortunately, most persons realize this and are willing and even anxious to

abide by established rules and customs. In cases where the sense of conformity is characteristic, the following methods of appeal can be used to create interest in accident prevention:

a *Standards.* The posting of safe-practice rules, codes, and approved procedures that are described as *standard* and *commonly accepted* by the majority of persons.

b *Comparison.* The violation of safe-practice rules can be treated as unpraiseworthy—an *exception* to a general rule.

c *System and regularity.* Regularity of procedures and systematically conducted safety work in general are of great importance where "conformity" is a predominating characteristic of employees. Such examples are:

Fixing hours for cleaning and oiling machines.
Definite and periodical practice of turning in defective tools.
Scheduled days for safety-committee meetings.
Emphasis on reporting minor injuries invariably to first aid.
Featuring steps determined by job safety analysis.

d *Leadership.* Selection of a leader who will set a good example in safe conduct for the purpose of influencing employees who readily "conform" to commonly approved practices. (See example under "Pride.")

e *Ridicule.* Inasmuch as the "conformist" fears ridicule, in certain cases it is advantageous to portray the violation of safe-practice rules not only as contrary to customary procedure but as silly or ridiculous.

Normal human beings desire to follow approved practices in all their activities, both in their work and in their private life. It is therefore not difficult, as a rule, to secure their active participation in a concerted movement for safety, provided they are shown that the safety movement has the approval and backing of those persons whose leadership and popularity are unquestioned.

Appealing to pride might better be thought of as providing the atmosphere of achievement for workers. Achievement is the top motivator on Herzberg's motivation scale. A job that provides a chance for individual achievement is a motivating job. Achievement is perhaps the most important thing to be built into safety programs.

7 *Rivalry.* Humans are highly competitive animals. Laboratory tests have shown convincingly that certain persons performing various simple tasks accomplish better results when competing with others than when working alone. Their interest in what they are doing seems to increase in proportion to the opportunity for demonstrating the superiority of their performance as compared with the performance of others. There are few persons in whom there is not at least a dormant sense of rivalry. Where this characteristic is prominent, it may be appealed to in several ways.

Outstanding no-accident records usually result from long-continued efforts on the part of both management and employees and not from intermittent campaigns or contests. Contests, however, have been proved a most satisfactory device for restoring flagging interest and providing a much-needed tonic for "pepping up" a lagging safety program, and for overcoming staleness, matter-of-factness, and even occasional downright indifference.

In appealing to the sense of rivalry there are few approaches, although these have many different applications:

a *Provide opportunity.* Rivalry requires opportunity to express itself, and such opportunity may be provided by the inauguration of safety contests or competitions. One department of a plant may compete with another, or the entire plant may compete with other plants that are entered in a national or statewide accident-prevention campaign. Opportunity may also be provided for the recognition of achievements by individual employees—perhaps by comparison with their own past records.

b *Set up objectives.* An objective can be established. It may be based on the number of work-hours worked with no lost-time accidents, on low severity rates, or on comparison with the past performance of comparable units.

c *Determine method of measurement.* Success depends in part on the adoption, in advance of the contest or competition, of a clearly stated formula or method of measurement. Those in common use are:

1 The number of chargeable accidents per million work-hours worked.

2 The number of chargeable accidents per 100,000 car miles or passenger miles (for vehicles).

Many variations of the above are often used. The definition of a chargeable accident may be modified to suit local circumstances. Frequency rates may be on a "per-person" or on a "unit-group-of-persons" basis. Cost or severity, often the degree of responsibility, and factors representative of the severity of exposure may be included as modifying elements in the method of measurement.

8 *Leadership.* The desire for leadership is strong in many persons and may be used to advantage in safety work. All workers who know correct safety procedures can exercise their liking for leadership by tactfully pointing out to careless or uninformed fellow employees the accident potentialities of their unsafe practices and by showing them the correct way to do things. Instructing seasoned workers to keep a watchful eye on the working habits of new employees appeals directly to their desire for leadership.

Employees in whom leadership is an outstanding trait may be interested in safety in the following ways:

a *Promotion.* The promotion of workers to supervisory or executive positions, because of safe performance coupled with other required characteristics, provides the opportunity to create interest in accident prevention through the inherent desire for leadership, and has the further advantage of associating safety with other phases of industrial work.

b *Additional responsibility in safety work.* The gratification of the desire for leadership may be satisfied by appointments on safety committees, by assignment of the task of accident investigation, by membership on first-aid teams, etc. All these provide opportunities for making speeches, giving testimony, relating experiences, and acting as chairperson of meetings, etc. Many opportunities of like nature exist in organized safety procedures.

This appeal also fits into modern behavioral thinking. Leadership appeals to the motivators of Herzberg of responsibility, achievement, promotability, and the work itself. Leadership is in fact an excellent appeal.

9 *Logic.* Some persons have a better faculty of reasoning than others and often take great pride in their ability to "see both sides of a question" and to arrive at conclusions that are logical and just. Such persons may be interested in safety to the point of correcting their own unsafe actions and taking a more constructive part in organized accident prevention if they are approached on the basis of facts and figures.

Facts are essential to the exercise of logic, and those of most importance when using the method of appeal under discussion deal with accident frequency and severity, causes, and remedies. Other facts may relate to the analogy between production faults and unsafe practices, also to skill as a factor in safety, and to the many obvious advantages of safe conduct in general. Statistics showing the good results that have been accomplished through safety efforts may be featured. Such statistics need not be elaborate or involved. It is preferable that they be easy to remember, that they be sufficiently impressive to arouse interest, and that they provide definite proof of results accomplished by specific measures.

If a plant already has a good accident record and wishes to maintain it, the presentation of accident facts should be designed to show how safety efforts have resulted in an improvement, as compared with other similar plants or in the industry as a whole.

A few methods of appeal to logic are listed here:

a *Basic philosophy of accident prevention.* Presentation of such basic principles as

The occurrence of an accident is evidence of "something gone wrong," of an error or a miscalculation, an instance of inefficiency or of fumbling and bungling in general.

Accident prevention is a worthwhile activity inasmuch as all its results are beneficial and desirable.

The majority of all accidents are of preventable types.

Unsafe workers penalize themselves and their families more than their employer.

It is part of the employee's unwritten contract with the employer that he or she obey the safe-practice rules of the shop.

The unsafe performance of one employee frequently endangers the safety of other employees.

b *Opportunity for prevention.* Discuss the foundation of a major injury (see Chap. 2, Sec. 4); the fact that the continual violation of a safe-practice rule inevitably leads to accident and injury and that such violations may be checked before the injury occurs.

c *Facts of accident occurrence.*

d *Provide data relating to accident frequency and severity, accident types and causes, and remedial measures.*

10 *Humanity.* In the earlier days of industrial safety effort, before it was clearly enough established that accident prevention was definitely "good business," many employers supported safety activity almost wholly because of humanitarian reasons. In those employers the humane characteristic was strongly developed. Such cooperation as they received from their supervisory staffs and from employees was also prompted largely by humanitarian motives. Where this most commendable trait exists, it may be used to great advantage in accident-prevention work, even though the sense of responsibility, fear of personal injury, personal gain, rivalry, logic, and other characteristics are relatively weak or wholly absent. Humanitarianism is a widely spread personal quality and should not be overlooked in the search for best ways and means of appealing to employees with regard to safe performance. Further, the humanitarian impulse is one of the easiest to enlist in the plant accident-prevention program.

However, the plan should be directed toward the development of such mental attitudes and habits as tend to make the individual *foresee* accident occurrence and either eliminate the causes or guard against them at all times. It is better to point out the danger of accident inherent in a given situation before a person is dangerously exposed than it is to shout "Look out!" in the split second before disaster occurs. Interest in safety may be aroused through a person's sense of humanity by methods similar to those discussed under the heading Self-preservation and also in the following ways:

a *Express progress in terms of life and limbs.* There is great personal satisfaction in having helped to prevent suffering in others. The humane person is stirred to a point of far greater cooperation if bare facts dealing with numbers and frequencies of accidents are supplemented or replaced by terms of human injury and suffering.

b *Save a life campaign.* Inasmuch as personal gain and commercial rivalry are not likely to be strongly developed motivating qualities in persons in whom the humanitarian quality is well developed, the tone of safety campaigns should feature the saving of lives and the prevention of suffering.

Conclusion

In bringing this discussion to a close it is appropriate to explain that no attempt has been made to express the principles of either analytic or introspective psychology in detail. The point of view is wholly that of the practical accident-prevention engineer, and the purpose has been to express in plain terms simple methods available to laypeople who have only the average understanding of human nature and of mental processes.

It will be observed that no attempt has been made to describe methods of creating or stimulating dormant or nonexistent personal characteristics, notwithstanding that this would be eminently desirable. It is clear that accidents are far less likely to occur among employees that at one and the same time are fearful of personal injury, desirous of material and personal gain, loyal, responsible, proud, logical, humane, etc., than among employees in whom only one or perhaps none of these characteristics is strongly developed.

However valuable the effort might be, the direct attempt to develop such personal qualities does not come within the scope of this discussion, which deals primarily with the principles and simplified application of modern industrial accident prevention. For this reason the treatment of the subject is confined to the *finding* and *utilization* of *existing* qualities rather than to their development.

Space limitation prohibits greater detail of methods of application. It should also be pointed out that many of the approaches that are listed under a specific personal characteristic serve equally well when appealing to other characteristics. For example, it is probably true that the suggested approach to a person who is strongly motivated by a sense of responsibility would likewise appeal to the logically minded person, and vice versa.

The characteristics listed herein are not necessarily all that may be appealed to in accident-prevention work. Other motivating characteristics and additional applications of them have been omitted. The effort is made to impress upon the minds of safety directors, supervisors, and foremen that the employees whom they control are human beings with emotions and ambitions that may readily be appealed to in furthering the work of accident prevention.

CREATING AND MAINTAINING THE INTEREST OF MANAGEMENT

Although the primary purpose of this chapter is to deal with methods of creating and maintaining the interest of supervisors and the employees whose work they direct, it is undoubtedly of interest to comment briefly on ways and means whereby *management* also may be induced to take an active interest in accident prevention.

The list of motivating personal characteristics is, of course, fully as applicable to top management as to the workers. However, the details of application are quite different. For example, the approach to management cannot be the same as to the workers for several good reasons, one being the absence of the same form of supervisory authority that management itself has when it deals with salaried and wage-earning workers. Another reason is that there is less opportunity to establish mutually satisfactory relations between management and the persons who attempt to influence management's attitude than exists between the workers and management.

The task of interesting management also differs from that of creating the interest of employees, because certain methods of appeal do not have the same force in one case as in the other. The general manager and the machine tender may have an equally keen sense of pride, yet the "no-accident" button which the machine tender cherishes as a token of her departmental safety record might not prove to be so much of an incentive in the case of the general manager. The latter would probably be far more proud of earning a lower insurance rate than her closest competitor. However, the fact remains that management and employees possess similar motivating personal characteristics and that these should be given consideration when approaching the task of creating and maintaining interest. A few examples applying wholly to management follow:

Example 1

Mechanization of industry, business depressions or booms, emergency situations, and labor troubles fortunately do not entirely destroy the sense of humanitarianism that should be and still is an inherent quality of American businesspeople.

The accident record of a large food-products plant had become increasingly worse over a period of years notwithstanding that tools and equipment were reasonably safe and that physical and personal conditions in general were above average.

The trouble was of the "underlying cause" nature. Specifically there was no coherently planned safety organization. This situation existed because the president of the organization, who had a dominating personality, strongly objected to "red tape" and clerical procedures and persisted in regarding organized safety work as nonessential paper work.

However, the president was even more strongly of a humane nature and had demonstrated this characteristic time and again. She contributed generously to local charities, had endowed a hospital, made personal deliveries of Christmas and Thanksgiving baskets, maintained a plant health and benefit association, and had frequently granted leaves of absence with full pay to deserving employees.

The successful approach in her case can be summed up by the culminating statement of the safety engineer as follows:

"One executive order from you. Ms. President, is all that is required to make life happier for more than 100 of your employees and their families. Without that order, these employees will be injured, some of them seriously, in the next 12 months just as more than 100 employees were injured during the past year."

In reply to the inquiry "What order do you refer to?" the engineer said:

"Order your general manager to place accident prevention immediately on an effective, practical, and systematic basis, in accordance with the proposed plan that is already in his possession."

The suggestion was accepted, the order was issued, and the results fully met the engineer's expectation.

Example 2

A men's clothing manufacturer became actively interested in safety when he was shown that his annual compensation insurance premium was more than 20 percent greater than that paid by a local competitor and that the occurrence of preventable accidents was responsible.

An interesting feature of this case was the method of approach that was used by the plant-safety engineer. This engineer knew that the manufacturer boasted of having a better plant than his rival, paying better wages, having more skillful employees, using higher-grade materials, and selling a better product at no greater cost to the purchaser.

The engineer therefore seized the first opportunity to show this employer that in one respect at least—the frequency and cost of accidents—his plant was not so good as that of his competitor.

Example 3

The interest of the chief executive of another industrial manufacturing concern was aroused when the accident problem was presented in terms that were in common use with regard to sales, production, and overhead cost. The entire plant product was made, sold, and shipped by the ton. The "ton" was a unit of measure of cost that was applied to light, heat, power, taxes, and profit. The output of machines was on a tonnage basis. The number of employees required was estimated by the expected tonnage to be produced. Practically every form of cost and activity, except as related to insurance and accidents, was compared with tonnage of product.

This situation provided an interesting and useful approach to safety. A brief was prepared, therefore, in which the increasing frequency and the cost of accidents were expressed as "so many accidents and so many dollars per ton." Violations of safe-practice rules likewise were compared with tons of product, and probable savings as a result of improved safety work were calculated on a similar basis. Incidentally, the new unit of measurement led to the discovery that accidents were more expensive than several items of overhead cost that had consistently received far greater attention.

Example 4

In an iron foundry the chief executive became interested in safety when the direct analogy between the causes of accidents and the causes of production faults was explained to him. In this plant excessive production cost had occurred because instructions that certain portions of the batch be *weighed* before placing it in the cupola were disregarded. Employees also disregarded instruction that ladles be heated and thoroughly dried before use to prevent explosion from contact of molten metal with damp ladles. Thus, quality and volume of production and also the plant safety record were affected by lack of control over the performance of employees. The practices in both instances could readily have been observed and corrected. When this analogy was made clear to the executive, he became an enthusiastic convert to the idea that management has the ability, the opportunity and the power to prevent accidents. He put his thoughts into practice with amazingly favorable results.

Example 5

The interest of management in another case was created by proving that accident occurrence had a detrimental effect on the morale of the employees. In this plant management set great store on the maintenance of sympathetic and friendly relations with its employees. Long-term service was encouraged and rewarded by paying high wages, by recognizing ability, and by granting an employee committee the privilege to participate in decisions that affected certain policies of the company. Group life insurance was provided on an attractive basis. Vacations with full pay were given, and in several other ways the lot of the worker was made agreeable.

Although the plant and its equipment were satisfactory as to guarding of machines and other mechanical hazards, the accident record nevertheless was regrettably high.

Specific careless and thoughtless acts of employees were responsible, and the management was loath to apply the obvious remedy because it feared that a more strict enforcement of rules would disturb morale.

In order that corrective action of the right kind be initiated it was necessary to point out a number of instances such as follow:

1 The employees in an entire department had to be sent home in the middle of the afternoon when they became "jittery" following the occurrence of a fatal accident in which an employee became caught in a revolving shaft pulley.
2 Employees who worked on assembling machines asked to be transferred to other work because of minor injuries to several members of their group.
3 Two workers fainted when they witnessed a freight elevator accident in which an employee raised a shaftway gate and attempted to board the ascending platform. Although no personal injury resulted, the shouting of the employee who slipped and hung precariously from the platform was alarming.

Thus management was convinced that accident occurrence seriously affected employee morale. Safe-practice rules were more strictly enforced with no unfavorable reaction, accident frequency decreased, and morale improved.

These five examples illustrate but a few of the many ways in which the interest of management may be aroused with regard to safety. Further exposition seems unnecessary for several reasons. One of these is that management in general is already interested. The expressed sentiment of many of the nation's highest business executives, federal and state public officials, the national, state, and municipal chambers of commerce, trade and labor organizations, technical societies, and representative individual persons is that accident prevention definitely is *good business*.

REFERENCES

Argyris, C., *Personality and Organization*, Harper, New York, 1957.
Berne, E., *Games People Play*, Grove Press, New York, 1964.
Blake, R., and J. Mouton, *The Managerial Grid*, Gulf, Houston, 1964.
Herzberg, F., *Work and the Nature of Man*, World, Cleveland, 1966.
Hovland, C., L. Janis, and H. Kelley, *Communication and Persuasion*, Yale University Press, New Haven, Conn., 1953.

Kollatt, D., and R. Blackwell, "Direction 1980," *Changing Life Styles*, Management Horizons, Columbus, Ohio, 1972.

Lawler, E., and L. Porter, *Managerial Attitudes and Performance*, Irwin-Dorsey, Homewood, Ill., 1968.

Likert, R., *The Human Organization*, McGraw-Hill, New York, 1967.

Mager, R., and P. Pipe, *Analyzing Performance Problems*, Fearon, Belmont, Calif., 1968.

Maslow, A., *Motivation and Personality*, Harper & Row, New York, 1954.

McGregor, D., *The Human Side of Enterprise*, McGraw-Hill, New York, 1960.

Petersen, D., *Safety Management—A Human Approach*, Aloray, Englewood, N.J., 1975.

Skinner, B., *Science and Human Behavior*, Macmillan, New York, 1963.

ELEVEN

A Motivation Model

In Chap. 10, much of what was discussed can be included under the heading of safety psychology. We reviewed a number of appeals that might be made to workers, and then took a look at a number of psychologists and their theories about the worker in industry. We'll continue to look at these in this chapter as we look at workers on the job and the kinds of influences upon them that determine whether or not they will work safely.

In this chapter, as in the last, our primary emphasis will be on workers, and we'll look at psychological theory as it applies to them primarily. In Chap. 13 we'll be looking at the role of supervisors, and in doing that we will also look at psychological theory as it might apply to them. In this chapter we'll primarily look at the model of accident causation and employee motivation presented earlier in Fig. 2-10 in Chap. 2. This model of worker performance is an attempt to depict the various motivational influences upon workers. These various motivational influences summed together determine what their performance will be—whether or not they will perform in a safe or an unsafe manner. Thus in this chapter we'll look at the various motivational influences depicted in the model (reproduced in Fig. 11-1).

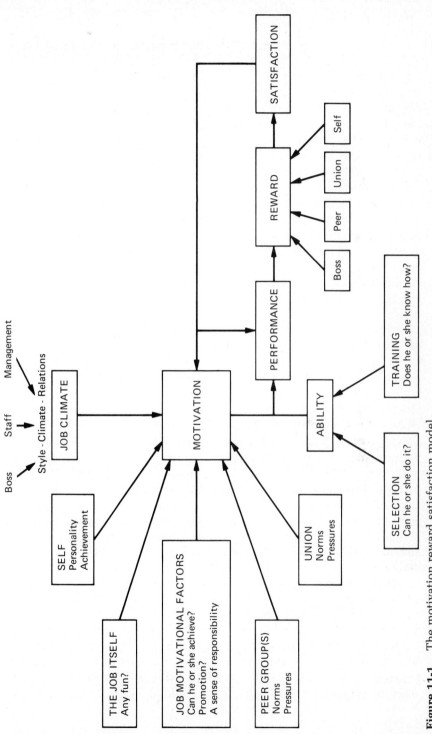

Figure 11-1. The motivation reward satisfaction model.

The motivational reward satisfaction model in Fig. 11-1 is a rather simple model attempting to describe the factors involved in determining employee performance. It says that performance (whether or not the employees perform in the manner we want them to) is dependent upon two things: (1) whether or not they have the ability to do the task (can they do it, and do they know what to do?) and (2) whether or not they are motivated to do it (which is a function of quite a number of organizational and personal things). If they know what to do (perform the task safely) and are so motivated to do it, they will perform.

Whether or not workers are able to perform (ability) is dependent upon our selection of them to be on the specific job and upon whether or not they have been trained. Motivation is somewhat more complex, being dependent upon such things as the pressures they live with daily (peer pressure, union pressure, other pressures), the motivational factors present on their job (do they turn them on or turn them off?), the job itself (does it have any meaning?), their own personalities and the job climate in which they live daily. Each of these will be briefly discussed in this chapter.

After performing the task safely, workers then experience certain rewards, which might be either positive or negative. Rewards of all kinds, coming from various places—from the boss, from the peer group, from the union, even from their own feelings (intrinsic rewards). These rewards determine how they feel about what they have done; they are either satisfied or dissatisfied with their performances, and that dictates whether or not they will continue to perform in that manner or try something else. This model is in harmony with a number of models of performance presented by a number of psychologists. It is notably close in makeup to the integrated model of the determinants of performance presented by Cummings and Schwab in their book *Performance in Organizations*. It is also similar to the model of supervisory performance presented by Lawler and Porter referred to earlier and explained in their book *Managerial Attitudes and Performance*. Both of these models are fairly well accepted in management thinking and in the literature.

In this chapter we'll look at each of the determinants in the model.

ABILITY DETERMINANTS

1 Selection

Figure 11-2 gives a broad outline of the process available to management in selecting employees. Ideally the process is based on job

> ### THE PROCESS OF SELECTION
>
> | *Biographical Data* | *Tests* |
> |---|---|
> | Application | Physical examination |
> | Interviews | Qualifications |
> | References | Defects |
> | Other sources: credit bureaus, | Skill tests |
> | schools, driving record, agencies | Job knowledge tests |
> | | Psychological tests |

Figure 11-2 The selection process.

standards which state that for a particular job a particular type of person is required. This is utopian, and the "exactly right" person for the job is seldom (if ever) found, but such a goal is still worth striving for. As the exhibit shows, we have two basic sources of information about an applicant: biographical data and test results.

The value and validity of most of these selection devices have been fairly well subjected to research in the field of occupational psychology, and that research leads us to these conclusions:

1 Accident proneness is real but only in a tiny percentage of people, and is probably a system of maladjustment. We probably cannot afford to find those people.
2 Accident susceptibles are an ever-shifting group. Finding them will not help much because next year the susceptibles will be a different group.
3 Job knowledge and skill tests can be administered effectively, provided the job criteria can be established in these areas.
4 Similarly, physical exams can be helpful, provided physical job criteria are available.
5 Interviews are generally invalid, not because the applicant is lying, but rather because our biases and stereotypes invalidate them.
6 Checks to previous employers (if we have established a good working relationship) will tell us an applicant's past history (assuming this is indicative of his or her future, which is not and probably cannot be proved).

There are some things that we can do to help select people that will perform as we wish them to (safely). However, it is a somewhat limited accident-prevention device. Usually we cannot tell ahead of time which employees will be safe or unsafe performers. While the past record might be a good indicator, it certainly is far from surefire. The accident-experienced applicant could be only a repeater and not be expected to remain in the accident-repeating group next year, or he or she could be actually prone (maladjusted) and continue to experience accidents. We simply have no way to determine this yet.

The research might also be encouraging from the standpoint that it seems to say that all but a very small percentage of the population are not prone—are no more likely to experience accidents than anyone else; thus if we properly deal with them in our company, they will not end up on our accident record. Any employee can in fact be a safe worker.

The second facet of selection is proper placement. Fitting the person to the job (or the job to the person). Placement attempts to put people on jobs that fit them physically and psychologically. This placement process is within our knowledge today. It does require proper job analysis and proper worker measurement. While within our knowledge, it is not always an easy task. It is often difficult, is usually expensive in terms of both time and money, and is currently often illegal as it must comply with the dictates of federal law regarding equal opportunity.

2 Training

Putting the training process in its simplest form, we can say that it has three steps:

1 Finding out where we are
2 Finding out where we want to be
3 Providing the difference

Unfortunately in safety training we usually spend our time almost totally in determining content and method. Theorists tell us to spend the bulk of our effort and concentration on 1 and 2 above. If we do a good job in analyzing these, 3 almost naturally falls into place. Theorists tell us that content is no more than the difference between 1 and 2 and that method of presentation is almost immaterial to the learning process.

Training is dealt with in some detail in Chap. 12.

MOTIVATION DETERMINANTS

1 Climate

There seems little question that the way workers see the company safety program strongly influences not only their behavior on the job but their ability to learn from and to respond to safety materials. Workers in different companies characterize their safety programs quite consistently in any given company. This was brought out in a study a number of years ago by Social Research Inc., made for Employers Insurance of Wausau. They found these kinds of general company types:

The Overzealous Company This is the kind of company where a great amount of safety equipment is required to be worn, machines are so guarded as to make them difficult to get near and work with, and the tone of the safety program is heavy. Such a company is likely to impose harsh punishment (a 3-day layoff, for example) for some minor infraction of safety procedures. There seem to be endless meetings, films, manuals, and preachments about safety, often not involving the worker directly in any personal way. In such companies response to safety education materials is not lively, and workers feel overexposed to it.

The Rewarding Company This is the kind of company that might offer prizes for safety records or for entering into safety slogan competitions. While the prizes are relatively small, there is a sense of competition for them. Employees in such companies feel that safety programs are important and that the company generally feels responsible about safety matters.

The Lively Company This is the kind of company that has a safety program which stimulates competition among the various plants, which offers plaques, which has boards to record the number of hours passed without accident, or which posts a continuing safety record at the plant entrance. These are companies who teach the workers to identify with safety goals, and the employees are proud of their record. Safety in these companies becomes one of the lively aspects of the job and is more than avoiding risk or accident; it is a concrete symbol and goal.

The Negligent Company This is the kind of company which seems to have programs only after the fact, that gets busy about safety only after a major accident happens. The workers here feel that the company does not really care, that it passes out safety equipment because it is current custom and to protect itself, and that it passes out safety information material in the same way.

This study further reported that the ideas about the friendliness of safety people, the manner in which safety is talked about and promoted, and the efficiency with which matters are attended to vary with the concept of the company as described above. In the overzealous company, the safety people usually seem loud and harsh, overly watchful, too quick to criticize. In plants where programs for safety are livelier and more rewarding, safety people are usually seen as friends, people much like the workers. In the more lax organizations, the employees complain there are not enough safety officials around and that they doubt reporting to them does much good since little happens. These conclusions are the result of detailed in-depth interviews of workers to assess the need for and worth of safety media. They do well describe types of climates common in industrial safety programs.

The safety program climate is but one aspect of what we might consider. Another is corporate climate in total. This is a major influence on the behavior of both managers and employees. According to Burt Scanlon, of Oklahoma University, one point of concern is that the employees' perception of the organization's climate and philosophy may be different from that which is intended. Two possible reasons may contribute to this difference in desired versus actual perception. First, perhaps not enough effort has been expended in communicating the guiding philosophy down the line, and the second factor may be a discrepancy between that which is professed and actual occurrences. The individual's closest point of contact with the organization is his or her immediate superior. If the superior's actions do not reflect the organizational philosophy, a perception discrepancy occurs.

Some of the basic climate requirements for maximum individual performance in an organization, according to Scanlon, are:

1 There must be central over-all goals or objectives toward which the organization is striving.
2 The objectives must be communicated down the line with the idea of getting commitment as to value, reasonableness, feasibility, etc.
3 Functional areas, departmental units and individuals must also have specific goals to obtain. These must be derived from the central goals, and interrelationships must be perceived.
4 The interdependency of all subunits within the organization in the accomplishment of results must be clearly established and a framework for interunit cooperation must exist.
5 Meaningful participation on the part of the individual should be the keynote. This means participation in the sense that the individual has a "real" part in determining his job objectives.
6 There must be freedom to work in the sense that a man has an opportunity to control and adjust his own performance without first

being exposed to authority and pressure from his superior. In addition, he must receive support and coaching from his superiors.

M. Scott Myers specifies a number of specific climate factors as necessary before motivation can occur:

The first of the climate factors is *expansion*. In a growth organization, it is hard to be a bad manager simply because, as the organization expands it offers increasing opportunity to achieve, to grow personally and professionally, to acquire a sense of responsibility.

Another climate factor is *delegation*. It is possible to encourage delegation by the manner in which you organize your business. Some of the desirable characteristics of the small organization can be retained within a large company by organizing the business into blocks, products, or families of products of income-producing services, or of staff functions.

A third climate factor is *innovation*. Among some 400 highly motivated managers, 73% said that their ideas were usually adopted. Of approximately 400 poorly motivated managers, 69% said their ideas were not used. Part of the reason the upper group is highly motivated is that they exist in a climate that is adjusted to change.

Fluid Communication is the fourth requirement for the climate, necessary to the job of getting the superior product or service out of the door. As organizations grow, a formalizing of communication usually follows which can actually result in a break-down of communication. In a small shop, Joe yells something across the room to Harry; information is exchanged, and that's all that's needed. But in our growing company we dictate memos, which are typed in duplicate, filling up cabinets and taking up floor space. Boys are hired to deliver the memos to distribution points, from which they go into baskets, tying managers to their desks reading and answering these communications. We are caught up in a growing mountain of paper work which worsens until some agile competitor comes along who doesn't have this paper problem and threatens us by giving superior service at a lower cost. We clean up our systems, discovering that we can throw out 90% of what we have in our files. And gradually they grow fat again. Business requires documentation. But the best way of communicating is using an informal system. Speak face to face and use the telephone wherever possible.

The fifth element of climate is *goal orientation*. This differs from authority orientation. When you ask a person why he is doing something and he says, "Because the boss told me to," that's authority orientation. But if, in answer to your question, he explains the job and what it will achieve for the organization—that's goal orientation.

The final organizational climate factor is *stability*. It means many things,

one of which is job security. They must feel assured that there will be a
job. This kind of stability is essential.

Rensis Likert, one of the most famous of the behavioral scientists,
also discusses climate to a large degree when he describes his "system
4" kind of company. He has isolated three variables which are
representative of his total concept of participative management. These
include (1) the use of supportive relationships by the manager, (2) the
use of group decision making and group methods of supervision, and
(3) the manager's performance goals. The manager's supportive relationship is shown by the degree that he shows of such things as:

1   Confidence and trust
2   Interest in the subordinate's future
3   Understanding of and the desire to help overcome problems
4   Training and helping the subordinate to perform better
5   Teaching subordinates how to solve problems rather than giving the
    answer
6   Giving support by making available the required physical resources
7   Communicating information that subordinates must know to do their jobs
    and also the information they wish to know so that they may identify more
    with the operation
8   Seeking out and attempting to use ideas and opinions
9   Approachability
10 Crediting and recognizing accomplishments

The use of group decision making and supervision means that the
group does not necessarily make all decisions. The emphasis here is on
the involvement of people in the decision-making process to the extent
that their perceptions of problems are sought, their ideas on alternative
solutions are cultivated, and their thoughts on implementing decisions
which have already been made are solicited. The participative process
can be applied on either an individual or a group basis. High-
performance goals imply that the superior is maximum-result-oriented.
Participative management involves the integration of people around
production. It does not mean that people take a back seat to production
at any cost. A climate with a potential reward of psychological
satisfaction and the achievement of it is tied directly to accomplishment on the job.

Likert measures the relationship of the above to productivity. He
states there is strong evidence to suggest that the organization which
exhibits a high degree of supportive relationships, and which utilizes
the principles of group decision making and supervision where there

are high-performance aspirations, has significantly higher levels of achievement.

Likert makes distinction between what he calls "casual," "intervening," and "end-results" variables. The casual variables refer to different management systems characteristics as follows:

System 1—explorative authoritative
System 2—benevolent authoritative
System 3—consultative
System 4—participative group

Likert explains each in detail in *The Human Organization*. The intervening variables, such as loyalty; performance goals of subordinates; degree of conflict versus cooperation; willingness to assist and help peers; feelings of pressure; attitude toward the company, job, and superior; and level of motivation are of key importance. The end-results variables refer to tangible items such as volume of sales and production, lower costs, higher quality, etc. (results).

Participation is one of the main ingredients in gaining employee commitment on an overall basis. It can lead to less need for the use of formal authority, power, discipline, threat, and pressure as a means of getting good job performance. Thus, participation and its resultant commitment become a positive substitute for pure authority. Commitment may be much harder to achieve initially, but in the long run it may prove much more effective.

Likert's description of participation, however, is really a discussion of climate. And an assessment of climate is fundamental to safety program success. Climate builds employee attitudes and acceptance of our program. It is a function of corporate climate, and our communications with employees depend to a large degree on our credibility. Climate, while hard to define, is important. What builds climate? Myers suggests six components, and Likert suggests ten. Scanlon spelled out six different ones. Each of us might be able to identify others from our experience. Anyone who has worked in an organization knows climate exists and knows its importance. There are companies with vastly differing climates, from permissiveness to tight control, from climates of freedom to climates of guardedness, from climates for creativity to climates that stifle creativity, from climates that allow you to work the hours that you choose to climates that are run by real, but nonexistent, time clocks, etc., etc., etc.—and climates that say safety is an important, integral part of the job to climates that say it is not.

Similarly we've all seen or experienced differing climates generated from safety programs as described earlier. The climates described are real. They are a product of both the corporate climate and the functioning safety program.

2 Self

The second area that determines a worker's level of motivation is the worker himself. Each of us is different, and some are more motivated to do a good job, or to perform safely, than others. It is a variable over which we have little control. We can, however, recognize the individual differences and manage accordingly. An example of the differences in people is shown by the research done on one specific type, the high achiever.

We happen to know a great deal more about one particular kind of motivation than we do about any other. Fortunately for students of industrial motivation this motive, achievement, has a more significant influence on the success or failure of industrial enterprises than any other motive we know about. By far the greatest volume of research, and probably the most sophisticated kinds of research, have been devoted to it.

The leading authority on the achievement motive seems to be David C. McClelland of Harvard University. He is the director of research on achievement motivation and economic development for projects funded by grants from a number of large foundations. According to McClelland, only about 10 percent of the United States population is strongly motivated for achievement.

According to McClelland, the most convincing sign of a strong achievement motive is the tendency of persons who are not being required to think about anything in particular, that is, when they are free to relax and let their minds just "idle," as it were, to think about ways to accomplish something difficult and significant.

McClelland's studies have identified three major characteristics of self-motivated achievers. These are "strategies" which achievers tend to follow throughout life, beginning at a surprisingly early age. They help to explain why achievers are likely to be successful; and they also indicate why supervisory tactics which may be appropriate for other kinds of people are often inappropriate when applied to persons with a strong achievement motive.

First, achievers like to set their own goals. They are nearly always trying to accomplish something. They are seldom content to just drift aimlessly and let life "happen to them." Further, they are quite selective about which goals they commit themselves to, and for this reason they are unlikely to automatically accept goals which other people, including their supervisor, select for them. Neither do they seek advice or help, except from experts or people who can provide needed skills or information. Achievers prefer to be as fully responsible for the attainment of their goals as possible. If they win, they want the credit, and if they lose, they accept the blame. Either way, they want the victory or the defeat to be unmistakably theirs.

Second, achievers tend to avoid the extremes of difficulty in selecting goals. They prefer moderate goals that are neither so easy that winning them would provide no satisfaction, nor so difficult that winning them would be more a matter of luck than ability. They gauge what is possible, and then select a goal that is as tough as they think they can make—the hardest possible challenge. This attitude keeps them continually straining their abilities to their realistic limits, but no further. Above all else, they want to win, and therefore they do not knowingly commit themselves to a goal that is probably too difficult to achieve.

Third, achievers prefer tasks which provide them with more or less immediate feedback, that is, measurements of how well they are progressing toward their goal. Because of the importance of the goal to them, they like to know how well they are doing at all times.

For achievers themselves, McClelland believes that many standard supervisory practices are inappropriate, and in some cases may even hinder their performance. Work goals should not be imposed on achievers; they not only want a voice in setting their own quotas, but are unlikely to set them lower than they think they can reach. Highly specific directions and controls are unnecessary; some general guidance and occasional follow-up will do.

High achievers are one example of a type of person with different internal motivations than the so-called average worker. There are others. The point is that individuals are different from one another, and this influences their level of motivation, and it also influences how they can be managed.

3   The Job Itself—Job Motivational Factors

There has been a great deal written in recent years, both pro and con, about job enrichment. The principle of job enrichment was mentioned in Chap. 10 as one of the solutions to the basic incongruence between the nature of mature people and the characteristics of organizations. Job enrichment is in one aspect the opposite of specialization; it is an attempt to give back to the employee the piece of the action that specialization took away. More important, it is an attempt to put some meaning into the job, and thus a sense of worth for the person on the job.

Here, we begin by looking at the job itself. In most cases, jobs either were not consciously "designed" at all or were designed primarily from the standpoint of efficiency and economy. To the extent that these steps have taken the challenge and opportunity for creativity out of a job, they have contributed to a demotivating effect. That is, apathy and minimum effort are the natural results of jobs that offer workers no more satisfaction than a paycheck and a decent place to work. These hygiene factors may keep workers from complaining, but they will not make them want to work harder or more efficiently. Offering still more hygiene, in the form of prizes or incentive payments, produces only a temporary effect. Herzberg views *job enrichment* as a means of introducing more effective motivation into jobs. He draws a clear distinction between the deliberate enlargement of responsibility, scope, and challenge (job enrichment) and the movement of an individual from job to job without necessarily increasing responsibility at all (job rotation). Herzberg has found job rotation unsatisfactory as a motivating tool, but has achieved impressive results with job enrichment. After an initial "adjustment" period, during which productivity temporarily declines, efficiency tends to rise well above previous levels.

Figure 11-3 lists the main dissatisfiers and motivators. Dissatisfiers seem to be such items as pay, benefits, company policies and administration, behavior of supervision, working conditions, and other factors that are generally peripheral to the task. Though traditionally thought of by management as motivators of people, these factors are actually more potent as dissatisfiers. High motivation does not result from their improvement. Dissatisfaction does result from their deterioration.

Job Dissatisfaction	Job Motivation
Company policies and administration	Achievement
Supervision	Recognition
Working conditions	Work itself
Interpersonal relations	Responsibility
Money, status, security	Professional growth

Figure 11-3 Motivators and dissatisfiers.

Motivators are such items as achievement, recognition, responsibility, growth, advancement, and other matters associated with the self-actualization of the individual on the job. Job satisfaction and high production are associated with these motivators, while disappointments and ineffectiveness are usually associated with the dissatisfiers. A challenging job which allows a feeling of achievement, responsibility, growth, advancement, enjoyment of work itself, and earned recognition motivates employees to work effectively. Factors which are peripheral to the job (work rules, lighting, coffee breaks, titles, seniority rights, wages, fringe benefits, and the like) dissatisfy workers. These could be considered tease items which will temporarily please but which will be long-range motivational failures. In other words, the factors in the work situation which motivate employees are different from the factors that dissatisfy employees. Motivation stems from the challenge of the job through such factors as achievement, responsibility, growth, advancement, work itself, and earned recognition. Dissatisfactions more often spring from factors peripheral to the job.

4 Peer Groups

Each employee is an individual, and also an integral part or a member of a group. Each manager must manage her crew as different individuals, but also as a group. As in chemistry, elements combine together to make other substances with entirely different properties; individuals combine together to produce a group which has entirely different properties. We have to recognize the group properties just as well as the individual's properties. A group has a distinct personality of its own. Each group makes its own decisions. The group sets its own work goals. These may be identical with management's goals, or they may be different. The group also sets its own safety standards, and it lives by *its* standards, regardless of what management's standards are, and regardless of what the OSHA standards say. A group is a number of people who interact or communicate regularly and who see themselves as a unit distinct from other collections of people.

A factor that often influences the safety of a person, and yet which is not always understood, is the problem sociologists call *group norms*. Group norms are really the informal laws that govern the way the people that belong to a group should and should not behave. Very often when members of a group are asked what their norms are, they cannot identify them and yet unconsciously their behavior is strangely influenced by them.

Group norms are the accepted attitudes about various things in the group situation. These include attitudes about how workers behave toward their boss, how they react to safety regulations, how they react to production quotas. It "codifies" their attitude about the company, manner of dress, and merit systems into recognized, accepted, and enforceable behavioral patterns. If a member of a group takes on a pattern of behavior or expresses an attitude that is in violation of that commonly accepted by the group, there are ways of punishing him to bring him back into line.

In an industrial organization, if the norms developed within the work group are favorable to safety, the group itself will encourage and even enforce safe practices much better than supervision can. (Hard hats in construction, for instance.) Group norms often develop which are against our safety rules. A group of workers might have an attitude that safety is for sissies. We do not really know why this type of philosophy becomes entrenched, but we know it often does. Often management's first response to such a situation is to pass a regulation which will force the person to violate the norm of his peers and follow management. If the group is a strong group (with a high degree of cohesiveness), the member will violate management's direction rather than run the risk of being shut out from his work group. We ought to understand this phenomenon, and our objective should be to find some way to change the group norm and get this phenomenon working for safety, rather than against it.

5 Unions

This may or may not be a major influence and determinant of worker behavior. And it may be an influence for either bad or good, depending upon the union stand on an issue, management's stand, their relationship, etc. It is mentioned here as it might be one of the factors—one of the determinants of behavior in safety for the worker.

REWARD

This refers to desirable outcomes for supervisors provided by themselves (intrinsic) or by others (extrinsic), either monetary or psychological. In safety, there is always an intrinsic reward, but perhaps not as often an extrinsic reward. Rewards are desirable states of affairs that persons receive from either their own thinking or the action of others (management).

The various rewards that a person might hope to obtain are the friendship of fellow workers, a promotion, a merit salary increase, an intrinsic feeling of accomplishment, etc. A given potential reward means different things to different individuals. For example, the friendship of peers (i.e., workers at the same job level) might be highly desired by one worker and be unimportant to another. A promotion might have very little positive value for one because he does not want to take on increased responsibilities, but for a middle manager in a large corporation, a promotion might be a reward of extremely high value.

Rewards then are first of all personal and individual. What is a reward to one person might be perceived as punishment by another. To properly motivate by administering rewards, the boss must know the person well enough to know what a reward is for that individual.

Secondly, as indicated in Fig. 11-1, rewards come from a number of different sources, from management, from the immediate supervisor, from the peer group, from the union structure, and from the person herself.

Thirdly, rewards can be either positive or negative. Positive rewards will tend to increase the likelihood of the individual performing in the same way again. Negative rewards, or no rewards, will decrease that likelihood.

And finally, the rewards most often thought of as positive, and valued, are those on the "motivation" scale as opposed to the "hygiene" scale. The model is based on a psychological theory known as the "expectancy x value theory." This states that people have expectancies or anticipations about future events. These take the form of beliefs concerning the likelihood that a particular act will be followed by a particular outcome. Such beliefs or expectancies could take values between 0 (no chance) and 1 (completely sure it will follow). Thus if a worker expects to get something favorable for performing a safety task every time she performs it, she will be quite likely to perform it. If she does not expect this, she will not perform it—she will do something else that will generate a reward. Her past experience tells her where the rewards are, both positive and negative. And this determines her current and future behaviors.

CONCLUSION

It is an understatement of considerable depth to say that the practical application of psychology to industrial problems lacks full appreciation by industry. Safety engineers, however, agree that psychology has a place of considerable importance in accident prevention. They are not psychologists, but their profession demands that they deal with persons.

In any event, there is no gainsaying the fact that psychology in accident prevention is of great importance. By means of its application the ideal may more rapidly be approached. Psychology has already proved its value in enough cases to establish it as a factor in the field of approved practice. Some of the more advanced research in accident-prevention psychology has been done in the areas of both industrial and motor-vehicle operations.

Psychological Research

Research by Hannaford had as its purpose investigation of the readily assumed relationship between the safety attitude of industrial workers and supervisors and actual accident experience on the job. In view of prior studies showing that the majority of accidents are caused directly by unsafe performance of persons, the establishment of such relationship is of great value. Quoting in part from conclusions by Hannaford:

1 "There is a positive and significant relationship between the safety attitudes of male industrial workers and their actual lost time accident experience over the five-year period studied." Stated another way, "As test scores worsen, the number of accidents experienced by the testees increases."
2 "There is a positive and significant relationship between the safety attitudes of male industrial supervisors and the actual lost time accident experience of male employees they supervised during the five-year period studied." Quoting further from the Hannaford thesis with regard to this conclusion, "As the test scores worsen, the accident experience increases."

There is a vast and practically virgin field of usefulness in applied safety psychology, and it is one that should receive more immediate and widespread consideration. By no means, however, would it be wise to act on the assumption that psychology is all that is required.

A fair analogy to the substitution of safety psychology for all other accident-prevention work may be taken from the circumstances attending the annual national celebration of the Declaration of Independence.

A fond parent who observes his offspring standing on the back porch on July 4 with an enormous cannon cracker in one hand and a lighted stick of punk in the other might reason that inherited instincts are at fault and might decide to provide a course of education to overcome them. He might even go so far as to conclude that the attitude of the citizenry in general regarding the manufacture and sale of explosives for noisemaking and fireworks is improper and should be changed. In consequence of this reasoning he might leave the situation on the back porch as described and proceed to organize a public welfare committee, arouse the interest of the town folk, pass resolutions, and assist in the enactment of legislation to prohibit the sale of fireworks to other than authorized persons. The child also might have inherited chance-taking and reckless tendencies, and if this should be shown by psychological research, a way would probably be found to combat them. If the majority of the children in the community possessed the same tendencies, grade-school curricula could well be extended to include studies designed to correct the situation.

Such action would be defensible on the ground that it would probably cure the situation eventually and that it would make the Fourth of July safer for the entire community, and further, that it not only would eliminate the dangerous cannon cracker and the equally hazardous torpedoes, Roman candles, and other kinds of fireworks, but would also teach children to exercise more common sense and caution. If such action *only* is followed, however, what becomes of the boy who is on the back porch with the cannon cracker in his hand?

In line with the simplified and practical approach to safety as advocated in this text, the boy is unnecessarily exposed to a dangerous object.

Immediate action is necessary, and none could be more effective than to do exactly as the fond parent did in this case, that is, snatch the explosive away from the child and then proceed as may be advisable in the effort to remove the underlying causes.

Industry presents a similar problem and one so serious that it constitutes an emergency that requires immediate action. Employees are exposed to mechanical hazards. No matter through whose fault these hazards exist, the best procedure indicates that they should be removed or guarded. Employees are continually being hurt because they do unsafe things or omit to take safety precautions. The immediate practical remedy must be based on knowledge of what these dangerous practices are and why they exist and on supervisory observation and control. If the reason for the unsafe acts, for example, should be misunderstanding of instruction, it is not always necessary to find out why the instruction is misunderstood.

A clearer explanation of the purpose of the rule or instruction may be all that is necessary. In other cases, as suggested in the text, investigation from the lay point of view might develop the cause "physical inability," poor eyesight for example. It might be found that strict adherence to the rule was awkward or even impossible. These are not psychological findings, but they are quite helpful and suggest effective remedies.

Psychological analysis is of great value in many cases, but since it is not *always* necessary to delve so deeply into the origin of accidents, it is clear that the immediate task is to determine and to correct the more obvious causes and subcauses; then if the problem does not yield to remedies based on these facts, to proceed as far as is practicable with psychological analysis.

In the meantime industrial management would be wise to give more serious consideration to the place of professional psychology in evaluating and controlling the safe and efficient performance of its employees.

REFERENCES

Cummings, L., and D. Schwab, *Performance in Organizations*, Scott, Foresman, Glenview, Ill., 1973.

Hannaford, E., *The Significance of Safety Attitudes in Industrial Accident Prevention*, doctoral dissertation, New York University, New York, 1957.

Herzberg, F., *Work and the Nature of Man*, World, Cleveland, 1966.

Lawler, E., and L. Porter, *Managerial Attitudes and Performance*, Irwin-Dorsey, Homewood, Ill., 1968.

Levy, S., and S. Greene, *The Effectiveness of Safety Education Materials*, Social Research Inc., Chicago, 1962.

Likert, R., *The Human Organization*, McGraw-Hill, New York, 1967.

Myers, M., *The Managers Role in Motivation*, presented at Western Printing & Lithographing Company, 1966.

McClelland, D., J. Atkinson, R. Clark, and R. Lowell, *The Achievement Motive*, Appleton-Century-Crofts, New York, 1953.

Petersen, D., *Techniques of Safety Management*, McGraw-Hill, New York, 1971.

Petersen, D., *Safety Management—A Human Approach*, Aloray, Englewood, N.J., 1975.

Scanlon, B., "Philosophy and Climate in the Organization," *ASTME Vectors*, vol. 5, pp. 30–35, 1969.

TWELVE

Safety Training

All accident-prevention work, whether or not it is educationally intended, is nevertheless educational in its effect upon the individual employee whom it necessarily involves. That this is true is clearly indicated by evidence that well-trained and careful workers may avoid injury on dangerous work and that untrained and careless workers may be injured under the safest possible conditions. Whether employers demand as their right that employees conduct their work safely or merely teach safety to the best of their ability and depend on the employee for personal initiative, the net effect is educational. The influence of the supervisor or foreman should not be overlooked. Whatever the form that safety education takes, it should be kept in mind that the supervisors are in the best position to convey its message to the individual worker and to interpret it for the worker in "shop" language. It can be said with justification that if the supervisors are thoroughly "sold" on safety, they in turn will "sell" the employees whose work they direct.

In shop practice, safety education is not specifically defined.

Ordinarily it refers to meetings and talks, personal contacts with authorities or teachers, the use of bulletins and posters and other reading matter, sound slides and motion pictures, first-aid instruction, and any oral or written instruction in avoiding hazards and cultivating safe methods of doing work.

LIMITATIONS OF EDUCATION AS APPLIED TO EMPLOYEES

Education, in the ordinary sense of the word, has its limitations when applied to employees. They may be talked to or talked at, they may read pages of printed matter and view pictures by the hundreds, and yet fail to apply the message to themselves. This is particularly true when educational efforts are general in nature—when they preach safety, its value, its necessity, and its other virtues, and fail to specify what behaviors are wanted. In this chapter we'll be talking more about safety training than about safety education. If there is a difference between the two, it is that training zeroes in on teaching precise wanted (safe) behaviors, whereas education tries to impart knowledges. The limitations mentioned above are primarily limitations of safety education, not safety training.

Specific safety training among *employees*, therefore, is largely a task for supervisors and foremen, who, by virtue of their authority and close daily contact, are in a position to convert safety generalities to the everyday safe-practice procedures that apply to individual tasks, machines, tools, and processes.

Not only should employees be taught that safety is worthwhile and that it is their duty to themselves, their families, the community, and their employers to avoid injury, but *they should also be told what specific dangers in their own line of work should be guarded against and what specific things they themselves may do to avoid injury.*

The most effective way to teach them these things is to utilize ordinary supervisory methods *just as they are already being used to teach the control of volume and quality of product.*

The safety training of employees by supervisors is so closely allied with the *art of supervision*, which in turn is so vital a factor in accident prevention, that Chap. 13 has been devoted to it.

The necessity for safety training in the industries is justified primarily because employees are not informed or because they do not know how to apply their knowledge. Thus, safety education is primarily the process of imparting knowledge of safe and unsafe mechanical conditions, safe and unsafe personal practice, and remedial measures.

In its broader sense, outside the field of industrial safety and especially with regard to classroom work, the process of education implies discipline of the mind and body, and it deals therefore with approaches and backgrounds and requires a continuity and regularity and a step-by-step planning that is often quite impractical in the case of the average plant or other industrial operation.

The industrial and traffic safety courses that are conducted in high schools and colleges, by safety councils, insurance companies, and the federal government are most commendable and valuable. Consistent, however, with the stated scope and objectives of this book, the treatment of such phases of education is omitted in favor of more direct safety educational work that is best described as imparting knowledge of pertinent accident facts and the exercise of normal supervisory procedures in the encouragement and enforcement of safe practices and conditions.

EMPLOYERS HAVE A RIGHT TO EXPECT THAT WAGE-EARNING EMPLOYEES WILL FOLLOW RULES OF SAFE PRACTICE

Employers, with fairness to all, are justified in expecting that employees will abide by safe-practice rules. With this as a primary principle, general educational safety work may be added, with the result that the ideal situation is approached.

The following cases describe commonly used safety educational methods:

Case 1

A Texas oil-refining plant made good use of the so-called conference type of safety education.

In addition to the regular functioning of the executive and supervisory safety committees, provision was made for periodical meetings of supervisors and key workers in each department. In accordance with a carefully planned program, subjects for discussion were selected in advance of each such meeting, and plans were made for the leadership of discussion and for encouraging participation by the individual members.

Three general topics were selected, namely:

The principles of industrial accident prevention
The major causes of departmental accidents
Corrective action

Prior to the initiation of the conference meetings, a comprehensive accident pamphlet was prepared. This included a statement covering the objectives, nature, and scope of the text and separate exhibits defining an accident and its several identifying and cause factors, practical methods of accident investigation, the responsibilities of and opportunities for supervisors in accident prevention, corrective action, and typical unsafe acts and mechanical hazards.

The first meetings were devoted to the principles of prevention work. The effort primarily was to obtain uniformity of thought, expression, and understanding as to terms and definitions, objectives, and methods.

Case 2

In addition to the adoption of more effective supervisory methods in the enforcement of rules for wearing goggles, the management of a foundry coincidentally inaugurated an educational program of considerable merit. Employees are known to imitate those whom they respect or like or whom they consider leaders, and advantage was taken of this common trait in the effort to bring about the more general use of goggles. Certain loyal key workers were selected, who were convinced that goggles could and should be worn, and were instructed to advocate the use of eye protectors by argument and by example but, of course, without the knowledge of the other employees that a prearranged plan was being followed.

After this plan had been put into effect, a group of workers in the grinding department, one of whom was an advocate of wearing eye protection, was observed to break up shortly before the noon luncheon period was over, and the following conversation was overheard:

"Come on this way, Jim, you're headed wrong. What the ———?"

"I'll be along in a jiffy, got to go to the toolroom first."

"What for?"

"My goggles are cracked. I can't work with them and I won't work without any at all, so I'm going to get another pair."

This worker was liked and respected by his associates, and as a direct result of his example the department in which he worked was the first to reach the 100 percent mark in the use of eye protection.

All executives and supervisors voluntarily adopted the practice of carrying goggles in their pockets and of wearing them when inspecting work where eye hazards existed and even when passing near the grinding, chipping, snagging, and sandblasting operations.

A great variety of additional approaches in shop safety education are applied successfully. A few examples follow:

Planned executive contacts with individual employees

Judicious use of praise for good records, expressed in bulletins and house organs which feature control of specific hazards

PLANNED TRAINING

In this chapter a sharp distinction has been made between the ordinary safety educational methods too generally employed in our industrial activities and planned training that is intended to teach recognition and avoidance of known accident causes. Both are good, but planned training is so far superior in all respects that there is no good reason for failure to adopt it, especially when the method of determining real accident causes may be applied so readily.

Planned training is merely that part of the total training process done systematically by the organization, as opposed to the portions of the training process occurring daily by supervisors with their people. Putting this planned training process into its simplest form, we can say that it consists of the organization going through this process:

1 Finding out where people are in terms of their current knowledges and skills
2 Finding out where people should be in terms of the behaviors required to perform the job—safely
3 Figuring out a systematic way to provide the difference

As you can see, the above puts a concentration in the training task of defining training needs. In safety training, we have historically put our concentration in other areas—on training method, on content, on other things. Learning theorists tell us to spend the bulk of our time in defining needs, for if we do a good job of this, all else falls easily into place.

There is a considerable amount of literature on defining training needs. While oversimplifying, it might clarify to say that the definition of training needs encompasses 1 and 2 above, 2 being a function of policy and 1 being the major planning job of the trainer or safety person. Management tells us, through its policy statements, where we want to be and what each person's function is once there. We must then find out how able our people are to fulfill these duties and responsibilities: Can they do it if their lives depend on it? Defining needs is, of course, the key point in the process. It provides the objective and sets the criteria for measurement. We would expect that any function this vital to the success of the training would be in widespread use among industry. In one study it was found that a careful and systematic investigation was conducted in only one out of ten companies. Very little training effort is based on any systematic appraisal of the training and development needs.

One of the better texts on industrial training is *Training in Business and Industry* by McGehee and Thayer. To these authors the assessment of training needs involves a three-part analysis. An analysis of the organization should uncover the resources and objectives of the company which relate to training. An analysis of the operations should define corporate jobs and tasks. The manpower analysis explores the human dimensions of attitudes, skills, and knowledge as they relate to the company and the employee's job.

Organization analysis involves a study of the entire organization: its objectives, its resources, the allocation of these resources in meeting its objectives, and the total environment of the organization. These things largely determine the training philosophy for the entire organization. The first step in organization analysis is to obtain a clear understanding of both short- and long-term goals. What is the company trying to achieve in safety in general and specifically by department. A second step in organization analysis is inventory of the company's attempts to meet goals through its human and physical resources. The final step is an analysis of the climate of the organization.

The climate of an organization is, in essence, a reflection of its members' attitudes toward various aspects of work, supervision, company procedures, goals and objectives, and membership in the organization. These attitudes are learned; they are a product of the members' experiences both within and outside the work environment. A training program may be designed to effect certain changes in the organizational climate, for instance, our safety training certainly hopes to influence the employees' attitudes toward safety.

Job analysis for training purposes involves a careful study of jobs within an organization in a further effort to define the specific content of training. It requires an orderly, systematic collection of data about the job. We are familiar with this through our job safety analysis procedures. Other ways are also available for job analysis.

1 Observations. (Is there obvious evidence of unsafe acts or poor methods? Are there incidences on the part of individuals or groups that reveal poor personnel relationships, emotionally charged attitudes, frustrations, lack of understanding, or personal limitations? Do these situations imply training needs?)
2 Management requests for training of employees
3 Interviews with supervisors and top management personnel to accumulate information about safety problems, as well as interviews with employees concerning safety

4 Group conferences with interdepartmental groups and safety advisory committees to discuss organizational objectives, major operational problems, plans for meeting objectives, and areas in which training could be of value

5 Comparative studies of safe versus unsafe employees to underline the bases for differentiating successful from unsuccessful performance

6 Questionnaire surveys

7 Tests or examinations of safety knowledge of current employees; analyses of safety sampling

8 Supervisors' reports on the safety performance of employees

9 Accident records

10 Actually performing the job

Manpower analysis focuses on individuals and their performance on the job as it relates to safety. The performance analysis chart shown in Fig. 12-1 helps define individual training needs. This performance analysis chart was devised by Dr. Phil Brereton of Illinois State University, and seems somewhat self-explanatory. The chart helps to identify whether or not the person does in fact have a training problem at all. Too often we jump to training as the solution to almost any problem. While training is highly important, it is the solution only when the person's problem is the result of a lack of knowledge or skill. If the problem is the result of a skill deficiency, training might be the answer (see the left branch of the chart).

Many times, however, the problem is not the result of a skill deficiency, in which case the right branch of the chart applies, leading to other solutions.

If training seems to be the answer, we must then begin to think about how the training should be accomplished. Should it be one to one performed by the supervisor? Should it be a classroom exercise? Should it be a practice learning experience on the job?

Much has been written about this. We seem to be finding today, however, that method of instruction is somewhat less important to the learning process than some other variables. First we might look at what we know about human learning behavior. There are more empirical data here than perhaps anywhere else in psychology. Here are some of the "knowns."

Motivation and Learning

Learning theorists generally agree that individuals will learn most efficiently if they are motivated toward some goal which is attainable by learning the subject matter presented. It is necessary that the

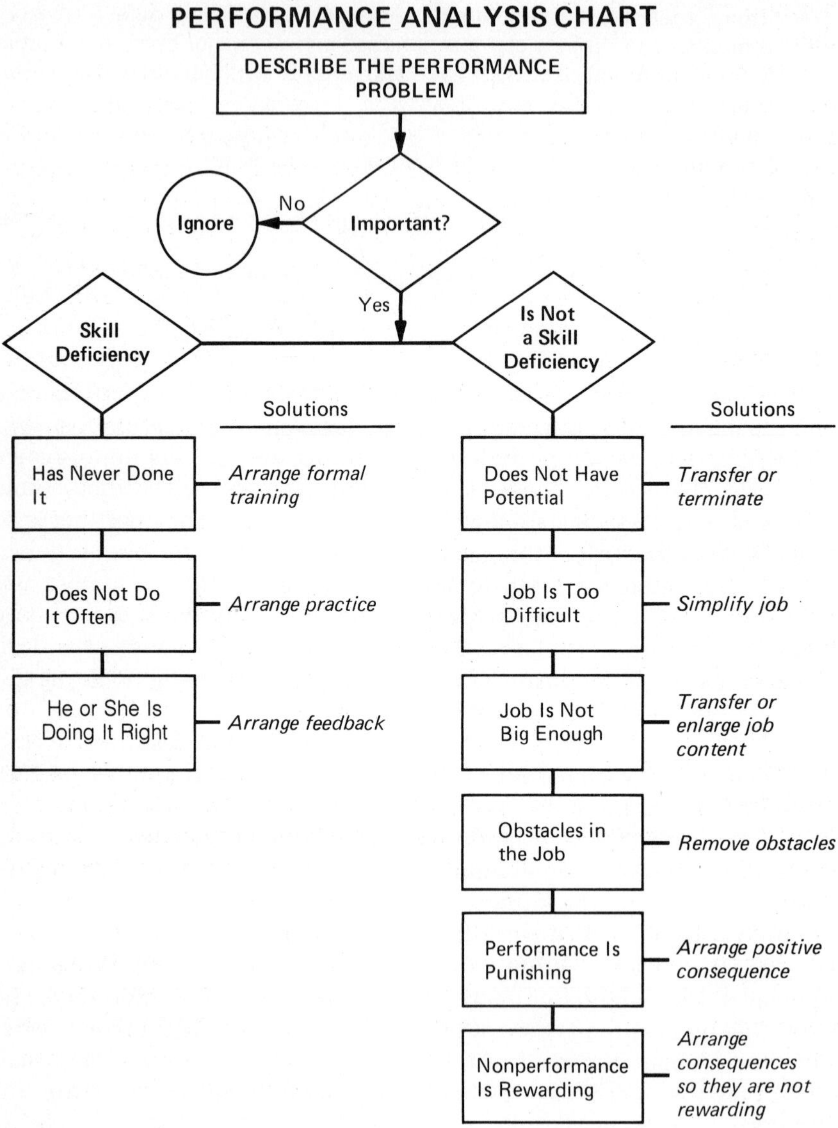

PERFORMANCE ANALYSIS CHART

```
┌─────────────────────────────┐
│ DESCRIBE THE PERFORMANCE    │
│          PROBLEM            │
└─────────────────────────────┘
```

         Ignore ◄── No ── Important?

                     Yes

Skill Deficiency		Is Not a Skill Deficiency	
	Solutions		Solutions
Has Never Done It	Arrange formal training	Does Not Have Potential	Transfer or terminate
Does Not Do It Often	Arrange practice	Job Is Too Difficult	Simplify job
He or She Is Doing It Right	Arrange feedback	Job Is Not Big Enough	Transfer or enlarge job content
		Obstacles in the Job	Remove obstacles
		Performance Is Punishing	Arrange positive consequence
		Nonperformance Is Rewarding	Arrange consequences so they are not rewarding

Figure 12-1 The Brereton model.

outcome must be desired and the behavior must appear to the learners to have some relation to achieving that outcome. If the behavior-outcome relationship is obscure and the learners are striving toward their desired goal, they may ignore attempts to teach them the new behavior and try other kinds of behavior which appear to be relevant to their goal.

In conducting a training course for supervisors on safety, for example, some people may feel that they have more important production problems to worry about and will spend training time thinking about them and complaining about being taken away from the job to learn a lot of nonsense. Or they may enjoy the opportunity to get together with the "gang" and swap stories. Still others may see the training class as an opportunity to show how much they already know and to strive for greater recognition in the eyes of the trainer and their fellow trainees. A few may see that training may aid them in their job. The behavior of people is oriented toward relevant goals, whether these goals are safety, increased recognition, production, or simply socialization. People attempt to achieve those goals which are salient at the moment, regardless of the trainer's intent.

Reinforcement and Learning

Any event which occurs so as to change the probability of a given response is said to be reinforcing. Positive rewards for certain behavior increase the probability that this behavior will occur again. Negative rewards decrease the probability. Actually, punishment seems to inhibit the occurrence of response rather than eliminate or extinguish it. Instead, the response which leads to the avoidance of punishment is reinforced. However, it should be emphasized that the role of punishment in learning is not completely clear yet. Often, failure to reinforce a response may have a better long-range effect than a reprimand from the trainer.

Whether or not an event is reinforcing will depend on the perceptions of the individual who is learning. What one person regards as a rewarding experience may be regarded by another as neutral or nonrewarding or even punishing. In general, however, these are various classes of reinforcers, food, status recognition, money, companionship, which are reinforcing to almost everyone at one time or another.

Practice and Learning

Individuals learn that which they do. They will acquire and maintain, in practicing a skill, those kinds of behavior which they perform and which are reinforced. Without practice they quickly lose that which they have learned. In safety training, then, the follow-up and practice are as important as the initial learning. Also, spaced practice (a little at a time) seems to be more effective than massed practice. This seems to be true both for learning rates and for retention. It is better to have a number of short sessions than one long one.

Feedback and Learning

Most experts state that telling trainees how they are doing is essential for good training. It is difficult for trainees to improve their performance unless they are given some knowledge of their performance. Are they performing correctly? If not, what is the nature of their errors? How can the errors be corrected? This is essential. Some theorists strongly emphasize immediate reinforcement to each bit learned. The success of programmed instruction is based primarily on this concept.

Meaningfulness

In general, meaningful material is learned and remembered better than material which is not meaningful. It is difficult to state just what is meant by the term "meaningfulness" because no universal measure of this important concept has been developed. For instance, new information in your own specialty is more easily learned than material in another specialty. In your own specialty, you are familiar with the meaning of many technical terms and have used these concepts in many situations; they are "your own." Delving into a new area requires the acquisition of new concepts and principles as well as the acquisition of the information being presented.

The implications of the rather consistent experimental findings on meaningfulness with respect both to original learning and to transfer to other situations are quite clear. In order to simplify the trainee's job, the material must be made as meaningful to him as possible. In reading through training courses or watching safety films, we often wonder toward what audience the course or film was being aimed. Too frequently it appears that the safety director is trying to write in terms which are most meaningful to herself or in terms which will impress her superiors or peers, ignoring the fact that the new trainee will not have the familiarity with terms, concepts, machines, and materials which will help him make sense out of what is being taught.

The concept of meaningfulness has implications not only for the way in which material is presented to the trainee, but also for the preparation of the material which the trainer must carry to others. The trainer must try to think in the trainee's terms, to put her material across with examples and language familiar to the worker, and also she must attempt to supply as many associations for new ideas and concepts as possible so that they become more meaningful.

Climate and Learning

The classroom environment makes a difference. Research has concluded that to encourage high rates of achievement in highly technical subjects, the environment must be challenging. To encourage achievement in nontechnical areas, classes should be socially cohesive and satisfying.

Research where anxiety and stress are built into the training situation shows that quantity of ideas increased; increases did not transfer to other tasks; there is a curvilinear relationship between anxiety and performance, with poor performance being associated with intermediate anxiety; and anxiety interferes with verbal ability.

These then are some of the insights into the behavior of people involved in learning. Whichever methods are chosen, they should take into account the things we do know from the research.

Within this framework, then, safety training programs should be devised. First, they must meet the organizational and individual needs of those persons to be trained. Next they must incorporate the principles that determine human learning in terms of motivation, reinforcement, practice, feedback, meaningfulness, and climate. Then method might be considered, and here the choice is almost endless. Here are some typical approaches:

1 On-the-job training
2 Job rotation
3 Coaching
4 Interning—apprenticing
5 Junior board
6 Classroom (off-the-job training)
7 Vestibule training
8 Lectures
9 Special study
10 Conference
11 Films, television
12 Case-study approaches

13 Role playing
14 Simulation (gaming, in-baskets, etc.)
15 Team training
16 Programmed instruction
17 Laboratory training (T-groups)

This is only a partial listing. Each has advantages and disadvantages. Method is important, although less crucial than the earlier mentioned considerations of needs and learning principles. The method should be somewhat congruent with the subject matter. Gaming is, for instance, a good way to teach problem-solving skills, but a poor way to teach factual information.

EVALUATING TRAINING RESULTS

The evaluation of training is not simple. We try to determine what changes in skill, knowledge, and attitude have taken place as a result of training. We also try to determine how these skills, knowledges and attitudes contribute to organizational objectives. McGehee and Thayer suggest evaluation measures in four broad categories: (1) objective-subjective, (2) direct-indirect, (3) intermediate-ultimate, and (4) specific-summary.

Objective-Subjective The major distinction between an objective and a subjective measure is its source. A measure is objective if it is derived from overt behavior. If the measure represents an opinion, a belief, or a judgment, it is subjective.

Direct-Indirect A measure is classified as direct if it measures the behavior of the individual or the results of his behavior. An indirect measure assesses the action of an individual whose behavior can be measured only by its influence on the actions of others. For instance, supervisory training effectiveness is usually measured indirectly as we read the results of the participant's (supervisor's) efforts for our determination of training effectiveness. As a result of our supervisory training program on safety, did his department have a better safety record?

Intermediate-Ultimate An intermediate measure might be the supervisor's test grade—an ultimate measure of the results back on the job.

Specific-Summary Somewhat related to the problem of intermediate and ultimate measures of training outcome is the problem of measures which are used as an index of successful performance of a specific phase of a job or as an index of the degree of performance of the total job against its potential contribution to organizational goals. Should we measure the effectiveness of our supervisory inspection training program by the number of code violations or by improvement in the department's accident record?

These four categories of measures are not mutually exclusive. A rating scale could be a subjective, intermediate, direct, and specific measure. It also could be a subjective, direct, ultimate, and summary measure. Regardless of their type, measures must have certain characteristics if they are to be used in studying training outcomes. These characteristics are relevance to job criteria and corporate goals, reliability (consistency), and freedom from bias.

Too often in safety training there simply is no measure of any kind to evaluate whether or not the training has accomplished its objectives—or if it has accomplished anything. There should be. Participants in the training process, regardless of organizational level, should have a measurable improvement in some skill or behavior, and management should insist that the training be evaluated in terms of improvement of the skill or in terms of measurement of behavior change.

CONTRIBUTIONS OF OUTSIDE AGENCIES

Under the heading General Safety Education may be included the excellent work done by insurance companies; local and national safety councils; schools; chambers of commerce; magazines and newspapers; national and state labor departments and industrial commissions; engineering, technical, and statistical organizations; and many other groups and individuals. The means and methods used include lectures and addresses, radio broadcasts, preparation and circulation of printed matter, inauguration of safety campaigns, sound slides, motion pictures, engineering research, preparation of safety codes, and legislation and enforcement procedure.

These activities, to be sure, do not provide an answer for industrial executives who want to know how to prevent accidents in their particular plants, and of necessity they are general in nature. They reach the public at large, however, including industrial workers who must eventually benefit to such an extent as to justify the time, effort, and expenditure involved.

MISCELLANEOUS EDUCATIONAL ACTIVITIES

In the foregoing examples some special phase of safety educational work has been featured. It is seldom, however, that success is achieved when only one of the many forms of safety education is utilized. Best results, rather, are achieved when, in addition to emphasizing one particular step, the program is made to include helpful corollary activities. Some of the more commonly used educational procedures are listed here:

1 Placing responsibility on supervisors for the investigation of accidents and for correcting unsafe practices and mechanical hazards
2 Safety meetings, especially conference-type meetings
3 Publicity—safety bulletins, posters, notices, special letters, payroll envelope inserts, slides and films, house organs, etc.
4 Safety plays
5 Instruction in first aid and resuscitation
6 Employee rule books
7 Hiring and training programs
8 Use of loyal employees in "setting a good example"
9 Safety messages on work orders, on correspondence, etc.
10 Contests
11 Job safety analysis
12 Featuring specific safe-practice rules
13 Attendance at safety conferences
14 Preparation of safety codes

REFERENCES

McGehee, W., and P. Thayer, *Training in Business and Industry*, Wiley, New York, 1961.

Petersen, D., *Safety Management—A Human Approach*, Aloray, Englewood, N.J., 1975.

THIRTEEN

Formula for Supervision

Industrial management, including its safety engineers, has long recognized the vital status of its key people, the supervisors and foremen.

It acknowledges its dependence on supervision for the direct control of worker performance and the maintenance of environmental conditions in so far as these affect the quality and volume of production and the safety of personnel.

In view of this general recognition and acknowledgment by industry of the status of its supervisors, it is no less than paradoxical that supervision is not being taught, to either practicing or prospective first-line leaders, as an art that is subject to specific definition and rules of application. This is not to say that no effort is being made along such lines. Indeed, in a great many cases, the larger and more progressive industrial establishments exercise meticulous care, first in the selection of supervisory material and then in the development of supervisory ability not only by practice and experience but also by means of class instruction.

Selection is, of course, a necessary and valuable first step in determining probable fitness. Practice and experience are also of great value, but selection is preliminary to instruction, and practice and experience are but long-term cut-and-try methods of acquiring information and necessary skills. *Direct instruction* in the art of supervision is the immediate need, and it is in this field that industry is weak.

TRAINEE SUPERVISORS ARE BEWILDERED

The great bulk of available text is largely descriptive of need and value, of hints and suggestions, of broad generalities, of the characteristics of persons, and of examples to illustrate these things. Morals or conclusions may be drawn from such examples, but when these are so voluminous, so varied, and so diversified in nature, the student or prospective supervisor may become confused and have difficulty in visualizing either the complete story or its basic structure.

An intelligent young man who had the advantage of an engineering-college education and who had just completed one of the most complete supervisory training courses known to the writer made the following statement:

> I now have my certificate, the blessing of the instructor, and the notice of my first appointment to a supervisory job. I have the material that was distributed. There's a stack of it over a foot high including a tome on psychology and human engineering. I also have a wealth of my own notes relating to what seemed to me to be the highlights of the course.
>
> All of this should make me confident of success, yet I am thoroughly bewildered and uncertain. It makes me think of the bit of cynicism that goes, "At last we have succeeded in converting absolute chaos into well-regimented confusion."
>
> I now have a set of rules, admonitions, or guides for successful supervisory performance. Here they are, at least most of those I have been impressed by:

1. Deal with people as human beings, not machines.
2. Lead, don't drive or push.
3. Get people to like and respect you, create loyalty, win cooperation, instill confidence, build morale, and make people feel that they *belong.*
4. Listen to grievances.
5. Give credit when due, and time it psychologically.
6. Explain changes in advance.
7. Give orders clearly and precisely.

8   Ask for opinions and suggestions.

9   Be patient and impartial, consistent, friendly, and courteous.

10 Display personal interest in the home life, the hobbies, avocations, recreations, and personal problems of your workers.

11 Don't argue or be dogmatic when you disagree.

12 Get to know *your own* personal characteristics so as to avoid irritating or antagonizing others.

13 Get to know the personal characteristics, likes, dislikes, whimsies, convictions, idiosyncrasies and motivating qualities, and fundamental instincts of your workers.

14 Same as above for your boss, so as to enable you to get along with him.

15 Recognize your responsibilities to both management and labor.

16 Run your department as a business.

17 Find out what the workers really want most.

18 Test your men to check attitude and ability.

19 Maintain a personal-history record of each employee.

20 Put the "team" and competitive spirit to work.

21 Learn to recognize symptoms of trouble.

22 Correct misdemeanors only when a person has cooled off.

23 Anticipate difficulties and remove obstacles in advance, plan ahead and organize.

24 Interest the workers in quality of production.

25 Keep in sound physical health and develop a saving sense of humor.

PRACTICING SUPERVISORS ARE CONFUSED

Continuing, the trainee said:

> There were many more ideas, but I just can't recall them now. They are all good, undoubtedly all are *essential*, but they are not related to each other in any *orderly way*. Nor have I been told of any framework or basic structure into which I can fit them.
>
> When I start on my supervisor's job tomorrow I know I'll immediately be confronted with the need of handling certain problems. One of them may be a minor or even a major grievance, someone may be unhappy, stubborn, or even insubordinate, errors and mistakes may have to be corrected.
>
> What I would like to know is *what* to do, *what* to say, and *how* to handle the problem. I agree with all 25 of the things on my list, but for the life of me I can't see how I can apply *all* of them at once, nor have I a clear idea which one of them best suits a particular problem.
>
> Suppose Jim Smith's work is way behind and I start with item (1) and decide to deal with Jim as a human being and not a machine. That's a fine idea to start with but I've still got to do or say something and take some definite action. Of course I can consider item (7) and *give orders clearly* and *precisely* to Jim to speed up his work. Ten to one that won't fix it. If

the mere giving of orders is all that is necessary, there wouldn't be very much to the job of supervision. So I look over the list and see that I *must not drive* Jim, must *give* him *due credit, get* him to *like me*, make him *feel that he belongs*, avoid *argument*, and so on. Meanwhile his work volume is down and the superintendent is "on my neck" for departmental output. The answer must be in the book somewhere, maybe it is item (25) about developing a *saving sense of humor*, but it still seems to be up to me to find the key pieces in the jigsaw puzzle in my own way.

These remarks of the trainee are typical. Libraries are full of material on various phases of supervision. There are books, pamphlets, articles, and lectures galore. They *describe, preach, prove need* and *value, philosophize*, and *generalize*, but do not tell how. What is needed is a simple formula—one that can be memorized, understood, and applied to *any* and *all* supervisory situations and problems. Pending the development of something better, such a formula is given herein.

FORMULA FOR SUPERVISION

Supervision is properly described as the *art* of *controlling* the *performance* of *persons* and, to the extent of authority vested in the supervisor, *controlling also* the *conditions* of their *environment*. Emphasis is placed on *control*. Thus it must be clear that in addition to *having* certain knowledges, abilities, and attitudes, the supervisor must *do* certain things. Therefore a workable formula must contain specific steps of supervisory *action* and at the same time be so devised as to indicate when, where, and how the background of knowledge, attitude, and ability may be used.

This background of knowledge, etc., may be termed the *foundation* of supervision as distinguished from its *structure*. The several factors in the formula are the members, especially the *frame* or *skeleton* of the structure itself. Incidentally, the formula should deal with *problems* of the supervisor, because solving problems is the true expression and the ultimate test of supervisory ability.

Formula for Solving Supervisory Problems

1 Collect and analyze data (identify the problem)
2 Select a remedy
3 Apply the remedy
4 Monitor results

These four steps are shown graphically in Fig. 13-1. They are also

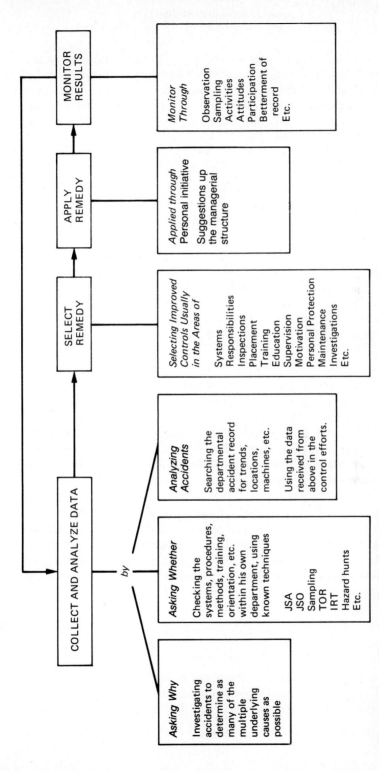

Figure 13-1 A flowchart for the supervisor.

explained in some detail in the figure. This figure is the same as that shown earlier in Chap. 3. It is an obvious adaptation of our flowchart for the supervisor shown in Fig. 3-2. Since most of these steps have been discussed in considerable detail in earlier chapters as they pertain to other levels of the organization, a short explanation here should suffice.

Collect and Analyze Data

As in the diagram for the safety professional in Fig. 3-1, the diagram for the supervisor contains three methods:

Asking Why   At the supervisory level asking why accidents have occurred or asking why acts and conditions happen is basically a process of investigating accidents for cause and of inspecting for hazard. Perhaps the key to obtaining good data here is whether or not the investigation leads past mere symptom to actual causes of the incident.

Asking Whether   Here the supervisors have a number of techniques available to them. The job is one of determining whether or not all is being done that should be done to control the situation that exists in their jurisdiction. They can do this by merely looking and asking questions. They also have many systematic techniques that they can use: job safety analysis, discussed earlier; job safety observation (JSO), a systematic method of observing work processed; safety sampling, a statistical tool to examine the acts of their people; technique of operations review (TOR), discussed earlier; the incident recall technique (IRT), an interviewing process to unearth potential accident causes; hazard hunts, a method of involving employees in finding things wrong; and others. While there is not time to cover each of these in detail here, the reader is referred to the new text *Safety Supervision*, which discusses each in detail.

Analyzing Accidents   Since at a department level there are relatively few accident data that can be analyzed, this is somewhat limited as an information source to the supervisors. They can observe and keep track of the accident and incident occurrences under their jurisdiction, which will provide some information, and they can look at company-wide data that they might receive from the corporate safety department.

These sources might give them some insight and direction for their selection of a remedy.

Select Remedy

Using the information available, and their analysis of that information, supervisors can then select remedies to the problems defined. Remedies within their control within the department usually can be categorized in areas of improving their own internal systems and procedures, better definition of responsibilities, better measurement of results (holding their employees accountable in better ways), more effective inspections, better placement of people, training, education, tighter (or looser) supervision, other motivational methods, use of personal protective equipment, improved maintenance, more detailed investigations, etc. All these remedies are directly within their control.

Or the remedy may be to recommend or suggest change in areas beyond their direct control: up the ladder, to maintenance, etc.

Apply Remedy

For those remedies directly under supervisory control, application is simple. The supervisors merely start doing it. For remedies outside their control, application is almost totally out of their hands in most cases.

MONITOR RESULTS

Supervisors should continually monitor the safety effectiveness of their departments. Similarly, their bosses and staff safety will also monitor the supervisors' effectiveness. Monitoring at the supervisory level can be done by a variety of ways, through simple observation, through safety sampling, through measurement and recording of safety activities being performed by people, by informally (and perhaps formally) measuring or getting a feel for the attitudes of the workforce, by the amount of participation of workers, and by the betterment eventually of the actual accident record of the department.

Of these, those measures that measure the activities being performed are better and more meaningful than measures of accidents. At this level of the organization it is too easy to assume good safety is being achieved merely because no accidents are happening. This could be merely luck, rather than results. The larger the department, the better accidents are as a measuring stick.

HOW SAFETY GETS ACCOMPLISHED

The following material from *Safety Supervision* seems relevant:

A good safety record does not just happen—it is the result of a number of people doing a number of things well. Taken as a whole these activities have been called a "safety program." Our industrial safety programs, then, are basically collections of recommended procedures and actions. These are typical components of safety programs:

Management's statement of policy

Corporate safety rules

Definitions of responsibility

Screening of employees

Placement of employees

Training of employees (orientation)

Ongoing training

Supervisory training

Motivational activities

Inspections

Investigations

Record keeping

Record analyses

First-aid training

Medical facilities

Others

WHO PERFORMS THE ACTIVITIES?

A number of people are involved in a "safety program."

Top management. Those who decide what they want done and then set direction or policy.

Middle management. The group in an organization that is located between the policymakers and the first-line supervisors.

Staff people. These are the various specialty people who have a function of assisting the top in setting policy, of working with the middle in some rather ill-defined ways, and of influencing the first-line supervisors in a number of ways to get them to want to do what the specialty people and management want done.

The supervisor. In the eyes of most safety professionals the first-line supervisor carries out the most critical functions. The second most critical functions are top management's.

A 1967 National Safety Council survey brought out a number of rather interesting points. The purpose of the survey was to determine which factors are considered most important to a comprehensive industrial safety program.

One hundred forty-eight industries took part in the survey by completing a questionnaire to rate the importance of various safety activities. A total of 78 activities in eight safety program areas were included. Industry people rated the importance of the eight major areas as well as the groups of activities within each area. The rank order of the major safety program areas is as follows:

Supervisory Participation
Top Management Participation
Engineering, Inspection, Maintenance
Middle Management Participation
Screening and Training of Employees
Records
Coordination by Safety Personnel
Motivational and Educational Techniques

The emphasis in the survey results is on supervisory and top management participation. This indicates that most people saw the supervisor as the crucial link directly affecting employee behavior. Top management must provide the initial push of a safety program, the supervisor must maintain program momentum daily, and middle management must participate to create the chain of communication and command.

Top Management   Other than the first-line supervisor, the man considered most important to any safety effort is the big boss. His role in the safety program is to:

• Issue and sign safety policy.
• Receive information regularly as to who is and who is not doing what is required in safety as determined by some set criteria of performance.
• Initiate positive and/or negative rewards for that performance.

Middle Management   The industrial safety survey also showed the importance of middle management participation. The list giving the rank order of the major safety program areas indicates that the role of the mid-manager is more important to safety success than the screening or training of employees, the record-keeping function, the coordination by the safety manager, and all motivational and educational techniques. In the survey the third most important item on the list of 78 was "middle management setting an example by behavior in accord with safety requirements."

The survey also spelled out what the role of the mid-manager might be in safety. It is his role to restate policy, to participate in safety meetings, to review employee safety performance, to establish checks to insure adherence to safety program goals, to set an example, to utilize safety performance as a measure of management capability, and to serve on investigating committees. In short, he is to be an active participant in the program by transforming the executive's abstract policy into supervisory action.

Safety Manager   What a safety manager does is crucial to the success of a safety program in any company. First of all, he should structure systems of measurement so that accountability can be fixed and rewards can be properly applied to the right people at the right time to reinforce the desired behavior. Some of these systems might incorporate estimated costing, sampling, rating effort, performance measurements, and others described later.

Secondly, the safety manager must be a programmer: a person who oversees those aspects of the safety program that are not completely under his control. He must make sure that safety is included in orientation, that safety training is provided where needed, that safety is a part of supervisory development, that things are done to help keep the organization's attention on safety, and that safety is included in employee selection, in the medical program, and in other areas.

Thirdly, he must fill the role of a technical resource, knowing how to investigate in depth, where to get technical data, what the standards are, and how to analyze new products, equipment, and problems.

Fourthly, he must fill the role of systems analyst, searching for reasons why accidents happen and whether or not proper controls are in effect.

The Supervisor   The major role of the safety program belongs to the first-line supervisor. Everything that everybody else does is worthless if the supervisor does not do his job. What is his job in the safety program? While this may vary from company to company there are four key tasks that belong to the supervisor in every safety program.

1   *Investigating* all accidents to determine underlying causes.
2   *Inspecting* his area routinely and regularly to uncover hazards.
3   *Coaching* (training) his people so they know how to work safely.
4   *Motivating* his people so they want to work safely.

The attitude of the majority of supervisors today lies somewhere between total acceptance and flat rejection of comprehensive accident prevention programs. Most typical is the organization in which line managers do not shirk this responsibility but do not fully accept it either and treat it as they would any of their defined production responsibilities. In most cases their "safety hat" is worn far less often than their "production hat," their "quality hat," their "cost control hat," their "methods improvement hat," etc. In most organizations, safety is not

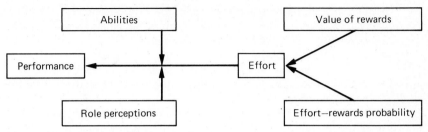

Figure 13-2   Factors affecting performance.

considered as being as important to the line manager as many, if not most, of the other duties that he performs.

What does a manager's attitude toward safety depend on? It depends on (1) his abilities, (2) his perception of his role, and (3) the efforts that he expends (see Fig. 13-2). All are important, and a manager will not turn in the kind of performance we want unless all three are taken into account.

EFFORT

There are two basic factors which determine how much effort a person puts into his job: (1) his opinion of the value of the rewards and (2) the connection he sees between his effort and those rewards. This is true of a manager's total job, as well as any one segment of it, such as safety.

The Value of Rewards   The manager looks at his work situation and asks himself, "What will be my reward if I expend effort and achieve a particular goal?" If he considers that the value of the reward which management will give him for achieving the goal is great enough, he will decide to expend the effort.

"Reward" here means much more than just financial reward. It includes all those things that motivate people: recognition, chance for advancement, ability to achieve, increased pay, etc. In fact, most research today into supervisory motivation indicates that the rewards of advancement and responsibility are the two greatest motivators.

In effect, the line manager looks at the rewards that management entices him with and makes a judgment whether those rewards are great enough. Chances are that he will decide they are if the rewards are in terms of advancement and additional responsibility, rather than some of the lesser enticements that management too often selects. In safety, just as in other areas, management's chosen rewards are too often small, too unimportant, to entice the line manager.

The Effort-Rewards Probability   If a manager wants to assess how probable it is that the rewards really depend upon his effort, he asks himself the following kinds of questions:

1 Will my efforts here actually obtain the results wanted—or are there factors involved beyond my control? (The latter seems a distinct possibility to him as he thinks about safety.)
2 Will I actually get that reward if I achieve the goal?
3 Will management reward me better for achieving other goals?
4 Will it reward the other manager (in promotion) because of seniority, regardless of my performance?
5 Is safety really that important to management, or are some other areas more crucial to it right now?
6 Can management really effectively measure my performance in safety, or can I let it slide a little without management knowing?
7 Can I show results better in safety, or in some other area?

The line manager asks himself these questions and others unconsciously before he determines how much effort he will expend on safety. He must get the right answers before he will decide to expend the needed effort for results.

In safety these kinds of questions are crucial. Often line managers decide that their personal goals are better achieved by expending efforts in other areas, and too often their analysis is correct because management *is* rewarding other areas more than safety.

Changing this situation is the single greatest task of the safety professional. Change can be achieved by better measurement of line safety performance and by offering better rewards for line achievement in safety.

ABILITY

Job performance does not depend simply on the effort that a manager expends. It depends also on the abilities he brings to his task. "Ability" here means both the inherent capabilities of the man and his specialized knowledge in the field of endeavor being performed in. In accident prevention, this means our ensuring through supervisory training that the line manager has sufficient safety knowledge to control his people and the conditions under which they work. In most industries lack of knowledge is not a problem. Usually, however, the line manager knows far more about safety than he applies. Many people agree that a manager can achieve remarkable results on his accident record merely by applying his management knowledge, even if he has little safety knowledge. If a manager does not have adequate safety knowledge, the problem is easily handled through training.

ROLE PERCEPTION

More important than ability is the area of role perception. The line manager's perception of his role in safety determines the direction in which he will apply his efforts. Lawler and Porter describe good role perception as that in which the manager's views concerning his placement of efforts correspond closely with the views of those who will be evaluating his performance.

In safety, role perception means: "Does he know management's desires on accident control?" and "Does he know what his duties are?"

In the area of role perception, we should be searching for answers to some questions about our organization and about each line manager in our organization. These questions should concern the content and effectiveness of management's policy on safety, the adequacy of supervisory training, the company safety procedures, the systems used to fix accountability, etc.

MOTIVATORS AND DISSATISFIERS

The four most important motivators are:

1 *Advancement.* Advancement seems to be the most important motivator for supervisors—considerably more so than for the employees.
2 *Responsibility.* Degree of responsibility is important, but less important than for the employees.
3 *Possibility of growth.* Possibility of growth is far more important for supervisors than for employees.
4 *Achievement.* Achievement is important, but not as much so as with employees.

To improve supervisory performance then, rewards built around these motivators will work best.

GETTING SUPERVISORY ACTION

Supervisors can be motivated to action by employees, but only after management has decided that such action is necessary. Following is a passage from *Safety Supervision*, which attempts to identify to top managers what they need to do to generate supervisory action in safety:

Most bosses haven't the foggiest idea whether or not their supervisors are doing anything about safety—and then they wonder why they are not getting good results. Your supervisors perform in those areas that are important to you—the only way they know what is important is by how you measure their performance.

When you do not measure their performance in safety, you are very clearly telling them that it is not important to you. As a result, it will not be important to them. They will not perform if they are not measured. Therefore, measure them.

Decide what you want your supervisors to do. Make it clear that you want them to carry out the four core tasks—investigating accidents for cause, inspecting for hazards, coaching, and motivating. If you wish to have them do other tasks also . . . make that clear. Have them tell you regularly what they have done. Have them send you a report . . . on the first of each month.

SET CRITERIA FOR ACCEPTABLE PERFORMANCE

Determine what level of performance in each of the four core areas is acceptable to you. For instance:

1 In investigating accidents for cause, acceptable performance might be:
   a Investigating all injuries serious enough to require the attention of a doctor.
   b Conducting investigations within 24 hours of the injury.
   c Finding at least five contributing causes for each injury.
   d Correcting at least two hazards for each.
   e Making at least two recommendations to management for preventing each accident.
2 In inspecting for hazards, acceptable performance might be:
   a Making four inspections each month.
   b Correcting at least two hazards on each inspection.
3 In coaching, acceptable performance might be:
   a Holding four meetings each month to coach the group.
   b Providing orientation each month for all new employees.
   c Contacting five employees individually each month.
4 In motivating, acceptable performance might be:
   a Using two types of safety materials each month.
   b Developing one structured attitude-building activity per man each month.

REWARD ACCORDINGLY

When the monthly reports come to you, look at them. Compare their numbers to your criteria. Occasionally spot-check to make sure their numbers are correct (see for instance if supervisor A has four inspection reports in his file). Then administer rewards—positive or negative.

If the supervisors meet the criteria, a number of kinds of positive rewards are possible. They might remain supervisors or stay in the running for advancement. You could give them compliments in memos, a bonus, congratulations in person, a monthly free dinner, green stamps, or a pat on the back. If the supervisors do not meet the criteria, these kinds of negative rewards are possible: not remaining a supervisor, not staying in the running for advancement or a bonus, not receiving a free dinner or green stamps, or receiving a chewing out from you in person, a zinger of a memo from you, or a kick in the _____.

Obviously, there are many other kinds of rewards you could use. You could even adopt the same reward structure you use for production or quality. Because most bosses are busy, the details can be worked out by your staff. But make sure the monthly reports go to you. (The staff can summarize them later.) And make sure the rewards come from you. No supervisor is so stupid that he would ignore your wishes—so safety is bound to improve if you take a direct personal interest.

REFERENCES

Planek, T., G. Driesen, and F. Vilardo, "Industrial Safety Study," *National Safety News*, August 1967.

Petersen, D., *Techniques of Safety Management*, 2d ed., McGraw-Hill, New York, 1978.

Petersen, D., *Safety Supervision*, Amacom, New York, 1976.

Porter, L., and E. Lawler, *Managerial Attitudes and Performance*, Irwin-Dorsey, Homewood, Ill., 1968.

PART THREE

Special Subjects

FOURTEEN

Law and the
Safety Professional

A ccording to a press release from the U.S. Department of Labor, Bureau of Labor Statistics, dated December 1, 1977:

> One in every 11 workers, on the average, in the private economy experienced a job-related injury or illness in 1976, according to a survey reported today by the Bureau of Labor Statistics of the U.S. Department of Labor.
>
> The survey, based on reports submitted during 1977 by employers in the private economy, shows an increase in the number of injuries and illnesses from 4.99 million in 1975 to 5.16 million in 1976. This rise of approximately 3 percent was proportionate to the rise in hours worked. Hence, the incidence rate at 9.2 injuries and illnesses per 100 full-time workers was virtually unchanged from 1975.[1]

Within the figures cited above by the Department of Labor, injuries and illnesses within the manufacturing and wholesale and retail trade categories accounted for 3.29 million in 1975 and 3.47 million in 1976—66 percent and 67 percent, respectively, of the total reported

[1]USDL–77–1031, Washington.

cases. These two industries are usually covered by the worker's compensation laws of the various states.

With the exception of the few small employers who may not be covered under these worker's compensation laws due to numerical exemptions, the injured employee typically had no recourse against the employer other than the right to collect worker's compensation benefits as provided by law. Since 1960, however, there have been an increasing number of cases brought against the employer's worker's compensation insurance carrier for the carrier's failure to make safety inspections or its failure to use due care in making safety inspections. There has also been an increase in the number of cases being brought into the courtroom against the safety director of the employer by a fellow employee alleging the same type of negligence. These two developments have caused a great deal of concern among safety professionals about just what their liability is to the injured worker. Each of these developments will be explored in some detail in the remainder of this chapter.

LIABILITY OF THE WORKER'S COMPENSATION INSURANCE COMPANY

In resolving the cases brought by the injured employee against the employer's insurance carrier, the courts have been confronted with two important questions: (1) Does the worker's compensation act of the state grant immunity to the insurer with regard to the making of safety inspections, and (2) does the insurer have a duty to the injured employee, and if so, has this duty been breached?

The worker's compensation laws of every state have three common elements: (1) there is no recourse for the injured employee at law against the employer—worker's compensation benefits are an exclusive remedy; (2) the employer must provide coverage for the benefits under the law through either private insurance companies, the state fund if one is operating in the state, or self-insurance; and (3) the employee retains the common law right to sue a third party (other than the employer) if the injury is due to the negligence of said third party.

The question of concern at this point is whether or not the employer's insurance carrier falls within that third element—the right of the injured employee to sue a third party if the injury is due to the negligence of the third party. Condition 4 of the Standard Workmen's Compensation and Employers' Liability Policy states:

The company and any rating authority having jurisdiction by law shall each be *permitted but not obligated* to inspect at any reasonable time the

workplaces, operations, machinery, and equipment covered by this policy.
[Emphasis supplied.]

This condition makes the inspection permissive; it places no duty on the insured to make inspections. The purpose of the inspection made on behalf of an insurance company is two-fold—to obtain information for the home office underwriter and to render a safety service to the employer in which specific recommendations are made to improve the workplace. In some cases, no inspections are made.

There are some inherent deficiencies in the inspections done by the insurance company representative. They are not too effective inasmuch as they are performed on a periodic rather than a day-to-day basis—perhaps as infrequent as yearly. The reports of these inspections to the employer take the form of recommendations and are not mandatory—the employer may follow the recommendations or not; however, if not followed, the insurance company has three options available—continue coverage, cancel coverage, or not renew coverage. Finally, many of the safety personnel of insurance companies are not graduates of a recognized collegiate curriculum in safety and their only safety training may be that offered by the insurance company. This may change over time.

Prior to the 1960 case in New Hampshire of *Smith v. American Employer's Insurance Company*,[2] the assumption was widely held that the insurance carrier enjoyed the same immunity from suit by the injured employee as was granted to the employer under the state's worker's compensation law. This assumption of carrier immunity was rejected by the court in the Smith decision.

In the Smith case, the employee was injured by an explosion of a compressed air tank which had been inspected as part of the regular monthly inspection performed by the insurance company's representative. The plaintiff alleged that the carrier had a duty to use due care in making these inspections to determine whether or not the plant was safe and to notify the employer of any dangers. It was further alleged that the insurance company was negligent in performing this duty.

The response of the insurance company was that there was no cause of action inasmuch as it shared in the employer's immunity from suit by an employee under the worker's compensation law of the state of New Hampshire.

The contention of the insurance company was denied by the New Hampshire Supreme Court, and it was held subject to third-party tort action. This decision was nullified the following year by the legislature

[2]102 N.H. 530, 163 A. 2d 564 (1960).

of New Hampshire when insurance companies were granted the same immunity enjoyed by the employer. Nine years after this legislative action, however, the New Hampshire Supreme Court again held to a decision in which an insurance carrier which undertakes to assist in accident-prevention activities of an employer may be liable to an injured employee as a negligent third party.[3]

There have been numerous cases involving the question of the liability of the insurance carrier when it conducts safety and loss prevention inspections. The one that has received the most attention is that of *Nelson v. Union Wire Rope Company*,[4] and it will be discussed.

Voluntary Inspections

As previously pointed out, an insurance carrier has no obligation to inspect the premises of its insured; however, if the insurance company voluntarily inspects and does so negligently, some courts have held that it shall be liable to a third party who can show that this negligence either caused or contributed to an injury. Justice Cardozo stated the legal basis for such actions when he stated: "It is ancient learning that one who assumes to act, even though gratuitously, may thereby become subject to the duty of acting carefully, if he acts at all."[5]

The Nelson case was a decision by the Illinois Supreme Court in which the Florida compensation act was interpreted. The court recognized in that case that "one who gratuitously renders services to another . . . is subject to liability for bodily harm caused to the other by his failure, while so doing, to exercise with reasonable care such competence and skill as he possesses."[6] In other words, if the insurer voluntarily undertakes to perform an inspection, a duty is owed to the employees to exercise reasonable care in so doing and that it was not essential for the injured employee to show proof of reliance on such inspection to recover for injuries.

It was this last part of the decision, the failure to require reliance by the injured party in order to recover, that has caused the most discussion about this decision. The question of liability for voluntary actions relies heavily on the interpretation of the RESTATEMENT (SECOND) OF TORTS, § 323 and § 324A, which reads as follows:

[3]*Corson v. Liberty Mutual Insurance Company*, 110 N.H. 210, 265 A. 2d 315 (1970).
[4]*Nelson v. Union Wire Rope Company*, 199 NE2d 769 (Ill. 1964).
[5]*Glanzer v. Shepard*, 135 N.E. 275, 276 (N.Y. 1922).
[6]*Nelson v. Union Wire Rope Company*, Op. cit., at p. 773.

Section 323. *Negligent Performance of Undertaking to Render Services.* One who undertakes, gratuitously or for consideration, to render services to another which he should recognize as necessary for the protection of the other's person or things, is subject to liability to the other for physical harm resulting from his failure to exercise reasonable care to perform his undertaking, if

a   his failure to exercise such care increases the risk of such harm, or
b   the harm is suffered because of the other's reliance upon the undertaking.

Section 324A. *Liability to Third Person for Negligent Performance of Undertaking.* One who undertakes, gratuitously or for consideration, to render services to another which he should recognize as necessary for the protection of a third person or his things, is subject to liability to the third person for physical harm resulting from his failure to exercise reasonable care to protect his undertaking, if

a   his failure to exercise reasonable care increases the risk of such harm, or
b   he has undertaken to perform a duty owed by the other to the third person, or
c   the harm is suffered because of reliance of the other or the third person upon the undertaking.

Most prior cases based the decision on § 323 which requires reliance on the inspections by the injured party. The Nelson case however, finds support in § 324A. In that section, the reliance requirement may be by either the party for whom the services are rendered or the injured third party. In another case involving safety inspections, albeit in a different field of insurance, the Fifth Circuit Court of Appeals held that reliance on the part of the decedent was not necessary, but reliance by the owners was sufficient.[7] In another case following the reasoning of the Nelson case and § 324A, the New Hampshire Supreme Court held against the insurance company.[8]

Another question raised by the Nelson case is that of misfeasance versus nonfeasance. Nonfeasance involves a defendant's failure to perform an undertaking; however, where the undertaking was negligently performed, misfeasance has occurred. The appellate court, in this case, held that the conduct was nonfeasance; however, the supreme court characterized the conduct as misfeasance and allowed

[7]*Hill v. United States Fidelity and Guaranty Company*, 428 F. 2d 112 (5 Cir. 1970).
[8]*Corson v. Liberty Mutual Insurance Company*, op. cit.

recovery. Many cases following the Nelson case have also held such actions on the part of an insurance company to be misfeasance.

In general, therefore, it can be said that the courts have held that voluntary inspections for safety and loss prevention on the part of the worker's compensation insurer can be the basis for a successful suit by either an employee or the heirs of a deceased employee; however, where statutory immunity has been enacted by a state legislature, and such immunity is extended to the insurer, it will be upheld by the courts. In a particular state, the question of insurer immunity depends upon the interpretation of the worker's compensation law in that particular state. In order for insurance companies to feel safe from suit due to safety inspections, state worker's compensation statutes should be clearly worded to the effect that the immunity of the employer extends to the insurer of said employer. In the majority of states, the courts will recognize such legislative statements.

Summary

The courts have taken different approaches to the question of liability of an insurance carrier to an injured person as a result of negligent safety inspections that are voluntary. In some cases, the courts have looked very closely at the issue of a duty to third persons to exercise due care. The point of contention in this line of cases is whether or not there was reliance upon the inspection by the injured party. If there is a trend to be developed from the cases, it would be that of allowing recovery from the insurer if there is reliance shown by either the injured party or the insured employer.

Other courts have based their decisions on such cases on the question of statutory immunity. A majority of the courts where this question has been raised either uphold a statutorily granted immunity to the insurance company standing in the shoes of the employer or agree to a legislative intent to do so. In a minority of cases the courts have refused the extension of such immunity.

Though it has been pointed out that in the majority of cases the insurance company is immune from action by an employee of one of its insureds, the insurance company may not be immune in an action by an outside party. Thus, an insurance company may be found liable to a widow of an employee or to a guest in a hotel which had been inspected by an insurance company.[9]

[9]*Johnson v. Aetna Casualty and Surety Company*, 348 F. Supp. 627 (M.D. Fla. 1972); and *Hill v. United States Fidelity and Guaranty Company*, op. cit.

Finally, it should be pointed out that the National Commission on State Workmen's Compensation Laws, in its report of July 1972, stated as its recommendation 2.19:

> We recommend that suits by employees against negligent third parties generally be permitted. Immunity from negligence actions should be extended to any third party performing the normal functions of the employer.[10]

Though the commission upholds the right of the injured employee to sue a negligent third party and, if successful, repay his employer or the employer's insurance company for some or all of the worker's compensation benefits received, it addresses itself to the problem area of third parties who perform a role normally performed by the employer—safety inspections as an example. The commission recommended that the immunity granted to employers from suit by injured employees should be extended to the insurance carrier of the employee. Their reasoning is that if the injured employees cannot sue the employer for negligently carrying out safety inspections, they should not be permitted to sue the insurance carrier that is carrying out that which is normally a function of the employer. As we have seen, a majority of the courts are taking a similar view.

LIABILITY UNDER CO-EMPLOYEES SUITS

An injured person is permitted, under worker's compensation laws, to sue a third party when the injuries are the result of the negligence of that person. The question that is naturally raised in this regard is that of who is the third party. As has been pointed out previously, most states do not recognize the worker's compensation insurance carrier as a third party insofar as safety inspections are concerned. A person injured by an automobile driven by another not employed by the same employer as the injured person may sue that individual as a third party. Further, manufacturers of machines and equipment that are shown to be defective or lacking in some way have been held responsible for injuries suffered from the use of that machine or equipment by a user's employee. However, one of the most troublesome areas of dispute is the liability of a co-employee for injuries suffered by a fellow employee.

[10]Government Printing Office, Washington, p. 52.

Statutory Immunity

A large number of states have enacted legislation that restricts the right of a suit being filed against a co-employee, including within that term the safety director. Typical of the language of such statutes is that of New York, in which it is stated:

> The right to *compensation* or benefits under this chapter, shall be the *exclusive remedy* to an employee, or in the case of death his dependents, when such employee is injured or killed by the negligence or wrong of *another in the same employ.*[11] [Emphasis supplied.]

Courts have held on similar statutes that a suit could not be brought against a co-employee unless it fell within certain statutory exceptions. Examples of such exceptions include willful or unprovoked physical aggression, gross negligence so as to amount to wanton neglect for the safety of others, or intoxication. As a general rule, an intentional act or one that is grossly negligent will remove the co-employee from the protection of the statute.

Where such statutory immunity has been enacted, the courts have upheld the constitutionality of such statutory language as being a valid exercise of a legislative police power.[12] In spite of such holdings, there have been numerous cases challenging such legislation, too numerous to mention here.

Why have the legislatures and courts taken the attitude that co-employees are immune from suit? One reason advanced is that the worker and his or her employer have each given up a right. The worker has given up the right to sue the employer for common law damages; the employer has given up the right to use the common law defenses and has accepted liability. Since the co-employee has also given up the common law rights, that co-employee can expect freedom from liability for injuries to another employee for which he or she may be at fault.

Another reason lies in the basic purpose of worker's compensation. The underlying theory considers industrial injuries as a cost of doing business and, as such, is passed on to the consumer in the pricing mechanism. Losses to employees caused by parties not associated with that particular employer are permitted to be tried by a court as it would not be equitable to have the consumer pay for the negligence of a person or firm which had no connection with the work being performed. If, however, a co-employee could be held responsible for an

[11]New York Workmen's Compensation Law, § 29(6).
[12]*Lowman v. Stafford*, 37 Cal. Reporter 681 (1964).

injury, that co-employee would bear the burden of the cost of the injury and not the consuming public.

A third reason is that such suits would duplicate, if successful, the worker's compensation responsibility with common law damages and expenses. This occurs due to the obligation—real or imagined—of the employer to provide defense for the employee causing the injury as well as paying the claim, if successful, or through the additional expense incurred by the employer by naming all the employees as insureds under the general liability insurance coverage of the organization.

As a result of the foregoing reasons, many states have passed legislation that restricts the right of suit against a co-employee for industrial accidents.

Judicial Immunity

Some states, on the other hand, merely prohibit suits against the employer and permit suits to be brought against third parties. The question raised by this type of statute is whether or not co-employees enjoy the immunity of the employer or are considered to be third parties. Some jurisdictions have granted judicial immunity to the co-employee on the same basis as the statutorily granted immunity of the employer. Various theories have been advanced by the courts to support this conclusion. A Kentucky court stated that only a person with no connection with the work being performed could be sued. An Idaho decision was based on the premise that the co-employee was the agent of the employer and enjoyed the same immunity as the employer. Thus, the co-employee was merged into the employer and was not considered to be a third party within the meaning of the law.

Such decisions, however, have been in the minority. With no statutory immunity present, the courts take the position that they will not presume that the legislature intended to deny anyone his or her common law rights. Therefore, actions against any third party have been permitted. It has been such judicial decisions that prompted states such as New York to enact legislation specifically providing immunity to co-employees.

Unless the co-employee had financial resources with which to pay or had coverage from some insurance policy, why should one employee sue a fellow employee for damages? It is feared that the increase in the number of lawsuits brought against co-employees is primarily due to the new techniques for finding insurance coverage to pay.

Insurance Coverage

In the usual case, the employer's worker's compensation insurance company has already provided benefits to the injured party and is not applicable to tort action. The employer's liability policy may or may not apply, depending on the definition of "insured."

The members of the safety department of a firm may be involved in a vehicular accident with a fellow employee; however, this would normally not be in the pursuit of their duties in safety and will not be discussed. Concern is limited to insurance protection available to safety personnel when conducting their usual function, that of safety inspections and training within the organization. The question of concern herein, therefore, is the liability insurance protection for the safety personnel within the organization while carrying out the safety function.

The first determination that must be made in such a situation is whether the co-employee (the safety person) is a named insured within the policy definition of the employer's policy. If the definition of named insured includes all employees while working within the scope of their duty, then the employer's insurance company will, in all likelihood, assume the defense of the co-employee under a reservation of rights. Further the carrier may institute a declaratory judgment action in order to determine that the policy exclusions provide neither coverage nor an obligation to defend the co-employee. The exclusion referred to is the fellow employee one which is found in most, if not all, standard liability policies. It may, however, be removed if requested by the risk manager and agreed to by the insurance company and would, in that situation, appear as an endorsement to the policy.

If, on the other hand, the policy is so worded as not to include employees as named insureds, the co-employee is usually notified by the employer's carrier that defense should be tendered to either his personal attorney or the insurer that carries his personal general liability insurance—usually the homeowner's policy. Unless endorsed to the contrary, the homeowner's coverage excludes activities arising out of business pursuits and defense will be denied. Where does this leave the employee in the safety department of a firm?

Two basic alternatives are available to such an employee. First, the safety engineer working for an organization may have a homeowner's insurance policy endorsed to provide coverage for business pursuits. As long as her work does not include the giving of professional services, coverage would be provided for both the defense of an action and the payment of a judgment within the policy limits.

The other alternative available is the purchase of personal coverage to protect against professional liability. This is the only recourse available to safety consultants in business for themselves.

SUMMARY

Two principal areas of concern to the safety professional have been discussed in this chapter. The first involved the safety person working as a loss control representative for an insurance company that does inspections for its insureds. The other looked at the question of the responsibility of the safety director and the employees of the safety department of an organization for safety inspections and other safety work done for the employer on behalf of the safety and health of co-employees.

As has been pointed out, the courts have taken varying positions on the question of whether a worker's compensation insurer is liable to an injured person, such injury allegedly due to the negligent safety inspection voluntarily carried out by the insurer. The major point of difference among the courts is that of reliance on the part of the injured party. If a trend is discernible, it would be toward allowing recovery by the injured party if there is reliance on the inspection shown by either the insured employer or the injured party. Some states have enacted legislation that grants the same immunity to common law damages as is granted to employers under the worker's compensation statutes. The majority of states, by either legislation or judicial decisions, have granted such immunity to insurance companies, particularly if such legislative intent can be shown. Note should be made of the fact, however, that such immunity does not extend to injured parties who are not employees of the insured. Thus, though an insurance company may be immune from suit by an employee of one of its insureds, no such immunity exists in a common law action brought by a nonemployee third party.

Co-employee suits, on the other hand, are becoming more common as a means to avoid the exclusive nature of the worker's compensation laws. Though such are not justified under the basic theory underlying worker's compensation, the trend of the judiciary is to permit such suits unless prohibited specifically by statute. This represents an interesting commentary on the current liability scene. There has been a great deal of discussion on the pros and cons of "no-fault" systems in such fields as automobile liability and product liability; however, worker's compensation laws were, after all, the first no-fault laws enacted in this country. Such a maze of complex and expensive litigation was never anticipated when these laws were enacted. At their inception, worker's compensation laws were thought to be adminstratively simple, inexpensive, and still equitable to the injured employee.

Does the individual safety professional have insurance protection in the event of such suits? If working for an insurance company as a loss control representative, there is little likelihood of personal suit inasmuch as it would be the insurance company that would normally be sued, not the individual. If, however, a suit is filed by a co-employee, the question becomes much more complex. There may or may not be coverage for monetary damages and defense costs. It depends on who is named in the policy as the insured and whether or not there is a fellow-employee exclusion present. As far as coverage under personal homeowner's insurance policies, the individual safety person should be sure that the business pursuits endorsement is added to the homeowner's policy. Even then the question may be raised about whether this is a professional service, which is an exclusion of this same policy.

As to the individual practicing safety consultant, insurance coverage is necessary the same as for anyone else offering professional services. Such coverage may be simple or difficult to obtain, much depending on the experience, record, and training and education of the individual and the insurance market at the time.

FIFTEEN

The Role of the Insurance Company

There have been various estimates of how much money insurance companies in the United States spend on the safety function. The *Report of the National Commission on State Workmen's Compensation Laws*[1] estimated that $35 million was spent annually by insurance companies on safety services. This, according to the report, represented 1.1 percent of the standard premiums collected for worker's compensation insurance.

Another figure commonly used is that 2 percent of worker's compensation premium is allocated to safety services. This figure is questionable, however, inasmuch as this service is not reported in the insurance expense exhibit that is submitted to the regulatory officials annually. For some individual insurance companies, no doubt, this would be a reasonable percentage. For others, however, it may be too high or too low. It may be a safe conjecture to say that safety services of an insurance company consume between 1 and 2 percent of worker's compensation premiums.

[1] Government Printing Office, Washington, July 1972, p. 93.

Assuming this estimate to be fairly accurate, its application to the premiums reported in the annual report for 1977 of the National Council on Compensation Insurance for the jurisdictions under its administrative bureaus develops an expenditure for safety services by insurance companies of between $28,980,994 and $57,961,988.

According to the Insurance Information Institute, the worker's compensation insurance premiums written in 1975 were $6 billion. If the 1 to 2 percent estimate mentioned above is applied to those premiums, then the insurance industry spent between $61 million and $122 million on safety services.

It would be safe to estimate, based upon the two sets of figures presented above, that in 1977 the insurance industry spent in excess of $100 million in the area of worker's compensation safety services.

This, however, does not tell the entire story insasmuch as a great deal of safety services are unrelated to worker's compensation insurance. For example, boiler inspections, traffic safety efforts, and building inspections, as well as safety research, are being conducted by many insurance companies.

The only way to obtain accurate statistics on the amount of premium dollars being spent on safety services would be to obtain internal accounting records of every insurance company and develop the expenditures for safety from these documents. Obviously, this is an impossibility!

The point of all these figures is to raise the rhetorical question of what benefits are being received from the expenditure of such a vast sum of money. Just as the amount of the expenditure is extremely difficult to determine, so are the benefits derived. Someone should be benefiting or the expenditures would not be made. Our role in this chapter is to look at what the insurance companies are doing in the area of industrial safety, omitting from this discussion the other safety areas of boiler and machinery, transportation, etc. We will leave it to the reader to determine whether the benefits from such an expenditure are substantial or not.

A BRIEF HISTORY

Insurance companies, as early as 1911 when the first worker's compensation statute was upheld as being constitutional, had created inspection departments to inspect manufacturing plants in order to determine hazards and establish rates for insurance. These functions can be looked upon as being primarily for the benefit of the underwriting department of the insurance company. Economic and humanitarian considerations made it desirable for them to go beyond the mere inspection for underwriting, and these inspectors began to insist upon corrections of the physical hazards detected. In return, the insured would receive a reduced schedule rate. This, coupled with the statutorily required worker's compensation insurance, made it economically better to stop injuries and clean up from the workplace recognizable hazards.

There are varying estimates of how many safety professionals are employed by insurance companies. These range from 29 to 70 percent. At any safety meeting, either local, statewide, regional, or national, one gets the impression, correctly or not, that the majority of those in attendance are employees of insurance companies. Thus, though the insurance industry did not begin the safety movement, it has become an increasingly important force in it.

OBJECTIVES OF A WORKER'S COMPENSATION PROGRAM

The *Report of the National Commission on State Workmen's Compensation Laws*[2] describes the five basic objectives of a modern worker's compensation program. These are as follows:

1 Broad coverage of employees and work-related injuries and diseases;
2 Substantial protection against interruption of income;
3 Provision of sufficient medical care and rehabilitation services;
4 Encouragement of safety; and
5 An effective system for delivery of the benefits and services.

Four of the five objectives are not of concern to us in this text; however, number four is, and will be discussed next from the viewpoint of the commission.

Encouragement of Safety

The way to encourage safety efforts by employers, according to the commission report, is to provide economic incentives. One of the roles of rating worker's compensation insurance, to be discussed in the next chapter, is to assign to the employer—through the experience modification factor—the cost of that organization's work-related injuries and diseases. This is further enhanced for those employers opting for one of

[2]Ibid., pp. 35–40.

the retrospective rating plans, in which each employer bears the losses of her organization within the minimum and maximum rates. This position is strengthened by the commission report, wherein it is stated, "In order to provide the most powerful direct incentives to safety, we believe in strengthening the concept of relating each employer's workmen's compensation costs to the benefits paid to his employees."[3] This is exactly what experience shows!

The commission report further suggests that safety effort can be indirectly encouraged through the worker's compensation mechanism by strengthening the competitive position of organizations with a superior safety record. If the costs of work-related injuries and diseases are allocated to the appropriate organization, then those organizations with higher worker's compensation insurance costs due to a poor safety record may have to increase prices. Inasmuch as a rational consumer will purchase goods at the lowest price, those firms that have higher costs due to work-related injuries and diseases will find either their sales decreasing or their profits decreasing if prices are maintained at a competitive level.

It appears that, according to the commission report, the principal method of encouraging safety is to make greater use of experience rating inasmuch as this rating procedure does charge to the individual employer the cost of work-related injuries and diseases.

WORKER'S COMPENSATION AND ACCIDENT-PREVENTION SERVICES

As mentioned earlier in this chapter, millions of dollars are spent by insurers in the area of accident-prevention services. This is of significant importance to small and medium-size firms which cannot afford to provide their own safety programs under a full-time safety director. On the other hand, however, it must be pointed out that it is extremely expensive for an insurance company to expend substantial sums of money on safety programs for small employers who pay relatively small worker's compensation insurance premiums. One of the recommendations of the commission is that small firms should pay a supplementary safety premium for worker's compensation insurance and that the carriers be *required* to spend this premium on safety efforts for the class of organizations which pay this supplementary premium. Several problems are apparent from that seemingly simple recommendation.

[3]Ibid., p. 39.

One of the problems is that the recommendation does not define what it considers to be a "small" firm. Is it to be based on annual sales, dollars of asset values, dollars of worker's compensation insurance premiums, or just what? What one person considers to be small may not be so to another. Some criteria must be established to recognize variations in size to which this supplementary premium will be added, and if such a program were to be placed in effect, it would appear that the logical organization to make such a determination would be the National Council on Compensation Insurance. Inasmuch as this supplementary premium is to be added to the worker's compensation insurance premium, it would be proper to add it to those employers who do not pay sufficient premium to qualify for experience rating.

The other problem is that of enforcement of the safety effort to be furnished these small employers by the insurance companies. Some insurance companies have built a reputation on their safety and loss prevention activities not only in the field of worker's compensation, but in other fields as well. All too many insurance companies, however, do little in the area of safety and loss prevention. How can one be sure adequate safety services are being provided? Further, there are many "small" businesses which do not want safety personnel coming into their operations and taking the time from the management and employees that is necessary to perform a proper safety service. In this regard, the commission suggested that the worker's compensation agency in each state audit the services being provided. With the thousands of small businesses in the various states, this audit function would be an almost impossible task and would merely create another bureaucracy at the state level that may cost more in tax dollars than it saves in the reduction of accidents. This is not to say that such an audit procedure is not possible; however, details of its operation for reporting services received and determining what are "adequate" safety services must be carefully worked out.

Adequacy of Loss Prevention Services

In order to answer the question of adequacy of loss prevention services, one must first determine the standards that are in current use for such services. Loss prevention services provided by the insurance industry are provided by employees referred to as "loss control representatives" by most insurance companies. These individuals are described by many insurance companies as professionals in the field of safety. According to a statement of the American Insurance Association, the basic duties of the safety professional include providing technical guidance and assistance as follows:

1 Identification of loss producing conditions and practices, and evaluation of the significance and the severity of these factors;
2 Development of improved loss control methods and programs, through study of operations;
3 Communication of this information to the employer's top management in such a manner as to stimulate appropriate action;
4 Assistance to the employer in implementing necessary loss control programs through counseling, provision of educational, technical, and statistical facilities; and
5 Continuing evaluation of the loss record and effectiveness of the loss control efforts, and proposal of necessary revisions or additional measures.[4]

Inasmuch as the insurance industry considers these loss control representatives to be professionals, it is not surprising that there is substantial similarity between the insurance industry description quoted above and the Scope and Functions of the Professional Safety Position developed by the American Society of Safety Engineers. Though there are specific differences between these two statements, the similarities are of more significance than the differences.

Dual Role of Loss Control Representative

It must be recognized that the loss control representative of an insurance company not only performs the safety services mentioned, but also serves the home office underwriter by inspecting the premises of an existing insured or an applicant for insurance. The loss control representative may be the only insurance company employee (as differentiated from the agent or broker) who sees the insured; therefore, the report that is submitted to the underwriter is of tremendous value.

The insurance coverage provided by worker's compensation insurance is applicable only to those accidents that arise out of and in the course of employment. Therefore, the loss control representative should be familiar with the type of operations that are to be checked and, further, should know the usual, more frequent, types of accidents that can arise from these operations. With this in mind, the loss control

[4]Pickens, John L., "Statement of the American Insurance Association before the National Commission on State Workmen's Compensation Laws," Dec. 13, 1971, pp. 3–4.

representative must communicate detailed information to the underwriter, who, in turn, must decide whether or not to either write or renew the coverage.

In a research project conducted by Daniel S. Otremba at the University of Arizona, it was reported that a majority of home office officials of insurance companies who responded to a questionnaire stated that they felt the primary function of the loss control representative was to secure underwriting information. Just under one-half of the representatives responding agreed that this was their primary function; however, the study disclosed that about one-half of the time of a representative was spent on securing underwriting information, the other one-half on performing safety services.

Is it inconsistent to say, on the one hand, that the loss control representative is a safety professional and, on the other hand, that the primary function of this representative is to secure underwriting information? It is not necessarily inconsistent, though the inconsistency depends on the use of the underwriting information made by the representative. For example, if the loss control representative, in securing underwriting information either for new business or for the renewal of existing business, discovers some safety "problems" which are reported to the home office underwriter, doing nothing about these "problems" would be inconsistent with the position of a safety professional; however, if this is merely the starting point for recommending solutions to those "problems," then it would not be inconsistent.

It must be remembered that the loss control representative of an insurance company has no authority over an insured to force the installation or improvement of a loss control program other than through the report sent to the underwriter. In other words, if the insured or potential insured is not cooperative, then the insurance company can merely deny coverage or not renew. The representative can only recommend certain actions take place. This, however, may not be too much different than many safety directors' positions inasmuch as they may not have the full support of top management in the implementation of their recommendations. In many cases, the supportive recommendations of the insurance company's loss control representative, along with those of the safety director, are all that is necessary to get action on their recommendations.

What, then, is the position of the loss control representative of an insurance company in the reduction of accidents covered by worker's compensation insurance? As pointed out, one of the functions of this representative, and in some cases the primary function, is to secure underwriting information for the home office underwriter of the insurer. The other function, and in some cases it may be the primary function, is to render professional safety services to insureds. These two functions are not necessarily opposed to each other, depending on how one makes use of the other. The rendering of professional safety services to insureds by the representative is only in an advisory capacity. The insured is paying for these services, through the mechanism of the insurance premium, and should be supportive of these recommendations just as if a safety consultant had been hired; however, this is all too often not the case. If a direct cost has been incurred in the hiring of a consultant, the advice so obtained is usually followed within the economic constraints of the firm. Whether or not the insured wants the services of the loss control representative, a cost is incurred which is hidden in the premium and may not be known to the insured, particularly the small business owner. Such "free" advice to the insured is handled, unfortunately, much the same as other "free" advice obtained from other sources.

Contractual Rights

The Standard Workmen's Compensation and Employers' Liability Policy, revised on March 15, 1975, *permits* the insurance company to inspect the "workplaces, operations, machinery and equipment" of the insured; however, it is not *obligated* to do so. After judicial decisions involving the liability of insurers for injuries that took place after inspections were performed at the workplace, the insurance policy was revised with the addition of the following statement:

> Neither the right to make inspections nor the making thereof nor any report thereon shall constitute an undertaking on behalf of or for the benefit of the insured or others, to determine or warrant that such workplaces, operations, machinery or equipment are safe or healthful, or are in compliance with any law, rule or regulation.[5]

The insurance company, therefore, has the right to make an inspection if it so desires inasmuch as the policy states that it shall be permitted to do so. It would appear to be a violation of a contractual condition to prohibit a loss control representative from a firm's insurer from gaining access to the premises, and, as such, could, though not necessarily would, be grounds for the insurance company to deny

---

[5]This is condition number 4 of the standard policy.

coverage from that point forward on the basis of a violation of a condition of the contract.

Thus, by contract, the insurer is permitted to inspect the premises and operations of the insured. There is, however, no obligation for the insurer to provide the insured a report of its findings.

Most of the insurance companies writing worker's compensation insurance do provide loss control inspections and reports to their insureds. These, however, are merely recommendations when passed on to the insured, and do not imply or warrant that the workplace complies with any law or regulation such as a standard promulgated under the Occupational Safety and Health Act of 1970.

Insurance Company Alternatives

With the increase in awareness of the need for loss prevention brought about by the Occupational Safety and Health Act of 1970 and the judicial decisions involving the liability of insurance companies for their loss prevention inspections, some insurance companies have seen fit to spin off the loss control services from the regular services offered by them and establish separate loss control service corporations. The services of these separate organizations may be either purchased by the insurer for its customers or purchased separately by organizations insured by other companies. What has happened is that the loss control services of the insurance company, traditionally provided to customers without charge, are becoming profit-making organizations in their own right. In this way, it is hoped that since there is a direct charge made for these services, the recommendations will be followed by the purchaser rather than merely gathering dust when there was no direct cost allocated to it.

OTHER LOSS PREVENTION SERVICES OF INSURANCE COMPANIES

Insurance companies have been very active for many years in the field of safety outside the scope of worker's compensation. Included in this would be such endeavors as boiler and machinery inspections, elevator inspections, highway and fleet safety, construction and fire safety, and training and education in all areas of safety.

Most of the premium dollars spent for boiler and machinery insurance find their way into uses other than paying losses. Less than one-half of the premium, as a matter of fact, is allocated to losses, and a substantial portion is allocated to the inspection function. Historically, the insurance companies have been the leaders in the field of boiler inspection.

As in the case of boilers, the insurance companies have been among the leaders in elevator safety, including escalators and other lifting devices. Due to the research done by some insurance companies, many times in conjunction with that of the manufacturer, safety devices have been developed to make these "means of transportation" safer for the public.

The interest of the insurance industry in highway safety is well known. Besides being supportive of many safety organizations in this field, the Insurance Institute for Highway Safety was an obvious outgrowth of this interest and not only serves the motoring public, but also the motor fleet operators.

A great deal of the fire safety developed to make buildings safer for human occupancy has been done by the insurance industry. The rating of property for fire insurance recognizes poor construction from the standpoint of life safety and fire safety, thus encouraging better construction via the pricing mechanism.

In all areas of commercial insurance, the safety and loss control departments of the various insurance companies have done a great deal in the area of education and training. Posters galore on all facets of safety and loss control are available from insurers. Training sessions for supervisory and higher levels of management are available from insurers, some on a regular basis, others on an as-requested basis. Most of these efforts, as would be expected, are available for their insureds and not for the general public.

Finally, the insurance industry, either through its own efforts or through financial support of others, has been quite active in the area of safety and loss prevention research. As mentioned previously, a great number of the safety devices found today have been developed by insurance companies. Underwriters Laboratories was established by the insurance industry, though it is self-supporting today.

Some people are very critical of the insurance industry for its lack of leadership in this area of safety; however, during the early years of the safety movement, the insurance industry was the principal source of safety information and help, and to a large extent, it remains so today.

WHY ARE INSURANCE COMPANIES INVOLVED IN SAFETY?

This is a legitimate question to ask of the insurance industry. It appears that there may be a conflict of interests when the insurance company spends vast sums of money on safety and loss prevention effort on behalf of its clients. Let us explore this briefly.

One of the definitions of insurance is that a company collects premiums from the many to pay for the losses of the few. In other words, if the actuary is correct, the total premiums collected from insureds should be sufficient to pay for their losses, cover the expenses of the insurance company, and have some left over for a profit. Therefore, the only risk taken by an insurance company is that the pricing of the product is correct.

It can be argued that the only reason that an insurance company is interested in safety and loss prevention is to reduce the amount of losses incurred by its customers so that there will be more profit left over. This may be true in some cases; however, one must remember that the insurance pricing mechanism is both competitive and regulated. In other words, a company is not free to set its premiums as high as it wants due to the competitive nature of the marketplace and the rate regulation found in many states. As a rule, rates are based on experience, and this is true for most, if not all, lines of property and liability insurance. There are, of course, problems in developing experience figures, but as a general rule, experience determines rates.

The use of experience to determine rates is a long-run calculation inasmuch as there is no immediacy to the impact of good experience. In most lines of property and liability insurance, 3 to 5 years of experience is required to determine a change in rate level. Thus, 1 or 2 years of good experience would not be reflected immediately by reduction of insurance rates. This, of course, does not hold true in those situations where retrospective rating plans are in effect.

There is, then, some truth to the statement often heard that insurance companies are in the business of safety and loss prevention in order to increase their profits. This is, however, a short-run view of the situation. In the long run, since experience determines rates, the rewards for good experience will be returned to the insured population through the reduction in rates.

Another aspect of this question is that of competition. Insurance is a highly competitive product, and anything that can be done to give a company a competitive edge over others will help that insurer gain an edge in the marketplace. There are some insurance companies that have established a valuable reputation due to their efforts in this area of safety and loss prevention. One often hears and sees the word "service" used by insurance companies, and the loss control services offered by insurers are known in the commercial lines marketplace. The insurance companies with the largest volume of business in worker's compensation insurance, for example, are those with a reputation for their loss control activities.

SUMMARY

The insurance industry has been a leader in the field of safety and loss control for many years. The first edition of this book was written by an employee of an insurance company. Prior to that writing, insurance company safety and loss control was devoted almost exclusively to the inspection of premises of insureds. After the development of the concept that accidents can be caused by unsafe acts as well as unsafe conditions, the insurance industry turned its efforts toward teaching individuals how to perform their functions in a safe way.

Insurance company loss control representatives and other insurance company safety personnel have been involved in other safety problems outside the area of worker's compensation. Personnel of insurance companies are working on problems of pollution and noise, as two examples of this type of work. Also, insurance companies have been leaders in such fields as boilers, elevators, fire, and highway safety.

The loss control representative in the field for an insurance company performs a dual role—inspector for the underwriting department and safety consultant. There are various opinions about which is the primary function; however, both are valuable. If the representative is primarily performing inspections on behalf of the underwriter, it can be questioned whether or not that person is truly a safety professional. However, if the services of this person are primarily for the benefit of the insured in helping to reduce losses, then a professional service is being rendered. One of the major problems, if not the major one, is that these services are merely consultative, and the results of the services are merely recommendations to the insured, to be followed or not as the insured sees fit. Of course, cancellation or nonrenewal of the insurance can occur due to failure of the insured to follow these recommendations; however, this seldom occurs unless a severe fault is determined to be present. Thus, these recommendations tend to gather dust, and the same recommendations are made year after year without any action taking place. If there is a fault to be found with the concept of safety and loss prevention as practiced by insurance companies, it would be the failure to enforce the recommendations of their own loss control representatives.

The role of the insurance company can be looked at, also, as a competitive tool. The leading companies in the writing of commercial (as opposed to personal) lines of insurance have had safety and loss control activities for many years, and have built an excellent reputation for themselves in this area. Insurance, after all, is a highly competitive industry, and anything that will give a company a competitive edge over others will be beneficial to it.

A strong safety and loss control department can also aid a company in reaching its corporate goal of profit. Since loss experience takes time to develop and find its way into the rate-making process, reduction of losses will lead to increased profit for the insurance company. This is, as mentioned, a short-run effect inasmuch as experience will dictate price in the long run.

The insurance companies have taken a leading role in the safety and loss control field. Perhaps they have not done as much as they should have or could have. Their role has been primarily with large insureds, who, if under enlightened management, may not need as much of this service due to the fact that they have a full-time safety director and staff. The type of business that really needs this service is the small business, and here it is a question of how much service can be provided for the small premium dollars received. This is the point brought out by the National Commission, and their recommendation was to charge an additional premium for obtaining this service.

Many insurance companies, cognizant of the recommendations of the National Commission, have set up separate loss control and safety consulting companies under the ownership of the insurance company. In this way, safety services can be purchased separately from insurance protection. Only time will tell if this will be the role of insurance companies in safety and loss prevention, or if it will remain as it has been—a service to insureds in which mere recommendations are made that may or may not be followed and an inspector on the site for the home office underwriter.

SIXTEEN

Risk Management

The word "risk" has intrigued scholars for a long time with no development of a simple, understandable, generally acceptable definition. Generally speaking, there are two views on just what "risk" is. One equates "risk" with uncertainty of loss, the other with chance of loss. The first can be viewed as a subjective view of the term, the latter an objective, mathematical view. The use of the term in this chapter will be that of uncertainty of loss; therefore, one way to look at risk management is the management of uncertainties.

A more specific definition of this term is that risk management involves the efforts of all levels of management to control accidental losses by the use of all available means. Thus, surprise losses are eliminated or, at least, reduced to an irreducible minimum. The accidental loss envisioned in this definition may be that of either an asset—including the lives of employees—or anticipated income. The objective of risk management, therefore, is the prevention of the serious impact on the organization's financial structure from unplanned or uncontrolled losses at the lowest long-term cost possible. As will be pointed out later in this chapter, one of the key factors supporting the function of risk management is that of safety and loss prevention.

THE FUNCTIONS OF RISK MANAGEMENT

The functions of risk management, regardless of the type of organization or the purpose of the organization, are as follows:

- Risk identification
- Risk measurement and analysis
- Risk-handling techniques or methods
- Risk-handling technique selection
- Continual monitoring

The person who usually has the responsibility for seeing that these functions are being performed is referred to as the "risk manager." There are, however, many organizations in which this responsibility is being carried out by the insurance manager or someone with a title other than risk manager.

The risk management function gets involved in three different areas—property and liability insurance, employee benefits, and safety and loss prevention. The safety function, though, is quite frequently found in the personnel department of an organization inasmuch as anything having to do with people—employees—comes within the purview of the personnel department, and, it might be added, the employee benefit function may also be found there. Nevertheless, it is our contention that anything having to do with risk and its control and management belongs in the risk management department. This will be discussed at greater length in this chapter.

Risk Identification

One cannot manage something he or she knows nothing about. Therefore, the primary function of the risk manager is to identify the risks the organization faces. It should be mentioned at this point that the risks—or uncertainties—that are of concern to the risk manager are what are referred to as "pure" risks as opposed to "speculative" risks. Pure risk is that type of risk that, if it occurs, only a loss can take place—no gain—whereas speculative risk involves either a gain or a loss. Whereas fire and automobile accidents are characteristic of the first, business decisions fall into the second type. The risk manager is not directly concerned with marketing a new product or entering a new territory except for the pure risks that may be involved. A merger per se is not of concern to the risk manager; the additional exposure to loss that will develop is of concern. Thus, the risk manager is concerned with all the uncertainties of a "pure" nature faced by the organization, whether insurable or not.

A variety of tools are available to the risk manager to help in this identification function. The best tool is the interviews with employees and inspections performed, preferably with the safety manager, in the workplace. Reading job descriptions or being told what goes on in the plant only informs the risk manager of what is *supposed* to be going on; only by seeing it firsthand does the risk manager know what *is* going on.

Other tools that can be aids in identification are a detailed questionnaire that is used to elicit information about the organization and its various components; records of past incidents and losses; financial statements, which can be used to arouse questions in the risk manager's mind; management contacts, which can be very helpful in keeping the risk manager informed of future plans and thoughts of top management; and finally, keeping in touch with others in a similar position to see how problems are dealt with in other organizations.

The identification process is never-ending. New exposures develop; new laws are passed or new judicial decisions are rendered that create new exposures; new products are brought to the market; new advertising material is developed; and new employees are added by the organization. Contacts must be maintained with the personnel director, legal counsel, production manager, and anyone else who may bring these exposures to the attention of the risk manager.

Risk Measurement and Analysis

Once the risk manager has identified as completely as possible the risks of the organization, the next function is to analyze and measure them. This involves the use of relatively simple arithmetic; however, it can get as complicated as one wants to make it.

What must be done in carrying out this function is to obtain the most complete records of past losses as is humanly possible to obtain. This can be a very difficult process, especially if the organization has not maintained such records itself. Such records should be developed going back 5 years, if possible. What are needed are both the number of losses and the dollar amount of the losses, broken down by dollar brackets. For example, how many worker's compensation claims, each year, were under $500 and what was the total amount incurred; how many, and for how much, between $500 and $1,000, etc.

From these data, properly displayed, one can develop a budget for losses in the future under the assumption that the future will hold true to the past, adjusted for inflation and other similar factors. From this determination of future the risk manager can make an intelligent decision as to which is the best of the alternative methods to handle this particular risk.

The laws of probability are one of the analytical tools available to the risk manager. This text is not the proper place to go into this complex subject, and the reader is referred to a good mathematics or statistics textbook for this subject.

The purpose of measuring and analyzing a specific risk is to enable the risk manager to determine the best method, in his or her judgment, for protecting the firm's assets and earning capacity from accidental loss.

Risk-Handling Techniques

The techniques for risk handling available to a risk manager will vary depending on the type of organization involved. There are, however, four basic methods available, and only the degree of applicability will vary between types of organizations. These four are:

- Avoidance
- Reduction
- Retention
- Transfer

If one were to take the first letter of each of the above-listed techniques, the acronym would spell ARRT, which is another way to emphasize that risk management, like many fields of management, is more an "art" than a "science."

Avoidance Technically, this is not a method for handling risk inasmuch as the risk is never present in the organization. It is a method, however, available to a risk manager who has proper input into top management to persuade them not to enter a particular venture due to the risk involved. In other words, if the comments of the risk manager are solicited by top management *before* a particular action is taken, additional risk on the part of the organization may be avoided. For example, by showing management the risk and cost involved in entering international trade, management may decide not to do so, thus avoiding the risk.

Reduction   This involves reducing both the hazard and the amount of loss—in other words, trying to reduce both the frequency and the severity of loss. It is in this connection that the risk manager relies very heavily on the safety manager, and more will be said on this in a later section. Regardless of what technique or techniques are adopted by the risk manager, reduction of frequency and severity of losses is an imperative part of the job.

Retention   This is many times referred to as "self-insurance." Retention is a risk management device in which part, but not necessarily all, of a particular risk is retained for the organization as opposed to the purchase of insurance. This is best illustrated through the use of a deductible in any insurance policy. The level of the deductible is set at whatever amount the firm feels financially comfortable with, and the risk of loss above that amount is transferred to the insurance company. In some cases, plate glass as an example, the entire risk is retained due to the financial ability of the firm to pay for those losses without the need for an insurance company. There is a common misconception that retention or self-insurance contemplates retaining all losses; however, this is not true—only those amounts of loss potential are retained that the firm can financially retain. Any amounts over that should be transferred—perhaps to an insurance company.

Transfer   Insurance is the most popular transfer device. For a consideration—the premium—the financial consequences of a loss are transferred to a third party—the insurance company. If the contingency insured against occurs, the insurance company will indemnify the insured within the conditions of the contract. There are, however, other transfer devices for handling risk. Without going into the details of each, some of them would be hold-harmless agreements, subcontracting, and hedging.

Risk-Handling Technique Selection

After the risk manager has examined all the alternatives available for each risk, a decision must be made about which one or combination will be used for each such identified risk. In some situations, the best alternative will "stare you in the face." In most, however, the best is not that obvious. Inasmuch as avoidance is not truly a risk-handling technique and risk reduction must be carried on regardless of the method selected, the risk manager really has only two alternatives: (1) retention or (2) transfer. A major factor in making the decision between these two, or the combination of them, is the financial ability of the organization. These financial guidelines are prepared by those individuals in financial management of the organization.

Continual Monitoring

Once decisions are reached for handling risks, they are not "cast in stone". Changes can be made, and should be made, as conditions change, both inside and outside the organization. The insurance marketplace may change in such a way that it becomes uneconomical to purchase a certain type of insurance. The financial condition of the firm may change so that it may be either wise or unwise to continue, increase, or decrease the firm's retention plan. Taxes may be changed by the Congress or by a ruling of the Internal Revenue Service so that the current risk-handling technique is no longer suitable for that organization.

Part of the continual monitoring involves record keeping so that trends can be developed. What are the frequency and severity of a certain type of loss situation? What is being done about controlling these losses? In other words, a very close liaison between the risk manager and the safety manager must be maintained and is a vital part of the continual monitoring of the risk management plan.

RELATIONSHIP OF RISK MANAGEMENT AND SAFETY MANAGEMENT

Much of what has been said heretofore in this chapter about risk management would be relevant to the safety professional. It is not, however, part of the necessary knowledge safety engineers use as a foundation of professionalism. It is admittedly difficult to define where the safety profession begins and ends, what knowledge and skills it encompasses, and what should be taught in a curriculum of professional safety education. The profession of risk management can have these same comments directed to it.

Both risk management and safety management can be viewed as

pretentious labels for familiar functions, or they can be viewed as new designations for evolving management disciplines trying to define themselves, and trying to define their role and their niche in the affairs of management.

Their activities interface with nearly every function and task of management. They interface vigorously with each other. Both deal with risk which comes in two styles. Entrepreneurial risk makes the wheels go. It offers hope of gain along with the chance of loss. It is the soul of change and is therefore called "dynamic" risk (or, as used earlier in this chapter, speculative risk). The other type is "static" risk—a risk, a hazard, an exposure, an event or occurrence in which there is no hope of gain. If the event occurs, loss is the certain result, never gain (referred to before as pure risk).

The risk manager deals with static or pure risk. So does the safety manager. Both deal with events that cause loss, never gain.

The risk manager is a specialist in *funding* the loss. Whatever happens, the risk manager tries to assure that loss can be absorbed in normal cash flow, or buffered by reserves, or transferred to others by legal devices including insurance. The safety manager tries to *reduce and control* losses. That person is a specialist in preventing loss. Skillful use of these two specialties, preventing and funding, is all that can be done to cope with static loss.

Both the risk manager and the safety manager share the task of identifying and evaluating exposures, but often they are looking at different exposures. The risk manager will be more cognizant of contingent risks, of disruption of cash flow, of earthquake and flood, and of those exposures which must be dealt with primarily by funding plans rather than by prevention and control. The safety manager, on the other hand, will be more cognizant of those exposures for which a technology exists for prevention and control. The two functions can sometimes be powerfully combined in the same person or staff, but they are more often managed as distinct functions.

These two functions should be in fundamental communication with each other. They deal with the same undesirable phenomenon, static or pure loss, approaching it from opposite directions with much merging and overlapping in the range of strategies and methods available to them. The funding approach employs financial skills and knowledge of law and insurance. The prevention approach employs safety technology and the skills of leadership in group endeavor. Each influences the activities and priorities of the other. The feasibility and cost of prevention must be balanced against the feasibility and cost of funding. And vice versa. Each possesses a distinct body of specialized knowledge and skill brought to bear on the risks of static loss. Risk management and safety management should be in fundamental communication with each other.

Such interface is best illustrated in Fig. 16-1.[1]

Lines of Communication[2]

Regardless of where in the organization the risk manager and the safety manager are located, there must be communication between these two positions. There have been numerous discussions about where these positions should be located in the organization as well as reporting

[1]Adapted from D. A. Weaver, "Interface: Risk Management and Safety Management," *Professional Safety*, June 1976.
[2]Adapted from Mary Finnell, "Should Safety Management Be Included in the Risk Management Function?" *Risk Management*, April 1978, pp. 38–43.

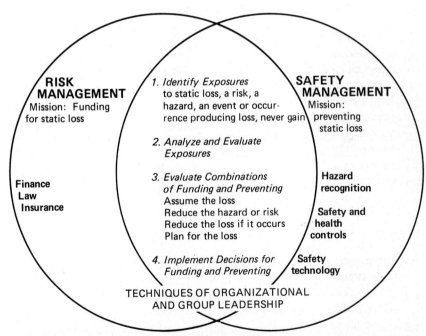

INTERFACE: Both Risk Management and Safety Management deal with static loss. Both interface with every function and operation in the organization. They interface vigorously with each other with much merging and overlapping in the range of strategies and methods available to them. They should be in fundamental communication with each other.

Figure 16-1.

responsibilities. Should the risk manager report to the safety manager or vice versa? Oftentimes the risk manager is found reporting to the financial executive and the safety manager reporting to the personnel executive, and unfortunately, there may be little liaison between the two. A good and effective safety and loss control program depends on good communication between these two positions, especially when they are located in different functional areas of the organization.

Based upon a questionnaire sent to the member corporations of the Risk and Insurance Management Society, Ms. Finnell received a response of slightly over 40 percent. Perhaps the results would have been different if the questionnaires had been distributed to members of of the American Society of Safety Engineers; however, the results that were obtained were of interest to those in both risk management and safety management.

Inasmuch as the risk management function is concerned with the problem of funding for loss, it was not too surprising to find that 42 percent of the respondents report to the financial function. Slightly under 70 percent of the respondents reported that their firm does have a safety manager, with the largest percentage (41 percent) reporting to personnel, employee benefits, and industrial relations departments. One of the surprising results of this survey was that 20 percent of the safety managers report to the risk manager, and this appears to be a trend. Where a firm did not have a safety manager, over one-half of the respondents reported that the risk manager performed the safety function.

It is of interest to note that while only 7 percent of the responding risk managers are located away from the corporate headquarters of their firm, 29 percent of the safety managers are so located. As a result of this location problem, only 23 percent of the risk managers have daily contact with their safety manager; however, over 60 percent of the risk managers are in contact with safety people once a week or more, and such contact is usually on a face-to-face basis. These communications usually dealt with safety and loss control subjects, as would be anticipated, such as worker safety, physical conditions, and frequency and/or severity rates. Risk managers responding found these contacts to be very useful, and just under one-half stated that they used all the information received.

Seventy-two percent of the responding risk managers felt that a full-time safety manager is needed, and another 17 percent said it could be a part-time job. Thus, 89 percent of the respondents felt the need to have the function of safety management carried out in their organization by either a full-time or part-time person. This gives credence to the importance of safety and loss control to the function of risk management and shows that risk managers in industry recognize this.

Almost one-half (48 percent) of the responding risk managers felt that the safety management function should be in their office or department. This corresponds with the thinking of the Editorial Advisory Panel of *Business Insurance*, in which 80 percent of the responding members of that panel agreed that safety and security should be included under the risk management function, and, further, just under 50 percent of the group surveyed had the safety and security responsibility as part of the risk management responsibility.[3]

SUMMARY

It was not until 1950 that a professional organization of risk managers—currently known as the Risk and Insurance Management Society—was formed. One can still look upon this as a developing area within management sciences. New ideas and concepts are being tested at various times, often due to conditions in the insurance marketplace, at other times due to innovative ideas of practitioners of the field.

The profession of safety is older than risk management. In some organizations recognition of the safety and loss control function preceded the recognition of risk management by many years. In others the opposite is true. Regardless of when either of these professions was founded, the reality is that now the two *must* work together as they never have before.

As has been pointed out previously in this chapter, the basic function of the risk manager is funding for losses that occur; the basic function of the safety manager is controlling and preventing losses from occurring.

These two important functions cannot go their own way without liaison and coordination. The funding needs of the risk manager can be reduced by the control and prevention activities of the safety manager. If the safety function does not operate as anticipated, the risk management function is there with the necessary funding.

Neither of these functions can claim a priority from management. Just as personnel is a recognized function of management, so should risk and safety be recognized. The two functions should be together in the corporate organization as a department inasmuch as both functions deal with risk, only in different ways. Since safety and loss control are

---

[3]Alt, Susan, "Safety Department Belongs in Risk Umbrella, Panel Agrees," *Business Insurance*, Apr. 4, 1977, pp. 1 and 29–30.

part of the process of risk management, whereas the opposite is, obviously, not true, it becomes apparent that the risk manager should head the department that includes safety and loss control.

Though there has been ample discussion of this point, there is a dearth of written material addressing itself to it. The study of Ms. Finnell is one of the first to determine what is going on in the business community rather than to just "talk" about it.

It was determined in that study that risk managers and safety managers are communicating in many areas of mutual concern. Their relationship is generally a good one. Risk managers themselves regard safety and loss control as a vital part of their function and expressed the idea that they could handle the funding function better if they also had the responsibility for the safety and loss control function. Conversely, the risk manager interested in safety and loss control can be a very powerful ally to the safety manager.

The manager of the safety function must become more aware of what the manager of the risk function is doing and how the work is being carried out. Similarly, the manager of risk must become vitally concerned with the safety and loss control function.

As Ms. Finnell stated in her concluding comments:

> But no matter where each function is located, safety managers and risk managers must realize their contributions to the organization are unique and vital. The effect on the company's stability and profitability by the two professionals working together will be much greater than each making an individual effort.[4]

[4]Finnell, op. cit., P. 43.

SEVENTEEN

Experience Rating

Since 1911, with the passage of the Workman's Compensation Law in the state of Wisconsin, every state has enacted a workman's compensation law, the last being the state of Mississippi in 1948. Though the laws vary on the number of employees necessary before the employer must comply with the requirements of the act, what employments are exempt from the act, and the benefits (in duration and amount) available under the act, the methodology of pricing the product is very similar among the states. The National Council of Compensation Insurance was formed in 1922 as an association of insurance carriers, and that organization gathers statistics and files rates on behalf of the insurance industry. As of the end of 1977, the National Council had 522 insurance companies affiliated with it.

The worker's compensation insurance system in 32 jurisdictions is administered through 15 National Council branch offices, as shown in Table 17-1. Administrative operations include policy review, preparation and distribution of intrastate individual risk experience ratings, and inspection of employer operations for proper classification assign-

Table 17-1   Administrative Bureau Operations, National Council on Compensation Insurance

Bureau	State
Arkansas	Arkansas
Florida	Florida
Illinois	Illinois
Indiana	Indiana
Kansas	Kansas
Mid-Atlantic	District of Columbia
	Maryland
Missouri	Missouri
Mountain States	Arizona
	Colorado
	Idaho
	New Mexico
	Utah
North Central	Iowa
	Nebraska
	South Dakota
Northeastern	Connecticut
	Maine
	New Hampshire
	Rhode Island
	Vermont
Northwestern	Alaska
	Montana
	Oregon
Oklahoma	Oklahoma
South Carolina	South Carolina
Southeastern (1)	Louisiana
Southeastern (2)	Alabama
	Georgia
	Kentucky
	Mississippi
	Tennessee

ment. Six states—Nevada, North Dakota, Ohio, Washington, West Virginia, and Wyoming—are monopolistic state fund states, and private insurance carriers cannot write worker's compensation insurance there. The remaining states have what are referred to as "independent bureaus."

RATE-MAKING PRINCIPLES

Rate making in worker's compensation insurance is prospective in its application. That is, it does not intend to recover any deficiencies of the past nor return any gains of the past. It is designed to provide rates for the future that meet the standards of the state rating law.

Rate making means the process of developing manual rates. In worker's compensation insurance, the rates are calculated for approximately 700 different classifications. These manual rates produce the final premiums for a large *percentage* of policyholders; however, the total *volume* of premiums for these policyholders is small inasmuch as that group is made up of predominantly small employers. The reason for this phenomenon is that the premium at manual rates is subject to modification for larger employers under one of several types of rating plans, which will be explained later in this chapter.

The intent in determining manual rates is to use past state experience to produce the expected premiums and losses that will occur during the period the new rates will be in force. Many actuarial techniques are employed to reach this point; however, it is not the intent of this text to get into the specific techniques employed. Rather, the overall approach will be looked at so that one can see how the safety and loss control function can have an impact on insurance costs in the area of worker's compensation insurance.

Rate-Level Change

The first step in this process is to determine the overall rate-level change for the individual state based on that state's own statistics. This overall change is distributed over three broad industry groups: (1) the manufacturing group, (2) the contracting group, and (3) the residual group labeled "all other industries." Once this overall change is determined for each of these three groups, it is distributed among the individual classifications in accordance with the volume and characteristics of the class experience.

In determining the overall rate-level change, the council uses two types of data. One is policy year data, which is a gathering of data for all policies written in a given 12-month period. These data are given a 50 percent weight in determining the overall rate-level change; the other 50 percent weight comes from using statewide calendar year experience, which is the experience for the 12 months ending either December 31 or June 30, whichever is more recent. A call is issued to the insurance industry by the council in early January and early July to get the latest 12 months' calendar year experience from these insurance carriers.

Calendar year experience differs from policy year experience in that the calendar year experience represents all transactions taking place within the time frame of the calendar year regardless of the policies from which those transactions were derived. Thus, an increase or decrease in the reserve on a particular claim that was received several years in the past would have an effect on the current calendar year figures. In policy year data, however, any changes in the reserves would be reflected in the data for the year in which the policy was written rather than the year the change took place.

Inasmuch as a rate-level change will affect future policies—those written after the effective date of the change—a projection factor is applied to the developed statistics. In this way, the past data are updated to recognize that the rates will have to be sufficient to cover future losses. Though this factor has been in use for some time in other lines of insurance, it was not until 1973 that the National Council of Compensation Insurance adopted it in the field of worker's compensation insurance rate making.

Once the overall rate-level change is determined for an individual state, the rate-level change for the approximately 700 classifications must be determined. The class experience is a function of the volume of data within the classification—referred to as "classification credibility" in the rate making. If a class is very big in a state, it will be fully credible for the change in its rate level. If a class is not very big, or has little or no experience, then the change in the rate level for that particular class will be the industry group average change.

Final Premium

As was mentioned earlier, though these manual rates produce the final premium for a large percentage of policyholders, the total volume of premiums for these policyholders is small. These manual rates are subject to modification for larger employers under one of several types of rating plans: the experience rating plan, which is mandatory whenever the annual premium exceeds $750, mandatory premium discounts that apply whenever the annual premium exceeds $1,000, and the optional retrospective rating plans. Retrospective rating plans and premium discounts are exclusive of each other. Due to the importance of experience rating plans, whether prospective or retrospective, the remainder of this chapter will be devoted to those methods and the role of safety and loss control in reducing costs of worker's compensation insurance.

EXPERIENCE RATING

The rating plan referred to as "experience rating" is, technically speaking, a prospective experience rating plan. Popular usage, however, is to call it merely an experience rate to differentiate it from the retrospective experience rate, to be discussed later.

The manual rates that were discussed earlier can be characterized by looking at them as the statewide average rate for a given classification regardless of the experience within a given entity. The only effect a particular organization can have on a manual rate is that its experience becomes mixed with that of other organizations in a similar group with similar occupational classifications; however, a few bad cases from other organizations can more than offset the good experience of a particular entity when these manual rates are determined. Therefore, larger organizations (those paying in excess of $750 per year in worker's compensation insurance premium) are mandatorily placed in an experience rating plan so that they can modify the statewide average experience by their own experience—good or bad.

Manual Premium

The beginning point in the calculation of the rate under an experience plan is the manual rate for each classification found in a given organization. That rate is multiplied by the number of $100 of payroll within that classification, and the result is the *manual premium*. Therefore:

$$\text{Manual rate} \times \frac{\text{payroll}}{\$100} = \text{manual premium}$$

Experience Modification Factor

The way in which the individual organization's experience is used to modify the statewide experience used to promulgate the manual rate is known as the "experience modification factor," or the "E mod," or just the "mod" factor. The mod factor is calculated for each qualified firm by the National Council, and as mentioned earlier, the qualification for its use is the production of $750 of premium for an average of 3 years, or $1,500 of premium for the last 2 years.

The experience period used for calculating the mod factor for an organization is the most distant 3 years of the most recent 4 years. In

EXPERIENCE MODIFICATION FACTOR TIME FRAME

1974	1975	1976	1977	1978	1979
1974	1975	1976	x	E	
	1975	1976	1977	x	E

x = year not used in the calculation.
E = year calculation is effective.

Figure 17-1   Experience modification factor time frame.

other words, in order to calculate the mod factor for an effective date in 1978, the experience period would be the years 1974, 1975, and 1976, inasmuch as the experience for 1977 would not be available for use in 1978. Figure 17-1 shows the years that would be included in the calculation of the experience modification factor for different effective years.

If experience data are not available for the full 3 years of the experience period, 1 or 2 years will be used if there is sufficient payroll to have the $1,500 rule applicable.

Three basic elements are used in the calculation of the experience modification factor: (1) frequency, (2) severity, and (3) size of risk. Each of these elements requires further inspection.

Frequency   The frequency of accidents reflects the safeness of a particular organization and is the most important of the three elements. Frequency of accidents has a far greater influence on the mod factor than either severity of losses or the size of the risk. This is so due to the fact that the formula used divides losses into two elements—the primary value and the actual value. Any loss up to $2,000 is placed in the formula at the full value of the loss. In other words, any loss up to $2,000 is considered a frequency loss and is charged 100 percent of its value in the formula. An insured with a large number of small losses (up to $2,000) can produce a very large primary charge in the modification formula.

Severity   For losses in excess of $2,000, the dollars are divided between the frequency portion and the severity portion of the formula, called "actual value" and "primary value" in the formula. The severity of the loss is of less importance than the frequency, however, since a loss in excess of $2,000 is limited to a maximum value. In Arizona, for example, the maximum chargeable loss for a single accident, single injury, is $78,000, and the maximum primary value for a single accident, single injury, is $10,000.

Size of Risk Small risks—those with less than $25,000 in expected losses—have their experience modification factor computed using only the primary values of the accidents. The larger the insured becomes, the more it will share in the severity portion of the formula. Thus, more of the actual cost of its own accidents will appear in the formula.

It can be seen from the method of calculating the experience modification factor that frequency control is the key to low net cost in experience rating, and this can only be accomplished by a sound safety and loss control program.

Standard Premium and Discount

The manual premium, already calculated, is then multiplied by the experience modification factor to arrive at the standard premium. Thus:

$$\text{Manual premium} \times \text{experience modification factor} = \text{standard premium}$$

If the standard premium is under $1,000, it then becomes the guaranteed cost for the year for the insured's worker's compensation insurance. If, however, the standard premium is in excess of $1,000, the insured is given a premium discount, the amount of which is determined by whether the insurer follows the stock or nonstock formula. The premium discount for a nonstock insurance company is lower than for a stock insurance company inasmuch as the former usually pays dividends. The premium discounts are shown in Fig. 17-2.

The standard premium less any applicable premium discount is the amount paid regardless of the number or amount of claims filed during the year, and is guaranteed for that year.

RETROSPECTIVE RATING

The foregoing has been an explanation of what is referred to as a "prospective experience rate," or simply an "experience rate." An optional plan available to larger employers is a retrospective rating plan, which is a type of experience rate. Under a retrospective plan (or merely retro), the insured pays the standard premium as a deposit premium at the beginning of the year. Based upon the insured's

PREMIUM DISCOUNTS		
Premium	Stock, %	Nonstock, %
First $1,000	0	0
Next $4,000	9.4	3.0
Next $95,000	14.7	6.0
Next $100,000	16.3	8.5

Figure 17-2   Premium discounts.

experience during the year, the insurance company will either pay a refund due to good experience or request additional premium due to poor experience.

The retro formulas are five in number: plans A, B, C, J, and D. The first four are considered to be tabular plans inasmuch as the maximum and minimum premiums are fixed by the plan rules, whereas plan D can have these factors negotiated between the insured and insurer and can also include other lines of coverage beside worker's compensation. Rather than go into the details of each of these plans which operate under the identical formula other than the specific minimum and maximum premium percentages, the basic retrospective rating formula will be examined.

The Formula and Its Components

The basic retrospective rating formula is as follows:

[(Standard premium × basic premium factor) + (incurred losses × loss conversion factor)] × tax multiplier = retrospective premium, subject to a maximum and minimum

The standard premium, as explained previously, is nothing more than the manual premium multiplied by the experience modification factor.

Basic Premium   The standard premium is then multiplied by the basic premium factor to obtain the basic premium. This basic premium factor provides an allowance for the insurance company expenses (other than the cost of adjusting losses) and what is known as an "insurance charge." Rather than go into the details of this insurance charge, suffice it to say that this will vary with the specific minimum and maximum percentages chosen. It takes into account the fact that the lower the minimum premium, the less chance there is for the insurance company to earn additional profits; also, the higher the maximum premium, the more losses that are covered by the premium.

Incurred Losses   Incurred losses include the losses paid in a particular year as well as those for which reserves stand on the books of the insurance company. The purpose of this figure is to reflect the total losses that took place in a given year, whether paid or not. Another way to look at incurred losses is from the accounting view that expenditures should be matched against revenue for identical periods of time. In other words, premiums should be sufficient to pay all losses for that period, whether the losses are actually paid within that period or at a later date.

Loss Conversion Factor   The loss conversion factor allows for the cost of adjusting and trying cases. For purposes of the formula, rather than cost-account for these expenses a factor is applied directly to the incurred losses, and the resultant figure is referred to as "converted losses."

Tax Multiplier   All states tax insurance companies on the premiums that they receive. Since the basic premium plus the converted losses is, in effect, the premium, taxes are applied to this sum. The percentage of tax applied by each state varies; therefore, the actual tax multiplier used will depend on the state in which the insured is located.

Excess Loss Limit   In some instances, an insured might want to add one additional factor to this formula, known as the "excess loss limit." This is an optional feature of retrospective rating plans and enables the insured, for an additional premium, to be assured that losses over a specific dollar amount will not be included in the rating formula. However, all losses are paid; those above the limitation are just not included when the formula is calculated.

THE ROLE OF SAFETY AND LOSS CONTROL

As one can see by examining the components of the various experience rating plans—whether prospective or retrospective—the basic element that differentiates between the various plans is the amount of incurred losses. The first effect of a sound safety and loss control program, as was pointed out, is in the calculation of the experience modification factor. Inasmuch as frequency is so much more important in this calculation than is severity, an employer can, to a certain extent, control the standard premium paid by having a good working safety and loss control program since one of the basic goals of such a program is to reduce the frequency of accidents. It is not our role at this point to go into detail on such a program; suffice it to point out the impact of such a program on worker's compensation insurance cost.

In a retrospective rating plan, regardless of the one selected, the role of a sound safety and loss control program becomes more obvious. The employer receives the benefits of such a program not only in the experience modification factor, but also in the retrospective rating formula portion that adds converted losses to the basic premium. Of course, the employer is subject to a minimum premium so that there will always be some cost to this coverage; however, this cost can be driven down to the minimum with a sound safety and loss control program.

One can measure the impact of a safety and loss control program in dollars and cents through its impact on the cost of worker's compensation insurance. This is one of the few, if not the only, areas in which such an accurate measure can be obtained. It is through the lowering of the experience modification factor and the reduction of incurred losses that such accurate measurement is available. One should not be misled, however, by rising costs of worker's compensation insurance in spite of a sound program, since this increased cost may be due to factors beyond the control of the safety manager—rising *manual* rates due to increased benefits and rising payroll which will be reflected in the manual *premium*. All that can be said in that situation is to think what the cost would have been without the safety and loss control program!

SUMMARY

Worker's compensation insurance costs are all that have been discussed in this chapter; however, safety and loss control personnel can have an effect on other insurance costs if they are brought into the planning stages of a new building or activity. The more safely a building is constructed—in the eyes of the Insurance Services Office, the rating organization of the insurance industry—the lower the cost to insure. This same rationale does not hold true in many other lines of insurance unless some form of experience rating is applied. There is more profit potential to the insured from a safety and loss control program in the field of worker's compensation insurance than in any other field of insurance. And if the work of the safety manager continued to effectively control losses, this profit could be realized year after year.

Not only does the insured realize a profit potential from the safety program in the experience modification factor calculation, but also, if in one of the retro plans, through lowered incurred losses. It is in this way that the safety and loss control department can really become a "profit center."

PART FOUR

Appendices

APPENDIX ONE

Background of Industrial Safety

By D. A. Weaver

In essence, a history of safety deals with the "organized" safety movement, and with the forces that brought an organized safety movement into being. It deals with institutions, for society deals with its problems through the use of its institutions and by the creation of new institutions.

The fundamental institution is government in the legislative and judicial processes. Its influence on safety can be traced back to Deuteronomy and to Caesar's legions which are reputed to have had a safety officer.

Legislation and judicial process concerning safety appeared early in the industrial revolution, and became massively significant as the twentieth century progressed. Private enterprise, business, the role of management, and the interface with the insurance industry all became key elements in the organization of the safety movement. Safety societies and safety institutions came into being in growing numbers and influence as technology created problems faster than they could be solved. As the 1970s drew to a close, organized labor had finally found an effective role at least in the health aspects of the workplace, and the nation's universities were at last beginning to build an interdisciplinary foundation for safety professionalism.

These are elements in the history of organized safety, the structuring of institutions to contain the nation's proliferating exposures. We can pick up the threads as mechanical industrialism was about to burst upon a naive world.

The early years of the nineteenth century saw the mechanization of industry rapidly gain momentum. Simultaneously with the increasing use of steam power, the handicrafts gradually declined. This didn't come about without a struggle, however, for as the textile factory system developed, there were many riots between the hand spinners and factory operators. Machines were even destroyed by hand workers, many of whom did all in their power to hinder plant operations. But merchandise was less costly to produce by machines, and eventually the hand workers gave up the losing struggle and took their places at the spinning machine and the power loom.

The births of industrial power and industrial safety were not simultaneous. The introduction of English workers to mechanized industry was accompanied by working and living conditions so bad as to defy adequate description. Debasement and social degradation came quickly in industrial centers. The population of Manchester grew to 200,000, though the city contained neither park nor playground. There was no system of water distribution, and workers were compelled after their day's work to go great distances for water and to wait in line with buckets. There were no schools, and living quarters were inadequate. Idiocy and bodily deformities were common. The death rate tripled.

EARLY CONDITIONS

These conditions were social. Although they were bad, plant conditions were worse. Factories were little more than shacks. Light, ventilation, and sanitation in those low-ceilinged, narrow-aisled structures were almost nonexistent. Rest rooms were unthought of. Two-thirds of the workers were women and children whose workday was from 12 to 14 hours. Machine guards were unknown. Occupational deaths and maimings were frequent.

Some governmental factory inspections were made in England as early as 1833, but it was not until nearly 1850 that actual improvements began to be made as a result of their recommendations. These efforts were the first attempts by government to improve industrial safety. As time went on, legislation shortened working hours, established a minimum working age for children, and made some improvement in sanitary and safety conditions such as providing for the fencing of mill gears and shafts. These beginnings of improved industrial conditions were a far cry from organized accident prevention as it is recognized today.

Although fatal and disabling accidents were common during early industrial days in England—as they have been in every country during similar periods—damages were seldom paid by plant operators. Lawmakers were slow to legislate for the greatest common good, for the doctrines of "negligence of the fellow servant" and "contributory negligence" were strongly entrenched throughout the world. It was not so long ago that American employers felt they were discharging their obligations to their employees when they gave a job as a sweeper or watchman to a person who had lost an eye, or paid the funeral expenses of a worker who was killed at his machine. Under the existing laws employers were usually not even compelled to do this.

The pioneer industrialists should not be judged too harshly for their attitude toward employee accidents. Many benefits, such as improved transportation and better lighting, heating, and plumbing, were being realized through the tremendous industrial advancement in the United States, and it was a firmly rooted belief, among employees as well as management, that a certain amount of human suffering and loss of life was necessary if this advancement were to continue. Accidents were accepted as an inherent part of industry. Employers were still unaware of the economic losses that accompany accidents.

And employees too, though they frequently resented the working conditions offered them, were not, in many instances, particularly interested in safety. Notable progress began to be made in plant safety only after plant management began to insist on safe working practices.

Plant management can be seen as the force of greed or as the creative thrust of the profit motive in a private enterprise system. As a force to organize safety it has operated both ways. It has been observed that ours is a society that can do anything if only we can find a way to make a profit at it. Accident prevention, or occupational safety and health, requires an investment of time, money, and effort. Who shall invest that time, money, and effort, and why? The organizing of the safety movement, still continuing today, consists largely in finding better answers to that question. The remarkable achievements of organized safety (and its remarkable oversights and failures) can be understood in that light.

In one way or another, the total society bears the cost of the efforts to prevent accidents as well as the cost of failure. The loss of property, and especially the loss of human potential, diminishes the whole of society. Both consciously and unconsciously, society allocates those costs in a web of legal, financial, and philosophical considerations. At first hand, the victim and the family bear the cost aided only by charitable and ethical considerations. Welfare can allocate the costs to the whole society in taxes. Legal liability, or tort action, can allocate the monetary costs back to an employer, or to an insurance company, and thus ultimately to an increase in the price of the product. Ideally the cost of the loss would devolve upon those persons who can best implement the remedial measures. They would then choose the cost of prevention in preference to the cost of the loss. It never happens in the absence of safety statesmanship, in the absence of organizing for safety. Successful organization for safety seeks to allocate costs so that it becomes desirable (profitable) for appropriate persons and institutions to focus time, money, and effort on effective prevention. The intervention of government by legislation is one way to focus safety effort, and Lowell, Massachusetts, provides an early example.

Lowell, Massachusetts, was one of the first industrial cities in the United States, manufacturing cotton cloth as early as 1822. The workers were mostly women and children from the surrounding farms, many ranging from 6 to 10 years old, who worked from 5 in the morning until 7 at night. How many girls' fingers and hands were lost in unguarded machinery no one will ever know. It was not long, however, before the increasing number of cotton mills in Massachusetts began to exceed the supply of farmers' daughters. Fortunately for the employers, a potato famine in Ireland greatly increased migration from that country, and many of the Irish settled in and around Boston. Labor again became plentiful for the mills, but with the influx of immigrant help the number of accidents soared. One result was the passage by the Massachusetts legislature in 1867 of a law requiring the appointment of factory inspectors. Two years later the first bureau of labor statistics in the United States was established. Coincidentally, in Germany acts were passed providing that all employers provide necessary appliances to safeguard the life and health of employees. At last, industry was learning that conservation of the human element is important. A few years later, Massachusetts, having discovered that long hours of activity produce fatigue and that fatigue causes accidents, passed the first enforceable law for a 10-hour maximum working day for women. In 1874, France passed a law providing special inspection service for workshops, and in 1877, Massachusetts compelled the guarding of dangerous moving machinery.

Catastrophe followed by legislation is a frequent pattern in the history of safety. Unfortunately, the theme of catastrophe followed by legislation can be kept current in the pages of the daily newspaper. As a historical device, the theme has been recorded in depth by the publication in 1975 of Brenda McCall's *Safety First at Last*, and by her subsequent articles published in the pages of *Professional Safety*, the journal of the American Society of Safety Engineers.

Legislation is one process by which government affects safety. Judicial process is another. Together they change the impetus for safety or create a new impetus, and impetus is defined as time, money, and effort. Safety technology for prevention goes begging until it becomes desirable (profitable) for some institution to devote time, money, and effort to its implementation, until an impetus for safety is created.

A history of organized safety is a study of the question "From whence stems the impetus for safety?" As the last quarter of the twentieth century got underway, legislation and judicial process had changed the answers to that question. The new answers and the process of change can be illustrated in four aspects of professional safety in the late 1970s.

1 *Occupational disease became the main concern of the workplace by the late 1970s.* Of course, occupational disease had always been part of the job of the safety professional, but it had been strangely muted. The main theme had been accidental injury. Then a new mechanism came into being, NIOSH, the National Institute for Occupational Safety and Health, created by the Occupational Safety and Health Act of 1970.

A wave of legislation dealing with safety and health broke over the nation in the 1960s and 1970s as will be discussed later. For the moment, it is enough to say that in a few years' time, NIOSH had redirected the thrust of safety in the workplace. The awareness of 14,000 accidental deaths in the workplace became the awareness of 100,000 deaths each year, the difference being deaths attributable to occupational disease. New chemicals and substances had been flooding into the workplace for years. The accustomed dust, mist, fumes, and vapors, and the "stuff in the blue barrel," took on an unperceived menace. Furthermore, it was not important to perceive the menace, to devote time, money, and effort in needless worry about things that didn't seem to hurt anyone, except for a few "allergic reactions." The effects of long-term exposure appeared as the effects of aging, or just bad health. Nonetheless, the evidence was there, often tragically obvious. NIOSH did not expose these things by a scientific breakthrough. It merely looked at the evidence. The question must be asked: What deficiencies in organized safety made it unimportant to look at the evidence, to focus on this problem? It's a question that must be pursued as this history proceeds.

2 *Products safety became part of the job of the safety professional in the workplace.* In a few years' time, in the late 1960s, a whole new aspect of the

safety profession came into being, replete with books, educational programs, techniques, publications in professional journals, and highly visible specialists. One side of the coin was labeled "product liability," who shall pay. The other side was labeled "product safety," who shall prevent.

The problem wasn't new. All that happened was a change in the rules, a change that revolutionized who shall pay for losses and who shall try to prevent them. People had long been hurt, injured, and killed by the products they bought in the marketplace—by tools that electrocuted their users, by toys that maimed or poisoned babies, by cars with engineering defects, by toxic substances presented to the unaware customer with no warning. The victim could sue, of course, but the legal rules precluded any optimism in collecting.

In this secure world, insurance companies included product liability in a casual clause in their general liability policies. It was not separately rated; it was just a generous throwaway since the victim couldn't collect, and it was remote that anyone would have to pay off. Safety journals occasionally reminded safety professionals that defective products really ought to come under their purview. Then the rules changed.

The rules were changed by lawyers and judges in court, in a series of decisions whose history is better written (and has been) by legal historians. Step by step victims found more and more opportunity to recover for the hurt they had suffered. In a swelling wave, the victims, real and imaginary, took full advantage of their opportunities. Insurance policies were hastily amended. Products insurance became a distinct coverage to be purchased at a dear price. Many manufacturers could find no insurance company to sell them products coverage at any price. It was an erratic revolution, extending over several decades and still going on in the late 1970s, with justice often being in the eye of the beholder.

Now that victims could recover, it became important to prevent their being hurt, A new specialty came into being, a new task for the safety professional, a change in the nature of the safety function operating in the workplace but reaching out to protect the public. An aspect of the organized safety movement had been "reorganized" so that it became desirable (profitable) to devote time, money, and effort to prevent a category of loss it had previously been profitable to neglect. Accountability had been fixed near the point of control. Safety should not be merely a technology. It must also be a study of how to implement safety technology in order to contain the nation's loss exposures.

3 *Labor unions became a force for workplace safety and health.* Labor had a dominant influence on the passage of the Occupational Safety and Health Act of 1970, an act which also created NIOSH. The act also triggered a reappraisal of the worker's compensation insurance system and the safety services of worker's compensation insurance carriers. Organized safety, the question of who shall put time, money, and effort into preventing, would never be the same.

The adversary relationship of unions and management operates effectively to influence legislation. It operates less effectively for safety in the workplace. It was management which was expected to put time, money, and effort into prevention, and management wanted control of how and how much.

Discipline and work rules sometimes embroiled a worker, and the union could hardly do other than support their person right or wrong. Safety issues could be used as a ploy to bargain for other benefits. Although some companies got cooperation in their efforts to prevent accidents, others felt that the unions were obstacles to safety. Difference in leadership was a factor, but not the key factor.

The key factor concerned "safety," a word which has come to mean accident prevention, in contrast to "health" which is applied to control of occupational disease exposures. The unions could hardly risk sharing responsibility for safety. To share responsibility was tantamount to an endorsement of the accident prevention program, and everyone knew there would still be accidents and injuries, that death would sometime strike, and that blame would readily be fixed with hindsight clarity. The key factor was that unions could not risk a shared responsibility for safety. They could not join with management in putting time, money, and effort into a shared responsibility and a shared blame. But they can share responsibility for health, and with the shift of emphasis to occupational disease they are doing so. Disease does not strike in an instant of time like an accident. Disease exposures are not dependent for control upon the worker's behavior, the simplistic unsafe act. Accidents result from a hazard in combination with other events, often improbable combinations of events. In contrast, pathogenic exposures can be measured and documented. Their results downstream in time can be predicted with no blame to workers, and control can be engineered at the expense of management.

In this context of events, unions can seek a shared responsibility for health in the workplace. The Oil, Chemical, and Atomic Workers, engulfed in the flood of new and untested substances in the workplace, raised an early cry of alarm. There were horrors—kepone, vinyl chloride, PBC, and others—which will become as archaic as "phossy jaw." Some management tended to respond with medical surveillance, assuring workers that they were being cared for, but denying them access to their own medical files. Bits of the fearful evidence were possessed by company doctors, government officials, industrial hygienists, labor leaders, and safety professionals. Effective response was slow and late, and only after horrors. In 1978, the AFL-CIO founded the Worker's Occupational Health and Safety Institute. It was billed as the "worker's equalizer." That "equalizer," it should be noted, was also slow and late.

The study of organized safety is a study of institutions, not just a study of safety technology. Organized safety tries to create reasons for an institution to engage in prevention, or it creates new institutions. All institutions operate under constraints, be it an insurance institution, a safety council or society, a government safety function, a standards or testing organization, or a labor union. The institutions seek to contain the nation's loss exposures, within the necessary constraints of serving themselves. The amazing thing is that no institution can commit itself solely to the welfare of the worker in the workplace, not even the unions.

Nonetheless, the reorganization of safety in the 1970s brought unions into play as an effective tool to implement safety and health technology in the workplace. It is likely to grow in effectiveness on the health aspects, more so than on safety. It should have a synergistic relationship with management, maintaining the adversary relationship which seems to be inevitable, but not necessarily antagonistic.

4 *The fourth aspect is the massive involvement of universities in the safety profession.* A chronology of safety may show the founding of Northwestern University Traffic Institute in 1936 and the founding of the Center for Safety Education at New York University in 1938. By the late 1940s, most major campuses had research or teaching activity related to safety and health—driver education, something in the engineering curriculum or in public health, or continuing education short courses and conferences. The American Society of Safety Engineers began the periodic reporting of progress in safety and health activities in the nation's universities. Steady growth and progress took place until the coming of OSHA in 1970; then safety became big news on and off campus.

The news on campus centered more and more on the involvement of NIOSH. NIOSH undertook a series of studies to determine what the deficiencies of safety and health were in the workplace. It looked at the overlooked evidence of occupational disease. It studied the activities of safety practitioners and outlined the courses and education needed, and developed training programs and materials. In the mid-1970s, NIOSH issued a little booklet entitled *Under-Graduate and Graduate Degree Programs in Occupational Safety and Health.* It gave information from nearly 80 universities offering degree programs in five major categories: occupational safety and health, occupational safety, industrial hygiene, occupational nursing, and occupational medicine. A companion bulletin showed 32 junior colleges, technical institutes, and universities offering associate degree programs in occupational safety and health. The expansion had been phenomenal. Users could also consult a NIOSH *Directory of Training and Education Resources in Occupational Safety and Health* for short courses, workshops, meetings, conferences, audiovisuals, and publications, as well as academic programs and courses.

Meanwhile, NIOSH set out to measure the nation's shortages for trained professionals in health and safety. Safety and health programs in the workplace suffered for lack of industrial physicians and nurses, for lack of safety engineers and industrial hygienists, and for lack of trained compliance officers for OSHA. It undertook to correct the deficiencies by the formation of Educational Resource Centers for Occupational Safety and

Health at selected universities. The first nine were organized in late 1977 and in 1978. For the first time, the resources of great universities were to be effectively harnessed in interdisciplinary programs of education. For the first time, a massive base for professionalism came into being.

NIOSH was a product of legislation. Indeed legislation was a necessary ingredient in the four great aspects described above. An overview of the role of safety legislation seems pertinent at this point.

LEGISLATION

A history of safety could record hundreds of acts of legislation as society sought to curtail the wave of accidents, each so simple to prevent by hindsight and often so unlikely or unpredictable by foresight. Some attacked definite physical and mechanical causes, or unsafe building or operating conditions, comparable to standards imposed by OSHA in 1970. Such laws have always been unpopular and difficult to enforce, and often irrelevant to perceived problems. No less so under OSHA. In fact, OSHA appears to be backing away from prevention of accidents in favor of prevention of occupational disease, for disease exposures can be successfully controlled by adherence to standards. The cost-effectiveness of such laws must be in doubt, but such legislation must be an experiment that a society must inevitably take.

It should be a high responsibility of the safety profession to assess such laws. In England the accretion of a century and a half of factory laws became a jungle which demeaned the safety profession and created a climate which caused accidents, until a thoroughgoing revision took place soon after 1970.

Other laws try to allocate the cost of accidents. Early employer liability laws sought to offer recourse to the injured worker, but the common law defenses retained by employers reduced their effect to futility. Not until worker's compensation laws could the worker hope to recover. Worker's compensation laws fixed the costs of accidents inevitably on the employer, thus producing a tremendous interest in prevention. Worker's compensation laws became a major foundation stone in the organized safety movement. It will remain so to the extent that it successfully allocates costs to the person or institution which can implement preventive measures. Its success, as will be discussed later, is not unflawed.

The term "employer" focuses accident prevention on the workplace. Actually, legislation extends to accidents anywhere they occur—at the lake, at home, on the highway, or wherever. Beyond accidents, there is also occupational disease, or stated in the positive, occupational health. Safety and health express the aspirations of society, the aspirations contained in the revolution of "consumerism" that swept the nation beginning with the early 1960s. Ralph Nader became the leader, the symbol, and the apostle for all those who felt that our society produced too many victims who had too little recourse in our laws. A wave of safety legislation ensued.

A partial list of legislation dealing with safety issues might include the Environmental Protection Act of 1969, the Transportation Safety Act of 1975, the Noise Control Act of 1972, the National Traffic and Motor Vehicle Safety Act of 1966, the Coal Mine Health and Safety Act of 1969, the Boat Safety Act of 1971, and the Consumer Product Safety Act of 1972. To safety professionals, the most notable and far reaching was the Occupational Safety and Health Act of 1970.

In terms of "organizing" for safety, changes took place in the patterns by which preventive action was brought to bear. The OSHA compliance officer changed the priority of attention given to safety in the workplace, although as the 1970s came to a close, the success of OSHA in preventing accidents was still being evaluated. The over-looked evidence of occupational disease came under the revealing eye of NIOSH. Safety services provided by insurance carriers came under scrutiny by the National Commission on Workmen's Compensation Insurance. Labor unions taught hazard recognition to their members. Standard-setting organizations reevaluated their work and their role in relation to the standard-setting authority of the federal government. Safety education in universities expanded. Businesspeople appointed people to safety positions as never before. The impetus for safety came from many new sources and directions, with time, money, and effort devoted to prevention, and from old sources with renewed vigor.

Massive intervention by government created a massive ferment in safety and health. The process left much to be criticized, and criticized it was. Nonetheless, behind it all lay a central truth. The intervention of government, legislative and judicial, must be an inherent part of organizing for safety. Not that government need "take over," but government creates the social mechanisms by which safety technology is implemented. It fixes accountability or allocates costs so that it becomes desirable or profitable for someone to devote time, money, and effort to prevent loss. Government intervention never commanded insurance companies to prevent accidents in the workplace. It simply created worker's compensation; it thereupon became profitable for insurance carriers to engage in loss prevention. In the absence of such mechanisms from which an impetus for safety can stem, safety technology is not applied. New disease hazards had been overlooked because they were not compensable. Defective autos were overlooked (until Nader), and driver training (which could be marketed) was the panacea for highway safety. Products were tested in the marketplace, and product safety was merely moralistic talk, because the victim had scant recourse. Organizing for safety means organizing someone, some institution, some organization to whom it is profitable or desirable to prevent loss. Ideally, for every hazard someone needs to control it for his own self-interest.

It is probable that every institution of society, and every discipline of learning, bears on safety problems in some way. Uncontrolled exposures fall between the cracks in their interaction. Every institution, every safety organization and professional society, could be analyzed to understand its role and its inherent constraints and limitations. An understanding of inherent constraints and limitations opens the opportunity for safety statesmanship, the opportunity to close the cracks. As an example, the insurance industry has been another massive institutional influence in organized safety.

INSURANCE

By the mid-1920s a new force was affecting the accident situation. Early employer liability laws had created a need for employers to purchase insurance to protect themselves against legal action brought by injured employees. Worker's compensation laws made such insurance protection imperative, even mandatory under the law. Insurance companies found it necessary to send inspectors to visit their new policyholders primarily to classify risks according to hazards so that proper rates might be charged. In fire and property coverages this process had long been going on.

As the inspectors became more competent in judging risks, their accumulating experience presented another role, to serve in an advisory capacity concerning the reduction of accident hazards. Thus, safety services were inaugurated by insurance carriers, and employers gradually became more aware of accident prevention as a means of reducing operating costs. For a long time, the main burden of accident prevention was shouldered by insurance companies, which, while protecting their own interests, were also rendering a valuable service to industry and to society.

For 50 or 60 years throughout the midcentury, the dominant and most visible impetus for safety stemmed from insurance companies.

This impetus had two forms. In one form, the insurance company protected its own interest by what might be called the underwriting mechanism, the business of assessing hazard and loss potential and pricing accordingly. In the other form, it was delivering a safety service to its customers or policyholders.

The first form, the underwriting mechanism, is almost invisible, but yet is a pervasive impetus for safety. Its effect is to make it profitable (at least to some extent) for the insurance purchaser to reduce hazards and thus reduce the insurance "rate." Its effect is most visible on fire and property hazards which are largely physical in nature, and hence can be priced and compared to reduced insurance costs. Planning for fire protection offers numerous ways to reduce hazard and to reduce insurance cost. This is the underwriting mechanism, pricing according to hazard and rewarding for reduction of hazard.

The underwriting mechanism demands a competent safety professional to evaluate exposures and to help the underwriter price insurance coverage correctly. Where it is shrewdly done it serves the whole of society. An early application of the principle to worker's compensation insurance took the form of a merit rating scheme that gave credit for superior plant physical condition and other factors. However, it proved less effective in controlling worker's compensation insurance costs because injury prevention is much less precise and predictable than fire prevention. Nonetheless, scheduled rating, as it came to be called, may have been an effective incentive to improve plant conditions in the early days of safety. As time went on, however, generous scheduled credits often went hand in hand with increasing frequency and cost of workplace injuries. As a result, scheduled rating dropped out of the worker's compensation insurance system and ended in its last holdout in California soon after 1970.

Scheduled rating is more than a historical oddity. It represents a self-serving competitive tinkering with the underwriting mechanism in a way which, nonetheless, broadly serves society's aspirations for safety. There are numerous pricing schemes in differing lines of coverage designed to reward success and penalize failure in preventing losses. The details are available to students of insurance. The point to be made here is that the underwriting mechanism is an important, though nearly invisible, aspect of organized safety, very much alive. It is a pervasive influence in determining who shall devote time, money, and effort to accident prevention, and how much. Its effect depends on success in allocating the costs of accidents back to the point of effective control, with sufficient rewards and penalties to make a difference.

Safety service to customers evolved out of the underwriting

mechanism. Insurance safety service became very visible, a competitive tool, the most frequent theme in insurance advertising. It became profitable for insurance companies to promote safety services to their customers. Many divorced their safety department from the underwriting department, signalizing their commitment to the goals of customer service while hazard recognition to guide the underwriter was reduced to a communications link. The impetus for safety stemming from insurance implied that workplace safety had been organized on a path of steady improvement, and steady improvement took place for about four decades before it faltered. By 1960, an identifiable condition labeled "delayed progress" had set in.

"Delayed progress" continued for 15 years before reorganization took place under the impact of legislation. During those years, and for a generation preceding them, the ubiquitous insurance safety engineer (inspector, consultant) was a major presence in the safety profession. The effectiveness of the presence had always been a lively topic of discussion. It culminated in OSHA, a proclamation that "delayed progress" could no longer be tolerated. Texas and Oregon passed laws attempting to demand effective safety services from insurers. The point is not to evaluate the effectiveness of insurance safety services. The point, more objectively, is to evaluate the mechanism.

Objectively, the mechanism produced more thousands of people working for safety than any other source. With few exceptions, there was no other way to get started in safety. An exception was New York University's Center For Safety Education, founded and funded by insurance interests. When a business decided it wanted a "safety expert," it tended to hire one from an insurance company, often the person who had been serving the business. Safety people took their insurance safety training into government jobs, or into teaching, or into the expanding safety associations, such as the National Safety Council. Until the "education revolution" of the 1970s the great majority of the nation's safety professionals got their start in an insurance company. As part of the organized safety movement, insurance companies more than anyone else trained people for safety work, and contributed their people to the whole of society as they were hired away.

One need not attribute this process to altruism nor to wisdom. It was a quite comfortable business arrangement. In the same comfortable fashion, insurance companies contributed leadership and support to just about any safety venture that came along. It was good public relations for their safety people to be visible in safety activities. After all, safety had become their most frequent advertising ploy. They could afford the time and the travel expense as a comfortable business arrangement, whereas the professional colleague in business often found the expense less comfortable to justify.

Needless to say, insurance had no monopoly. In all the web of organized safety, great leadership also came from business, government, the National Safety Council, associations, and professional societies, especially engineering societies. Nonetheless, insurance time, money, and effort infused them all. The prevention of loss, if successful, served underwriting profits, while the visibility of safety service served broad company goals whether or not prevention was successful. Objectively, it can be seen as an important aspect of organized safety. It seems reasonable to assume that it was also infused with its share of vision and wisdom.

The Insurance Institute for Highway Safety may be an example of vision. The escalating deaths, injuries, and costs of highway crashes were obviously beyond any service that could be rendered by the insurance industry. In fact, the whole problem, though subject to ameliorative countermeasures, was basically not readily subject to control. In this climate, many insurance companies sponsored research. Piecemeal, much money was spent on bits and dabs of the problem according to the vision and opinions of different executives. A pooling of resources was needed, a vision beyond immediate competitive advantage. The Insurance Institute for Highway Safety, founded in 1959, was the result of that vision.

Though the insurance presence in organized safety was infused with its share of vision, it also suffered from institutional blindness. It was a business arangement, and what didn't serve the business didn't get done. All institutions survive by functioning within their nature; all are limited. Without finger pointing, the notable achievements and the notable oversights of organized safety can be understood, and perhaps corrected, by considering the constraints inherent in the system. With considerable objectivity, the constraints on the insurance role in organized safety might be enumerated.

1 *If it wasn't compensable, insurers ignored it.* "It" could be anything that had not yet been written into worker's compensation laws. This is quite natural and not a moral fault. Nonetheless, insurers share with others the failure to see the mounting evidence of catastrophe in the flood of new toxic substances that flowed untested into the workplace. The blindness lay in the sweeping advertising, and probably the unconscious belief, that worker's compensation carriers were watching out for safety in the workplace. They weren't; they were controlling their own losses, not the losses suffered by workers, especially losses by noncompensable injuries and disease.

2 *Insurers saw the absence of product liability rather than the absence of product safety.* In this they were no worse and no better than the rest of organized safety. The rebellion against unsafe products took place outside the professional ranks of organized safety.

3 *Insurers were limited by a conflict of interest.* Vigorous leadership to update compensation laws, or to extend coverage to more workers or to new diseases, obviously served the insurers at the expense of their customers. Worker's compensation is designed to protect the employer, not the employee. In the years of "delayed progress," it can be said that the insurance safety role, the safety profession itself, and indeed the whole of the organized safety movement were working for the employer not the employee.

4 *The underwriting mechanism put the problem of "industrial back" beyond the reach of effective study.* The upright posture of human beings puts "backache" beyond the reach of simplistic causes and cures. But when backache could be attributed to bending, lifting, reaching, pushing, or pulling at work, it became compensable. It became part of safety engineering, which had only simplistic causes and cures for this complex medical problem. Proper lifting and excessive use of x-rays became the standard safety preachment, along with modifications to ease the workload. The underwriter paid the predictable flow of claims, was daunted occasionally by a jumbo back case, but included it all in the rate, neither better nor worse off in the long run no matter what happened. The whole complex problem became entangled in a system in which it paid no one to seek fundamental answers.

5 *The worker's compensation system failed to motivate smaller business to prevent employee injuries.* Before the passage of worker's compensation laws, employers could escape liability. When liability for workplace injuries became inescapably fixed, it was promptly buffered by worker's compensation insurance. The system sustained the injured worker and assured her compensation and medical care. The possibility of catastrophe for the employer was smoothed into a predictable expense for insurance. It was a better system, but it failed to organize effective implementation of accident prevention in smaller businesses. Experience was to show that as an enterprise grew and increased in number of employees, its accident record got worse before it got better.

6 *Insurers lavished safety service on large accounts which needed it least.* Retaining a large account was big business, and large accounts demanded every service they wanted. They had, however, expert staff of their own and were fully capable of defining their own needs and seeking whatever expertise they needed. Insurers, nonetheless, competed with service to these large accounts. All this is natural and inevitable except that the organization of safety worked in reverse. Those who needed it least got the most. Those who needed it most got less and less as they diminished in size, and the amount of the premium got less and less. Those of smallest size got little or no effective safety assistance or incentive from insurers.

7 *Insurance safety service stretched over many goals other than accident prevention.* It served goals of acquisition and retention of desired accounts; it served goals of image and public relations. All this was quite proper and good business, but it wasn't accident prevention. Of all the safety resources and manpower available to the nation, the greater part worked for insurance companies. It was all called "safety," but the greater part of the time, money, and effort served purposes other than safety.

8 *Loss control became synonomous with accident prevention.* Loss for the insurance company, for example, can be controlled by simply increasing the premium. The difference is not a mere play on words. Loss control focused on loss ratios and underwriting data to monitor how the insurance company was making out. Loss control monitors the flow of money. It measures nothing about the success or failure of the services offered to customers. Accident prevention, in contrast, monitors whether those services actually reduce the frequency and severity of accidents and injuries. Insurers never defined accident prevention to be their role, and hence never developed measures of their effectiveness in society. Except for anecdotal evidence, they are hard put to explain themselves or to justify their absorption of the major share of the nation's safety workforce and resources.

These observations are not meant to denigrate insurance companies or their safety services. They are meant to show the constraints within which insurance companies make their contribution to organized safety. All the elements in the organized safety movement are limited by constraints of their own. The National Safety Council, for example, survives by serving its members and by developing for them what they will buy. It can hardly do otherwise. Although the insurance mechanism has been dissected here, the constraints of any institutional input can be similarly dissected. Organized safety requires safety statesmanship, a task beyond faultfinding. Its elements are institutions. Their interaction determines what safety technology will be implemented. Their interaction is the key to understanding why we know how to prevent much more than we succeed in preventing.

The present mingles with the past and the future. Awesome technological change is with us, behind us, and in the future. The safety profession in all its variety has notably served society, even while major categories of loss continue to fall between the cracks. The American Society of Safety Engineers continues two decades of work to define the scope and functions of the safety professional, and to join with universities in establishing an educational foundation for professionalism in safety. The effort will surely continue to the end of the century.

Some of the ferment of change in safety can be captured in the "CSP" designation. Initiative from the American Society of Safety Engineers led to the establishment in 1969 of a Board of Certified Safety Professionals. Their mission was to certify safety professionals and to escalate the level of competence. It began with the ASSE as the sole professional society on the board. As safety sophistication increased, the Systems Safety Society (founded in 1964) became a member of the board, as did also the Human Factors Society (founded in 1957). By the mid-1970s, the CSP designation had become a notable force in escalating the standards of professionalism in safety, with an escalating future of its own.

The impelling growth of the National Safety Management Society (founded in 1968) dramatized the new demands being placed on the safety professional. Ideas, theories, and special groups rose to meet the demands. Suddenly, the year 2001 was quite close, with the hope of safety more deeply rooted in its technology, with implementation guided by assured techniques of management, and with the institutions of society more wisely structured by safety statesmanship.

APPENDIX TWO

Major Hazards and OSHA Checklist

The following checklist covers approximately 90 percent of OSHA's general industry standards, and is a good checklist in general of hazards. It is alphabetized by subject. The checklist was originally prepared for the September 1974 issue of *Job Safety and Health*, a publication of OSHA, and is now available in an OSHA pamphlet. It is not intended to be a substitute for either the standards published in the *Federal Register* or the more comprehensive hazard checklists that a company might prepare for its own internal uses.

1 ABRASIVE BLASTING

  a Blast cleaning nozzles shall be equipped with an operating valve which must be held open manually (deadman control). A support shall be provided on which the nozzle may be mounted when not in use.

  b The concentration of respirable dust or fumes in the breathing zone of the abrasive-blasting operator or any other worker shall be below the levels specified in 1910.93.

  c Blast-cleaning enclosures shall be exhaust ventilated in such a way that a continuous inward flow of air will be maintained at all openings in the enclosure during the blasting operation.

d The air for abrasive-blasting respirators shall be free of harmful quantities of contaminants.

2 ABRASIVE GRINDING

a Abrasive wheels shall be used only on machines provided with safety guards, with the following exceptions:

- Wheels used for internal work while within the work being ground;
- Mounted wheels, used in portable operations, two inches and smaller in diameter; and
- Type 16, 17, 18, 18R, and 19 cones, plugs, and threaded hole pot balls where the work offers protection.

b All abrasive wheel bench and stand grinders shall be provided with safety guards which cover the spindle ends, nut, and flange, except:

- Safety guards on all operations where the work provides a suitable measure of protection to the operator may be so constructed that the spindle end, nut, and outer flange are exposed;
- Where the nature of the work is such as to entirely cover the side of the wheel, the side covers of the guard may be omitted; and
- The spindle end, nut, and outer flange may be exposed on machines designed as portable saws.

c An adjustable work rest of rigid construction shall be used to support the work on fixed base, offhand grinding machines. Work rests shall be kept adjusted closely to the wheel with a maximum opening of ⅛ inch. The work rest shall be securely clamped after each adjustment. The adjustment shall not be made with the wheel in motion.

d Every establishment performing dry grinding shall provide suitable hood or enclosures that are connected to exhaust systems to control airborne contaminants.

e Machines designed for a fixed location shall be securely anchored to prevent walking or moving.

3 ACCIDENT RECORDKEEPING REQUIREMENTS

a Within 48 hours after its occurrence, an employment accident which is fatal to one or more employees or which results in the hospitalization of five or more employees shall be reported by the employer, either orally or in writing, to the nearest OSHA Area Director.

b Records as prescribed in the Recordkeeping Requirements booklet shall be kept for all accidents that result in a fatality, hospitalization, lost workdays, medical treatment, job transfer or termination, or loss of consciousness.

4 AIR RECEIVERS, COMPRESSED

a Air receivers should be supported with sufficient clearance to permit a complete external inspection and to avoid corrosion of external surfaces.

b Air receivers shall be installed so that drains, handholes, and manholes are easily accessible.

c Every air receiver shall be equipped with an indicating pressure gauge so located as to be readily visible, and with one or more spring loaded safety valves.

5 AIR TOOLS

a For portable tools, a tool retainer shall be installed on each piece of utilization equipment, which, without such a retainer, may eject the tool.

b Hose and hose connections used for conducting compressed air to utilization equipment shall be designed for the pressure and service to which they are subjected.

6 AISLES AND PASSAGEWAYS

a Where mechanical handling equipment is used, sufficient safe clearance shall be allowed for aisles at loading docks, through doorways, and whenever turns or passage must be made.

b Aisles and passageways shall be kept clear and in good repair with no obstructions across or in aisles that could create hazards.

c Permanent aisles and passageways shall be appropriately marked.

7 BELT SANDING MACHINES (WOODWORKING)

a Belt sanding machines shall be provided with guards at each nip point where the sanding belt runs onto a pulley.

b The unused run of the sanding belt shall be guarded against accidental contact.

8 BOILERS

Boilers are not covered by present OSHA standards. These are good practice procedures and might be incorporated into future OSHA standards.

a Boiler inspection and approval, on an annual basis, by a recognized boiler inspection service is satisfactory evidence of acceptable installation and maintenance.

b A valid boiler inspection certificate, bearing the signature of the authorized inspector and the date of the last inspection, shall be conspicuously posted.

c All boilers shall be equipped with an approved means of determining the water level such as water column, gauge glass, or try cocks. Gauge glasses and water columns shall be guarded to prevent breakage.

9 CALENDARS, MILLS, AND ROLLS

a A safety trip-type bar, rod, or cable to activate an emergency stop switch shall be installed on calendars, rolls, or mills to prevent persons or parts of the body from being caught between the rolls.

b A fixed guard across the front and one across the back of the mill, approximately 40 inches vertically above the working level and 20 inches horizontally from the crown face of the roll, should be used where applicable.

10 CHAINS, CABLES, ROPES, ETC. (OVERHEAD AND GANTRY CRANES)

a Chains, cables, ropes, slings, etc., shall be inspected daily, and defective gear shall be removed and repaired or replaced.

b Hoist chains and hoist ropes shall be free from kinks or twists and shall not be wrapped around the load.

c All U-bolt wire rope clips or hoist ropes shall be installed so that the U-bolt is in contact with the dead end (short or nonload carrying end) of the rope. Clips shall be installed in accordance with the clip manufacturer's recommendation. All nuts or newly installed clips shall be retightened after one hour of use.

11 CHIP GUARDS

Protective shields and barriers shall be provided, in operations involving cleaning with compressed air, to protect personnel against flying chips or other such hazards.

12 CHLORINATED HYDROCARBONS

a Carbon tetrachloride or other chlorinated (halogenated) hydrocarbons shall not be used where the airborne concentration exceeds the Threshold Limit Value (TLV) listed.

b Degreasing or other cleaning operations involving chlorinated hydrocarbons shall be so located that vapors from these operations will not reach or be drawn into the atmosphere surrounding any welding operations.

13 COMPRESSED AIR, USE OF

Compressed air used for cleaning purposes shall not exceed 30 psi and then only with effective chip guarding and personal protective equipment.

14 CONE PULLEYS (MECHANICAL POWER TRANSMISSION EQUIPMENT)

The cone belt and pulleys shall be equipped with a belt shifter so constructed as to adequately guard the nip point of the belt and pulley. If the frame of the belt shifter does not adequately guard the nip point of the belt and pulley, the nip point shall be further protected by means of a vertical guard placed in front of the pulley and extending at least to the top of the largest step of the cone.

15 CONVEYORS

a Conveyors installed within seven feet of the floor or walkway shall be provided with cross-overs at aisles or other passageways.
b Where conveyors seven feet or more above the floor pass over working areas, aisles, or thoroughfares, suitable guards shall be provided to protect personnel from the hazard of falling materials.
c Open hoppers and chutes shall be guarded by standard railings and toeboards or by some other comparable safety device.

16 CRANES AND HOISTS (OVERHEAD AND GANTRY)

a All functional operating mechanisms, air and hydraulic systems, chains, rope slings, hooks, and other lifting equipment shall be inspected daily.
b Complete inspection of the crane shall be performed at intervals depending on its activity, severity of service, and environment.
c An overhead crane shall have stops at the limit of travel of the trolley, bridge and trolley bumpers or equivalent automatic services, and rail sweeps on the bridge trucks.
d The rated load of the crane shall be plainly marked on each side of the crane, and if the crane has more than one hoisting unit, each hoist shall have its rated load marked on it or its load block, and this marking shall be clearly legible from the ground or floor.

17 CYLINDERS, COMPRESSED GAS, USED IN WELDING

a Compressed gas cylinders shall be kept away from excessive heat, shall not be stored where they might be damaged or knocked over by passing or falling objects, and shall be stored at least 20 feet away from highly combustible materials.

b Where a cylinder is designed to accept a valve protection cap, caps shall be in place except when the cylinder is in use or is connected for use.

c Acetylene cylinders shall be stored in a vertical, valve-end-up position only.

d Oxygen cylinders, in storage shall be separated from fuel-gas cylinders or combustible materials (especially oil or grease) a minimum distance of 20 feet or by a non-combustible barrier at least five feet high having a fire-resistance rating of at least ½ hour.

18 DIP TANKS CONTAINING FLAMMABLE OR COMBUSTIBLE LIQUID

a Dip tanks of over 150 gallons capacity, or 10 square feet in liquid surface area, shall be equipped with a properly trapped overflow pipe leading to a safe location outside the buildings.

b There shall be no open flames, spark producing devices, or heated surfaces having a temperature sufficient to ignite vapors in or within 20 feet of any vapor area. Electrical wiring and equipment in any vapor area shall be of the explosion-proof type. There shall be no electrical equipment in the vicinity of dip tanks, associated drain boards, or drying operations which are subject to splashing or dripping.

c All dip tanks, except hardening and tempering tanks, exceeding 150 gallons liquid capacity or having a liquid surface area exceeding four square feet shall be protected with at least one of the following automatic extinguishing facilities: water spray system, foam system, carbon dioxide system, dry chemical system, or automatic dip tank cover. This provision shall apply to hardening and tempering tanks having a liquid surface area of 25 square feet or more or a capacity of 500 gallons or more.

19 DOCKBOARDS

a Dockboards shall be strong enough to carry the load imposed on them.

b Portable dockboards shall be anchored or equipped with devices which will prevent their slipping. They shall have handholds or other effective means to allow safe handling.

c Positive means shall be provided to prevent railroad cars from being moved while dockboards are in position.

20 DRAINS FOR FLAMMABLE AND COMBUSTIBLE LIQUIDS

a Emergency drainage systems shall be provided to direct flammable liquid leakage and fire protection water to a safe location.

b Emergency drainage systems for flammable liquids, if connected to public sewers or discharged into public waterways, shall be equipped with traps or separators.

21 DRILL PRESSES

The V-belt drive of all drill presses, including the usual front and rear pulleys, shall be guarded to protect the operator from contact or breakage.

22 DRINKING WATER

a Potable water shall be provided in all places of employment.
b The nozzle of a drinking fountain shall be set at such an angle that the jet of water will not splash back down on the nozzle, and the end of the nozzle shall be protected by a guard to prevent a person's mouth or nose from coming in contact with the nozzle.
c Portable drinking water dispensers shall be designed and serviced to ensure sanitary conditions, shall be capable of being closed, and shall have a tap. Unused disposable cups shall be kept in a sanitary container, and a receptacle shall be provided for used cups. The common drinking cup is prohibited.

23 ELEVATOR

Elevators are not covered by present OSHA standards. These are good practice procedures and might be incorporated into future OSHA standards.
a All elevators shall be inspected annually by a competent inspection service or inspector.
b All hoistway openings shall be protected by doors or gates that are interlocked with the controls, so that the car cannot be started until all gates or doors are closed, and so that gates or doors cannot be opened when the car is not at the landing.

24 ELECTRICAL INSTALLATIONS

Every new electrical installation and all new utilization equipment installed after March 15, 1972, and every replacement, modification, repair, or rehabilitation after March 15, 1972, of any part of any electrical installation or utilization equipment installed before March 15, 1972, shall be installed or made and maintained in accordance with the provisions of the 1971 National Electrical Code, NFPA 70—1971; ANSI C1—1971 (Rev. of 1968).

25 EMERGENCY FLUSHING, EYES AND BODY

Where the eyes or body of any person may be exposed to injurious corrosive materials, suitable facilities for quick drenching or flushing of the eyes and body shall be provided within the work area for immediate emergency use.

26 EXITS

a Every building designed for human occupancy shall be provided with exits sufficient to permit the prompt escape of occupants in case of emergency.

b Where occupants may be endangered by the blocking of any single egress due to fire or smoke, there shall be at least two means of egress remote from each other.

c Exits and the way of approach and travel from exits shall be maintained so that they are unobstructed and are accessible at all times.

d All exits shall discharge directly to the street or other open space that gives safe access to a public way.

e Exit doors serving more than 50 people, or at high hazard areas, shall swing in the direction of travel.

f Exits shall be marked by readily visible, illuminated exit signs. Exit signs shall be distinctive in color and provide contrast with surroundings. The word "EXIT" shall be of plainly legible letters, not less than six inches high.

27 EXPLOSIVES AND BLASTING AGENTS

a All explosives shall be kept in approved magazines.

b Stored packages of explosives shall be laid flat with top side up. Black powder, when stored in magazines with other explosives, shall be stored separately.

c Smoking, matches, open flames, spark-producing devices, and firearms (except firearms carried by guards) shall not be permitted inside of or within 50 feet of magazines. The land surrounding a magazine shall be kept clear of all combustible materials for a distance of at least 25 feet. Combustible materials shall not be stored within 50 feet of magazines.

28 EYE AND FACE PROTECTION

a Protective eye and face equipment shall be required where there is a reasonable probability of injury that can be prevented by such equipment.

b Eye and face protection equipment shall be in compliance with ANSI Z87.1—1968, Practice for Occupational and Educational Eye and Face Protection.

29 FAN BLADES

When the periphery of the blades of a fan is less than seven feet above the floor or working level, the blades shall be guarded. The guard shall have openings no longer than ½ inch. The use of concentric rings with space between them, not exceeding ½ inch, is acceptable, provided they are adequately supported.

30 FIRE DOORS

Fire doors are not completely covered by present OSHA standards. These are good practice procedures and might be incorporated into future OSHA standards.

a Fire doors shall not be blocked or tied in an open position.

b Fusible links for fire doors shall be located so as to properly function in case of fire and shall not be painted.

c Three three-inch diameter vent holes, cut through the metal only, are required for tin-clad fire doors up to nine feet in height. (An additional vent hole is required for door from nine feet to 12 feet four inches in height.) The metal covering around the opening shall be secured with small nails, and the exposed wood thoroughly painted.

d A closing device shall be installed on every fire door except elevator and power-operated dumbwaiter doors equipped with electric contacts or interlocks.

31 FIRE PROTECTION

a Portable fire extinguishers suitable to the conditions and hazards involved shall be provided and maintained in an effective operating condition.

b Portable fire extinguishers shall be conspicuously located and mounted where they will be readily accessible. Extinguishers shall not be obstructed or obscured from view.

c Portable fire extinguishers shall be given maintenance service at least once a year with a durable tag securely attached to show the maintenance or recharge date.

d In storage areas, clearance between sprinkler system defectors and top of storage varies with the type of storage. For combustible material stored over 15 feet but not more than 21 feet high in solid piles, or over 12 feet but not more than 21 feet high in piles that contain horizontal channels, the minimum clearance shall be 36 inches. The minimum clearance for smaller piles or for noncombustible materials shall be 18 inches.

32 FLAMMABLE LIQUIDS INCIDENTAL TO PRINCIPAL BUSINESS

a Flammable liquids shall be kept in covered containers when not actually in use.

b The quantity of flammable or combustible liquid that may be located outside of an inside storage room or storage cabinet in any one fire area of a building shall not exceed:

- 25 gallons of Class IA liquids in containers;
- 120 gallons of Class IB, IC, II, or III liquids in containers; or
- 660 gallons of Class IB, IC, II, or III in a single portable tank.

c Flammable and combustible liquids shall be drawn from or transferred into containers within a building only through a closed piping system, from safety cans, by means of a device drawing through the top, or by gravity through an approved self-closing valve. Transferring by means of air pressure shall be prohibited.

d Inside storage rooms for flammable and combustible liquids shall be of fire resistive construction, have self-closing fire doors at all openings, four-inch sills or depressed floors, a ventilation system that provides at least six air changes within the room per hour, and in areas used for storage of Class I liquids, electrical wiring approved for use in hazardous locations.

e Outside storage areas shall be graded in such a manner to divert spills away from buildings or other exposures, or be surrounded with curbs or dikes at least six inches high with appropriate drainage to a safe location for accumulated liquids. The area shall be protected against tampering or trespassing, where necessary, and shall be kept free of weeds, debris, and other combustible material not necessary to the storage.

f Areas where flammable liquids with flashpoints below 100 degrees F are used shall be ventilated at a rate of not less than one cubic-foot-per-minute per square foot of solid floor area.

33 FLOORS, GENERAL CONDITIONS

a All floor surfaces shall be kept clean, dry, and free from protruding nails, splinters, loose boards, holes, or projections.

b Where wet processes are used, drainage shall be maintained, and false floors, platforms, mats, or other dry standing places should be provided where practicable.

34 FLOOR LOADING LIMIT

In buildings used for mercantile, business, industrial, or storage purposes, all floors shall be posted to show maximum safe floor loads.

35 FLOOR OPENINGS, HATCHWAYS, OPEN SIDES, ETC.

a Floor openings requiring access by personnel, such as stairway openings and ladderway openings, shall be guarded by a standard railing on all exposed sides except at the access point. Access to ladderway openings shall be further guarded so that a person cannot walk directly into the opening. Other floor openings shall be guarded by a suitable covering, and further guarded when the covering is removed by a removable standard railing, or shall be constantly attended by someone. Skylight openings shall be guarded by a standard skylight screen or fixed standard railing on all four sides.

b Open-sided floors, platforms, etc., four feet or more above the adjacent floor or ground level shall be guarded by a standard railing on all open sides, except where there is an entrance to a ramp, stairway, or fixed ladder.

36 FOOT PROTECTION

Safety-toe footwear shall meet the requirements of ANSI Z41.1—1967, Standard for Men's Safety-toe Footwear.

37 FORKLIFT TRUCKS

a All new forklift trucks acquired and used after February 15, 1972, shall comply with ANSI B56.1—1969, Power Industrial Trucks, Part II. Approved trucks shall bear a label indicating approval.

b High lift rider trucks shall be equipped with a substantial overhead guard unless operating conditions do not permit.

c Fork trucks shall be equipped with a vertical load backrest extension when the type of load presents a hazard to the operator.

d The brakes of highway trucks shall be set and wheel chocks placed under the rear wheels to prevent the trucks from rolling while they are boarded with forklift trucks.

e Wheel stops or other recognized protection shall be provided to prevent railroad cars from moving while they are boarded with forklift trucks.

38 GENERAL DUTY CLAUSE

Hazardous conditions or practices not covered in an OSHA standard may be covered under Section 5(a)(1) of the Act which states: "Each employer shall furnish to each of his employees employment and a place of employment which are free from recognized hazards that are causing or are likely to cause death or serious physical harm to his employees."

39 GUARDS, CONSTRUCTION OF

Guards for mechanical power transmission equipment shall be made of metal, except that wood guards may be used in the woodworking and chemical industries, in industries where atmospheric conditions would rapidly deteriorate metal guards, or where temperature extremes make metal guards undesirable.

40 HEAD PROTECTION

Head protective equipment shall meet the requirements of ANSI Z89.1—1967, Requirements for Industrial Head Protection.

41 HAND TOOLS

Each employer shall be responsible for the safe condition of tools and equipment used by employees, including tools and equipment which may be furnished by employees.

42 HOOKS, CRANES, AND HOISTS, ETC. (SEE CRANES AND HOISTS, NO. 16)

43 HOUSEKEEPING

All places of employment, passageways, storerooms, and service rooms shall be kept clean and orderly and in a sanitary condition.

44 JOINTERS (WOODWORKING)

a Each hand-fed planer and jointer with a horizontal head shall be equipped with a cylindrical cutting head. The opening in the table shall be kept as small as possible.

b Each hand-fed jointer with a horizontal cutting head shall have an automatic guard which will cover the section of the head on the working side of the fence or gauge.

c A jointer guard shall automatically adjust itself to cover the unused portion of the head and shall remain in contact with the material at all times.

d Each hand-fed jointer horizontal cutting head shall have a guard which will cover the section of the head back of the gauge or fence.

45 LADDERS, FIXED

a All fixed ladders shall be designed for a minimum concentrated live load of 200 pounds.

b All rungs shall have a minimum diameter of ¾ inch, if metal, or if the ladder is constructed of metal rungs embedded in concrete and exposed to a corrosive atmosphere, the rungs shall have a minimum diameter of one inch. Wooden ladders shall have rungs with a minimum diameter of 1⅛ inch. All rungs shall be spaced uniformly, not more than 12 inches apart, and shall have a minimum clear length of 16 inches.

c Metal ladders shall be painted or treated to resist corrosion or rusting when the location demands.

d Cages, wells, or ladder safety devices for ladders affixed to towers, watertanks, or chimneys shall be provided on all ladders more than 20 feet long. Landing platforms shall be provided each 30 feet of length, except where no cage is provided, landing platforms shall be provided for every 20 feet of length.

e Tops of cages on fixed ladders shall extend 42 feet above top of landing, unless other acceptable protection is provided, and the bottom of the cage shall be not less than seven feet nor more than eight feet above the base of the ladder.

f The side rails of through or side-step ladder extensions shall extend 3½ feet above parapets and landings. For through ladder extensions, the rungs shall be omitted from the extension and shall have not less than 18 nor more than 24 inches clearance between rails. For side-step or offset fixed ladder sections, at landings, the side rails and rungs shall be carried to the next regular rung beyond or above the 3½ feet minimum.

46 LADDERS, PORTABLE

a The maximum length for portable wood ladders shall be: step-ladders 20 feet; single straight ladders 30 feet; two section extension ladders 60 feet; sectional ladders—60 feet; trestle ladders—20 feet; platform step-ladders—20 feet; painter's step-ladders—12 feet; and mason's ladders—40 feet.

b   The maximum length for portable metal ladders shall be: single straight ladders—30 feet; two section extension ladders—48 feet; over two section extension ladders—60 feet; step-ladders—20 feet; trestle ladders—20 feet; and platform step-ladders—20 feet.

c   Step-ladders shall be equipped with a metal spreader or locking device of sufficient size and strength to securely hold the front and back sections in open position.

d   Ladders shall be maintained in good condition, and defective ladders shall be withdrawn from service.

e   Non-self-supporting ladders shall be erected on a sound base at a 4-1 pitch and placed to prevent slipping.

f   The top of a ladder used to gain access to a roof should extend at least three feet above the point of contact.

g   Wooden ladders should be kept coated with a suitable protective material.

h   In general industrial use, portable metal ladders may be used in areas containing electrical circuits, if proper safety measures are taken.

47   LIGHTING

Lighting is not covered by OSHA standards for general industry. These are good practice procedures and might be incorporated into future OSHA standards:

Adequate illumination, depending upon the seeing tasks involved, shall be provided and distributed to all areas in accordance with ANSI Standard A11.1. (Requirements vary widely, but a good rule-of-thumb is 20 to 30 foot candles for services, and 50 to 100 foot candles for tasks).

48   LUNCHROOMS

a   Employees shall not consume food or beverages in toilet rooms or in any area exposed to a toxic material.

b   Covered receptacles corrosion resistant to disposable material shall be provided in lunch areas for disposal of waste food. The cover may be omitted where sanitary conditions can be maintained without the use of a cover.

49   MACHINE GUARDING

One or more methods of machine guarding shall be provided to protect the operator and other employees in the machine area from hazards such as those created by point of operation, in-going nip points, rotating parts, and flying chips or sparks.

50 MACHINERY, FIXED

Machines designed for a fixed location shall be securely anchored to prevent walking or moving.

51 MATS, INSULATING

Where motors or controllers operating at more than 150 volts to ground are grounded against accidental contact only by location, and where adjustment or other attendance may be necessary during operations, suitable insulating mats or platforms shall be provided.

52 MEDICAL SERVICES AND FIRST AID

a The employer shall ensure the ready availability of medical personnel for advice and consultation on matters of plant health.

b When a medical facility for treatment of injured employees is not available in near proximity to the workplace, a person or persons shall be trained to render first aid.

c First aid supplies approved by the consulting physician shall be readily available.

53 NOISE EXPOSURE

a Protection against the effects of occupational noise exposure shall be provided when the sound levels exceed those shown in Table G-16 of the Safety and Health Standards. Feasible engineering and/or administrative controls shall be utilized to keep exposure below the allowable limit.

b When engineering or administrative controls fail to reduce the noise level to within the levels of Table G-16 of the Safety and Health Standards, personal protective equipment shall be provided and used to reduce the noise to an acceptable level.

c Exposure to impulsive or impact noise should not exceed 140 dB peak sound pressure level.

d In all cases, where the sound levels exceed the values shown in Table G-16 of the Safety and Health Standards, a continuing, effective hearing conservation program shall be administered.

e *Table G-16 Permissible Noise Exposures*

Duration per Day, Hours	Sound Level dB(A) Slow Response
8	90
6	92
4	95
3	97
2	100
1½	102
1	105
½	110
¼ or less	115

54 PERSONAL PROTECTIVE EQUIPMENT

a Proper personal protective equipment, including shields and barriers, shall be provided, used, and maintained in a sanitary and reliable condition where there is a hazard from processes or environment that may cause injury or illness to the employee.

b Where employees furnish their own personal protective equipment, the employer shall be responsible to assure its adequacy and to ensure that the equipment is properly maintained and in a sanitary condition.

55 PORTABLE ELECTRIC TOOLS (See HAND TOOLS, NO. 41)

56 POWER TRANSMISSION, MECHANICAL

a All belts, pulleys, chains, flywheels, shafting and shaft projections, or other rotating or reciprocating parts within seven feet of the floor or working platform shall be effectively guarded.

b Belts, pulleys, and shafting located in rooms used exclusively for power transmission apparatus need not be guarded when the following requirements are met:

- The basement, tower, or room occupied by transmission equipment is locked against unauthorized entrance;
- The vertical clearance in passageways between the floor and power transmission beams, ceiling, or any other objects is not less than five feet six inches;
- The intensity of illumination conforms to the requirements of ANSI A11.1—1965 (R 1970);

- The footing is dry, firm, and level;
- The route followed by the oiler is protected in such a manner as to prevent accident.

57 PRESSURE VESSELS, PORTABLE UNFIRED

Pressure vessels are not covered by present OSHA standards. These are good operating procedures and might be incorporated into future OSHA standards.

a All portable unfired pressure vessels should be designed and constructed to meet the Standards of the American Society of Mechanical Engineers Boiler and Pressure Vessel Code, Section VIII.

b Portable unfired pressure vessels not built to code should be examined quarterly by a competent person and subjected yearly to a hydrostatic pressure test of 1½-times the working pressure of the vessel. Records of such examination and tests should be maintained.

c Relief valves on pressure vessels should be set to the safe working pressure of the vessel, or to the lowest safe working pressure of the system, whichever is lower.

58 PUNCH PRESSES

a It shall be the responsibility of the employer to provide and ensure the usage of "point-of-operation guards" or properly applied and adjusted point-of-operation devices on every operation performed on a mechanical power press. This requirement shall not apply when the point-of-operation opening is ¼ inch or less.

b A substantial guard shall be placed over the treadle of foot-operated presses.

c Pedal counterweights, if provided on foot-operated presses, shall have the path of the travel of the weight enclosed.

59 RADIATION

a Employers shall be responsible for proper controls to prevent all employees from being exposed to ionizing radiation in excess of acceptable limits.

b Each radiation area shall be conspicuously posted with appropriate signs.

c Every employer shall maintain records of the radiation exposure of all employees for whom personnel monitoring is requested.

60 RAILINGS

a A standard railing shall consist of top rail, intermediate rail, and posts, and shall have a vertical height of 42 inches from upper surface of top rail to floor, platform, etc.

b A railing for open-sided floors, platforms, and runways shall have a toeboard whenever persons can pass beneath the open side, or there is moving machinery, or there is equipment which could be struck by falling materials.

c Railings shall be of such construction that the complete structure shall be capable of withstanding a load of at least 200 pounds in any direction on any point on the top rail.

d A stair railing shall be of construction similar to a standard railing, but the vertical height shall be not more than 34 inches nor less than 30 inches from upper surface of top rail to surface of tread in line with face of riser at forward edge of tread.

61 RAIL SWEEPS (See CRANES AND HOISTS, No. 16)

62 REVOLVING DRUMS

Revolving drums, barrels, or containers shall be guarded by an interlocked enclosure that will prevent the drum, etc., from revolving unless the guard enclosure is in place.

63 SAWS, BAND (WOODWORKING)

a All portions of band saw blades shall be enclosed or guarded except for the working portion of the blade between the bottom of the guide rolls and the table.

b Band saw wheels shall be fully encased. The outside periphery of the enclosure shall be solid. The front and back shall be either solid or wire mesh or perforated metal.

64 SAWS, PORTABLE CIRCULAR

All portable power-driven circular saws having a blade diameter greater than two inches shall be equipped with guards above and below the base plate or shoe. The lower guards shall cover the saw to the depth of the teeth, except for the minimum arc required to permit the base plate to be tilted for bevel cuts, and shall automatically return to the covering position when the blade is withdrawn from the work. This provision does not apply to circular saws used in the meat industry for meat cutting purposes.

65 SAWS, RADIAL (WOODWORKING)

a Radial saws shall be constructed so that the upper hood shall completely enclose the upper portion of the blade down to a point that will include the end of the saw arbor. The upper hood shall be constructed in such a manner and of such material that it will protect the operator from flying splinters, broken saw teeth, etc., and will deflect sawdust away from the operator. The sides of the lower exposed portion of the blade shall be guarded to the full diameter of the blade by a device that will automatically adjust itself to the thickness of the stock and remain in contact with stock being cut to give maximum protection possible for the operation being performed.

b Radial saws used for ripping shall have non-kickback fingers or dogs.

c Radial saws shall be installed so that the cutting head will return to the starting position when released by the operator.

66 SAWS, SWING OR SLIDING CUT-OFF (WOODWORKING)

a All swing or sliding cut-off saws shall be provided with a hood that will completely enclose the upper half of the saw.

b Limit stops shall be provided to prevent swing or sliding type cut-off saws from extending beyond the front or back edges of the table.

c Each swing or sliding cut-off saw shall be provided with an effective device to return the saw automatically to the back of the table when released at any point of its travel.

d Inverted sawing or sliding cut-off saws shall be provided with a hood that will cover the part of the saw that protrudes above the top of the table or material being cut.

67 SAWS, TABLE (WOODWORKING)

a Circular table saws shall have hoods over the portion of the saw above the table, so mounted that the hood will automatically adjust itself to the thickness of and remain in contact with the material being cut.

b Circular table saws shall have a spreader aligned with the blade, spaced no more than ½ inch behind the largest blade mounted in the saw. The provision of a spreader in connection with grooving, dadoing, or rabbitting is not required.

c Circular table saws used for ripping shall have non-kickback fingers or dogs.

d Feed rolls and blades of self-feed circular saws shall be protected by a hood or guard to prevent the hands of the operator from coming in contact with the in-running rolls at any point.

68 SCAFFOLDS

a All scaffolds and their supports shall be capable of supporting the load they are designed to carry with a factor of at least four.

b All planking shall be scaffold grade, as recognized by grading rules for the species of wood used. The maximum permissible spans for 2 x 9 inch or wider planks are shown in the following table:

	Full Thickness Undressed Lumber			Nominal Thickness Lumber	
Working load (p.s.f.)	25	50	75	25	50
Permissible span (ft.)	10	8	6	8	6

The maximum permissible span for 1¼ x 9 inch or wider plank of full thickness is four feet, with medium loading of 50 p.s.f.

c Scaffold planks shall extend over their end supports not less than six inches nor more than 18 inches.

d Scaffold planking shall be overlapped a minimum of 12 inches or secured from movement.

e Railings and toeboards shall be installed on all open sides and ends of platforms more than 10 feet above the floor except for scaffolds covering an entire interior floor with no exposure to floor openings or needle-beam scaffolds used in structural iron work. There shall be a screen with maximum ½ inch openings between the toeboard and the top rail where persons are required to pass or work under the scaffold.

69 SPRAY FINISHING OPERATIONS

a All spray finishing shall be conducted in spray booths or spray rooms.

b Spray booths shall be substantially constructed of steel, not thinner than No. 18 U.S. gauge, securely and rigidly supported, or of concrete or masonry; except that aluminum or other substantial noncombustible material may be used for intermittent or low volume spraying. Spray booths shall be designed to sweep air currents toward the exhaust outlet.

c There shall be no open flame or spark-producing equipment in any spraying areas nor within 20 feet thereof, unless separated by a partition.

d Electrical wiring and equipment not subject to deposits of combustible residues but located in a spraying area shall be of an explosion-proof type approved for Class I, group D locations or for Class I, Division 1, Hazardous Locations. Electrical wiring, motors, and other equipment outside of but within 20 feet of any spraying area, and not separated therefrom by partitions, shall not produce sparks under normal operating conditions and shall otherwise conform to the provisions for Class I, Division 2, Hazardous Locations.

e All spraying areas shall be provided with mechanical ventilation adequate to remove flammable vapors, mists, or powders to a safe location and to confine and control combustible residues so that life is not endangered.

f Electric motors driving exhaust fans shall not be placed inside flammable materials spray booths or ducts. Belts or pulleys within the booth or duct shall be thoroughly enclosed.

g The quantity of flammable or combustible liquid kept in the vicinity of spraying operations shall be the minimum required for operations and should ordinarily not exceed a supply for one day or one shift.

h Conspicuous "NO SMOKING" signs shall be posted at all flammable materials spraying areas and storage rooms.

70   STAIRS, FIXED INDUSTRIAL

a Standard railings shall be provided on the open sides of all exposed stairways. Handrails shall be provided on at least one side of closed stairways, preferably on the right side descending.

b Stairs shall be constructed so that rise height and tread width is uniform throughout.

c Fixed stairways shall have a minimum width of 22 inches.

71   STATIONARY ELECTRICAL DEVICES

All stationary electrically powered equipment, tools, and devices, located within reach of a person who can make contact with any grounded surface or object, shall be grounded.

72   STORAGE

a All storage shall be stacked, blocked, interlocked, and limited in height so that it is secure against sliding or collapse.

b Storage areas shall be kept free from accumulation of materials that constitute hazards or pest harborage. Vegetation control will be exercised when necessary.

c Where mechanical handling equipment is used, sufficient safe clearance shall be allowed for aisles, at loading docks, and through doorways.

73 TANKS, OPEN-SURFACE

When ventilation is used to control potential exposures to employees, it shall be adequate to reduce the concentration of the air contaminated to the degree that a hazard to employees does not exist.

74 TOEBOARDS

a Railings protecting floor openings, platforms, scaffolds, etc., shall be equipped with toeboards whenever persons can pass beneath the open side, there is moving machinery, or there is equipment which could be struck by falling material.

b A standard toeboard shall be at least four inches in height and may be of any substantial material, either solid or open, with openings not to exceed one inch in greatest dimension.

75 TOILETS

a Every place of employment shall be provided with adequate toilet facilities which are separate for each sex. Water closets shall be provided according to the following: 1-15 persons, one facility; 16-35 persons, two facilities; 36-55 persons, three fac'lities; 56-80 persons, four facilities; 81-110 persons, five facilities; 111 to 150 persons, six facilities; over 150 persons, one for each additional 40 persons.

b Each water closet shall occupy a separate compartment which should be equipped with a door, latch, and clothes hangers.

c The requirements of (a) and (b) above do not apply to mobile crews or normally unattended locations, as long as employees working at these locations have transportation immediately available to nearby toilet facilities.

d Adequate washing facilities shall be provided in every toilet room or be adjacent thereto.

e Covered receptacles shall be kept in all toilet rooms used by women.

76 TOXIC VAPORS, GASES, MISTS, AND DUSTS

a Exposure to toxic vapors, gases, mists, or dusts at a concentration above the Threshold Limit Values, contained or referred to in Safety and Health Standards, shall be avoided.

b To achieve compliance with paragraph (a), administrative or engineering controls must first be determined and implemented whenever feasible. When such controls are not feasible to achieve full compliance, protective equipment or any other protective measures shall be used to keep the exposure of employees to air contaminants within the limits prescribed. Any equipment and/or technical measures used for this purpose must be approved for each particular use by a competent industrial hygienist or other technically qualified person.

77 TRASH

Trash and rubbish shall be collected and removed in such a manner as to avoid creating a menace to health and as often as necessary to maintain good sanitary conditions.

78 WASHING FACILITIES

a Adequate washing facilities shall be provided in every place of employment and maintained in a sanitary condition. For industrial establishments, at least one lavatory with adequate hot and cold water shall be provided for every 10 employees up to 100 persons, and one lavatory for each 15 persons over 100.

b A suitable cleansing agent, individual hand towels or other approved apparatus for drying hands, and receptacles for disposing of hand towels, shall be provided at washing facilities.

79 WELDING (See also CYLINDERS, COMPRESSED GAS, NO. 17)

a Arc welding equipment shall be chosen for safe application to the work and shall be installed properly. Workmen designated to operate welding equipment shall have been properly instructed and qualified to operate it.

b Mechanical ventilation shall be provided when welding or cutting:

- beryllium, cadmium, lead, zinc, or mercury;
- fluxes, metal coatings, or other material containing fluorine compounds;
- where there is less than 10,000 cubic feet per welder;
- in confined spaces.

c Proper shielding and eye protection to prevent exposure of personnel from welding hazards shall be provided.

d Proper precautions (isolating welding and cutting, removing fire hazards from the vicinity, providing a fire watch, etc.) for fire prevention shall be taken in areas where welding or other "hot work" is being done.

e Work and electrode lead cables shall be frequently inspected. Cables with damaged insulation or exposed bare conductors shall be replaced.

80 WOODWORKING MACHINERY

a All woodworking machinery such as table saws, swing saws, radial saws, band saws, jointers, tenoning machines, boring and mortising machines, shapers, planers, lathes, sanders, veneer cutters, and other miscellaneous woodworking machinery shall be effectively guarded to protect the operator and other employees from hazards inherent to their operation.

b A power control device shall be provided on each machine to make it possible for the operator to cut off the power from each machine, without leaving his position, at the point of operation.

c Power controls and operating controls should be located within easy reach of the operator while he is at his regular work location, making it unnecessary for him to reach over the cutter to make adjustments. This does not apply to constant pressure controls used only for setup purposes.

d Each operating treadle shall be protected against unexpected or accidental tripping.

APPENDIX THREE

OSHA—The Supervisor's Responsibility

The supervisor's role in OSHA compliance is a major role. The supervisor first plays a large part in your company's initial efforts to get into compliance with the standards, and then plays an even larger part in the ongoing efforts needed to keep in compliance.

In this section an excellent publication is reprinted, called "A Foreman & Supervisor's Guide to Occupational Safety and Health Act." It was published originally by the FMC Corporation, and authored by Harold Hodnick of their insurance department. It was later reprinted by the Ohio Manufacturer's Association of Columbus. It seems an excellent summary of the law for the line supervisor.

A SUPERVISOR'S GUIDE TO OSHA

This guide has been prepared to assist foremen and supervisors of Machinery Group plants to initiate compliance with safety and health standards now in effect under the Occupational Safety and Health Law.

This guide in no way pre-empts any standards encompassed by the Law. It places priorities on those standards felt to be of prime and general importance.

Readers of this guide are encouraged to become totally familiar with the complete OSHA Regulations and to comply as quickly as possible with all applicable standards.

PENALTIES

Penalties for violation are severe. For example:

- a single violation—up to $1,000.
- willful violation—up to $10,000.
- a violation not corrected in the allotted time—$1,000 per day for each day the violation continues.
- employees death as a result of willful violation—up to $10,000 and a jail sentence of up to six months.
- false statement or report—up to $10,000 and six months in jail.
- violating the posting requirement (notice to tell your employees about their rights and obligations under the Law and to spell out the provisions of the applicable standard)—up to $1,000.
- for assaulting, resisting, opposing, intimidating or interfering with a Compliance Officer—up to $5,000 fine and three years in jail (if with a dangerous weapon, up to $10,000 fine and ten years in jail).

PROVISIONS OF THE LAW

The Occupational Safety and Health Act of 1970 became effective April 28, 1971. This law is tough. It provides for fines and even imprisonment. It requires the employer to provide a safe place to work and places heavy responsibilities on the foremen and supervisor.

The law specifically requires that:

- "Each employer shall furnish to each of his employees employment and a place of employment which are free from recognized hazards that are causing or are likely to cause death or serious physical harm to his employees."
- "Each employer shall comply with Occupational Safety and Health Standards promulgated under this Act."
- "Each employee shall comply with the Occupational Safety and Health Standards and all rules, regulations and orders issued pursuant to this Act which is applicable to his own actions and conduct."

Therefore, as a foreman and supervisor you cannot allow unsafe acts to continue. You cannot permit poor housekeeping, defective or unguarded machinery, poor ventilation, insufficient light or extreme noise. A foreman cannot treat safety training and compliance with safety rules as a secondary responsibility.

The law designates the U. S. Department of Labor as the enforcement agency and authorized it to issue standards. Department of Labor (Compliance Officers) are authorized to enter places of business to examine work conditions, machinery, devices, equipment, material and practices. They may enter for a general inspection or to investigate a specific complaint.

The law permits any employee or any union representative to submit a complaint charging unsafe conditions or lax safety practices to the Department of Labor Area Director. The Department of Labor is obligated to dispatch unannounced an inspector, if "reasonable" grounds for a violation appear to exist.

A foreman or supervisor cannot discriminate against an employee who asks for an inspection, nor can he discuss or discipline the complainant in any way.

Management must allow an employee or an union representative to accompany the Compliance Officer. A management representative is also permitted. If a violation is found, a "citation" describing the nature of the violation, assessed penalty and prescribed time in which it must be remedied will be issued. Abatement times are mutually agreed to between the Compliance Officer and management. If the violation is of an extremely hazardous nature, use of a particular machine or process may be forbidden or the operation can be shut down. In some cases, a court order could shut down an entire plant. The Law provides for hearings and appeals registered by either management or employees.

Injury and illness records must be kept at each work place and such records must be available to the Compliance Officer on request.

Regulations encourage foremen and supervisors to conduct periodic inspections of their operations. Such action would show that an operation is moving ahead on a voluntary basis toward getting the job done.

OSHA RULES AND REGULATIONS—STANDARDS

Walking and Working Surfaces. This section of the standards sets forth the rules and regulations concerning minimum requirements for the purchase, construction, operation, inspection and maintenance of both permanent and temporary facilities used for access to/or serving as working or walking surfaces.

OSHA	FOREMAN SHOULD

A General

All places of employment, passageways, storerooms and service rooms shall be kept clean and orderly and in a sanitary condition.

Floors will be clean and dry.

Floors will be free from protruding nails, splinters, holes and loose boards.

When mechanical handling equipment is used—sufficient safe clearance should be allowed.

Make sure that your area is clean and neat at all times.

Insist that all stock is to be stacked neatly and all waste materials be placed in containers and removed when needed.

Make sure that floors are swept when needed and all spills (oil, water, etc.) wiped up immediately.

Inspect the floors, walls and pillars for loose wires; nails, etc.

Make sure that the aisles where hand trucks are used are free from any materials that may cause a hazard.

B Floor & Wall Openings

Temporary openings may have temporary barricades.

Permanent openings should have permanent guard rails.

Hand rails are needed in all permanent steps.

Toe boards needed in all stairs above 4'.

Any stairway with four or more risers shall have hand railings.

All wall openings, a cut-out area at least 30"H×18"W, must have protection to make sure that employees do not fall into them. A wall opening could be a ventilation shaft, trash chute, etc.

Make sure wherever you have an open hole in a floor located in your area, it is guarded. If the opening is temporary a barricade must be placed around it—if it is permanent you must have a railing constructed around it.

Make sure that the hand rails on all steps in the plant are in good order.

Make sure that all wall openings are guarded so that no employee may accidentally fall into them.

C Portable Wood Ladders

The standards for sectional ladders are a maximum of 60' in

It's management's responsibility to make sure that the ladders you

OSHA

length when extended; rung size of at least 1-⅝"; the sides of the ladder having a depth of 3-¾".

Care and inspection of wooden ladders. Management:

1 shall maintain all ladders in good condition.
2 shall replace safety feet if defective.
3 should store all ladders so they are easily inspected.
4 should make sure that wood ladders are stored out of the elements.
5 should make sure that ladders are stored to prevent sags.
6 should paint all wooden ladders.
7 should dispose of defective ladders.

FOREMAN SHOULD

use are of an approved type, however, you are responsible for making sure they:

1 are used correctly. (The right ladder for the right job.)
2 are stored properly.
3 are not broken or cracked.
4 have their "safety feet" intact.
5 are free of grease.
6 look neat and clean.

D Portable Metal Ladders

Sectional ladders may be 48' to 60' long but must have an overlap of 5' in the middle.

Ladders (platform type) may be no longer than 20'.

The three most important things for the supervisor to check in the use of metal ladders are:

1 Are the "safety feet" in place?
2 Are the rungs clean and free from slippery materials?
3 Are there any bends in the ladder that may weaken it?

You must insist that the floor is clean and free from oil, etc., where the ladder will be placed.

A platform ladder may never be leaned against anything—it must be opened before use.

OSHA   *FOREMAN SHOULD*

E   Scaffolding

It is the management's responsibility to provide proper scaffolding.

Once the scaffolding is in place, you must make sure that the platform is used the way it was intended to be used and that no "horseplay" happens on or around the scaffolding.

You should make sure that barricades are placed around all scaffolding to protect people from walking under it.

The regulations are far more detailed than you have read above, but if you have a special question concerning a walking surface or climbing apparatus, check with your plant safety representative for proper and safe procedure.

Occupational Health and Environmental Controls. This section is probably the most complex and confusing of the standards because it deals primarily with toxicity levels, noise levels and other specific data about materials that may cause occupational diseases or other hazards that may affect the employee, either temporarily or permanently.

This section of the code relates heavily to many toxic or explosive substances such as hydrogen, acetylene and combustible liquids (vapors). Attention should be brought to the safety/personnel manager if you are using types of material that you think may be hazardous in order that he can be sure that this section of the standard is being properly fulfilled.

OSHA   *FOREMAN SHOULD*

A   Ventilation

Exhaust ventilation systems must be provided to produce maximum protection against air born dusts, flammable or explosive vapors, or potentially toxic substances.

If the environment cannot be controlled by engineering, the employee must be provided with personal protective equipment.

Make sure that all ventilation systems are spot checked periodically to see if they're working and the filters are clean.

Check all dust collectors and change filters frequently; all hoses must be in good repair.

Enforce the rule that respirators of an approved type are to be worn while in the environment where

OSHA

FOREMAN SHOULD

spraying, blasting, etc. take place when required.

If there seems to be a high accumulation of a gas or an undesirable odor in your area—check it out. Report all gas or liquid leaks as soon as they are detected.

Know the characteristics of the solvents, cleaners, etc. in your department.

Be sure that all employees in your area are instructed as to what to do in case of an emergency, especially if they are working with materials that may explode. Your subordinates should know:

1 which fire extinguisher to use and how to use it.
2 areas of egress (how to get out of his work place)
3 who to call and how to alert them in case of a fire.

You should make sure that all solvents, etc. are in approved type cans with self closing lids and properly labeled.

Hazardous Material. In conjunction with the environmental control, this section is primarily devoted to the materials that will deflagrate or burn, giving off toxic or other undesirable effects and their containers; not the toxic qualities of them.

The section about compressed gases is a special section. The supervisor should be aware of what the properties of the gases are, how they will act under pressure and stress, in heat, etc.

OSHA

This section specifically deals with:

FOREMAN SHOULD

Be familiar with the dangers of all potentially dangerous materials used in your area. (Especially those under pressure.)

OSHA

1 Gas cylinders and compressed gases
2 Safety valves
3 Acetylene
4 Hydrogen
5 Oxygen
6 Nitrous oxide
7 Flammable and combustible liquids
8 Dip tanks
9 Explosives and blasting
10 Ammunition nitral agents
11 L. P. gases
12 Anhydrous ammonia

FOREMAN SHOULD

Be sure that all compressed gas cylinders are secured in place with a chain or cable and cannot tip over.

Check each container and pipe line in your area for cracks, dents or any other signs of weakness in the containers.

Teach your subordinates to respect the materials they use—not fear it.

Make sure that each container is properly marked.

Personal Protective Equipment. This section deals with most equipment needed to protect employees from potential hazards that cannot be engineered out of an operation.

OSHA

The section covers: 1) head, eye and face protection; 2) respiratory protive devices; 3) foot protection; and 4) protective clothing for protection against electricity.

It is the responsibility of management to purchase proper protective equipment for the people in the plant.

FOREMAN SHOULD

It is the responsibility of the supervisor to:

Make sure that everyone wears the equipment in his area.

Inform the Personnel/Safety Manager of the original or additional need for glasses, shoes, aprons, masks, etc. in your area. (In some locations this is done through the nurse.).

Make sure that if the employee buys his own equipment the Safety Manager checks it to see if it is of an approved type.

Make sure that:

1 Glasses are not scratched, broken, etc.
2 The equipment is comfortable (man will not wear otherwise).
3 All equipment is clean and neat.

4 The protection equipment is in operable condition, masks disinfected, etc.

5 Repair is done properly on protective equipment.

General Environmental Controls. This section is devoted to protect the worker from unsafe and unhealthy conditions arising from the general environment.

OSHA

Such topics as: 1) sanitation; 2) color codes; 3) signs and tags showing physical hazards are some of the general items covered.

The general classifications for color coding are:

Red—fire protection equipment and apparatus—danger—stop.

Orange—designated for dangerous parts of machines.

Yellow—caution (against striking against or getting caught in).

Green—safety and first aid equipment.

Blue—warning against storing or moving of equipment under repair.

Purple—radiation.

Black and White—traffic and housekeeping.

No food is allowed to be eaten in the areas where toxic materials are present.

Many demands are placed upon management to have a clean and healthy environment for its employees. Some of the areas covered are:

1 Good housekeeping
2 Proper waste disposal

FOREMAN SHOULD

Keep your area clean and neat at all times.

Set the example of keeping the work area, lunch room and toilet facilities clean.

Know that no spitting (specifically) is allowed: except in approved containers.

All refuse shall be removed at least daily, more if needed.

Rodent and insect control must be practiced—if pests are in the work area the supervisor should alert the management.

No "common drinking cups" are allowed.

The supervisor should check the toilet facilities once daily; if cleaning is needed, maintenance should be contacted.

All signs (warning—watch your step, etc.) must be readable.

You should be aware of the general color coding and what each means.

3 Adequate rodent, insect control
4 Proper toilet facilities
5 Etc.

Medical and First Aid.   This section deals with the basics of first aid and the administration of first aid in the plant.

OSHA

Management must be certain that the health of the employees will be adequately protected by providing available emergency medical facilities.

FOREMAN SHOULD

It is your responsibility to make sure that if one of your subordinates gets hurt (regardless of how minor) he goes to first aid and gets attention as soon after the incident as possible. (Even the most minor incidents, slivers, foreign bodies in eyes, etc., may be serious injuries if not taken care of immediately.)

If the employee is available, he should be questioned as to how the incident could be prevented.

Fire Protection.   This section is to insure that protection is provided to fight small fires only and also to clarify what type, number and spacing of fire extinguisher, maintenance and testing of equipment and systems are in a plant.

OSHA

Covered in this section is the use and maintenance of:

1 portable extinguishers
2 standpipe and hose systems
3 sprinkler systems; and
4 the storage of materials

FOREMAN SHOULD

The supervisor should be thoroughly checked out and able to operate all fire fighting devices in his department.

The foreman should be aware of the different types of fires that may occur in his area and instruct his employees (at least every 6 months) how to fight all types of fires in the area.

The supervisor should spot check the extinguishers in the area periodically to make sure that they have not been tampered with and that the content of the extinguisher has not been used.

Compressed Gas and Air Equipment. This section deals with problems or potential hazards that may occur from the rupture or explosion of pressure vessels, which include cylinders, portable tanks and compressed air receivers. Testing, inspection of tanks and the codes dealing with the safety relief valves are cited.

OSHA

Management's responsibility dictated by this section covers requirements for inspection testing and examination of cylinders and the installation of safety relief devices on cylinders, tanks and air receivers.

FOREMAN SHOULD

Report all:

1 Dents in cylinders.
2 Cuts or gouges in cylinders.
3 Corrosion or pitting of the cylinder.
4 Cylinder necks that have "hairline" cracks.
5 Cylinders that are not stored in dry, clean area.

Make sure that employees do not use compressed air to clean their skin or clothing.

Know that no more than 30 P. S. I. (at the nozzle head) is allowed for the cleaning of machines or parts.

Material Handling and Storage. This section deals with the potential dangers involved with the manual and mechanical handling of materials and the storage of items.

OSHA

Industrial truck, mechanical equipment, overhead and gantry cranes, derricks, etc., are covered in this section.

OSHA has definite specific standards that are applicable to each of the aforementioned devices.

Management should provide safe equipment with which their employees are expected to work.

FOREMAN SHOULD

Make sure that all aisles are clear where lift trucks are to be used.

See that the storage of materials should not in itself create a hazard.

Insist that all open pits, vats, ditches, etc., be guarded.

Know that some of the regulations for power trucks are:

1 Overhead guards must be in place at all times with "high lift" trucks.

2 Only trained-authorized opera-tors can operate lifts.

3 No man may stand or pass under a suspended load.

4 No riders are allowed.

5 No passing of other trucks (unless other truck has stopped).

6 No speeding or horseplay.

7 No fuel oil spills should be visible on truck.

8 Trucks should be clean.

9 If trucks seem excessively noisy, they should be reported.

Request that all defective tools be removed from service.

Inspect portable grinding wheels regularly.

See that power tools should be unplugged when the operator is not there and that all hand tools are stored in a safe place.

Machinery and Machine Guarding. This section insures that protection is provided for every machine at the point of operation and/or power transmission.

Most machines and grinders can be made safe through engineering which may protect the employee against hazards.

If you think that a machine should be guarded better than it already is, you should contact your safety engineer/personnel manager. (All moving parts should be covered.)

OSHA

Woodworking machinery, machine guarding, abrasive wheels, mills and calendars in rubber and plastics, power press, hot metal stamping and mechanical transmission apparatus are covered in this section.

FOREMAN SHOULD

All existing guards should be in place at all times when a machine is in operation.

Machines should be "locked out" when work is being done on them.

Hand and Portable Powered Tools and Equipment. This section deals with the protection of the employee at the point of operation for various types of hand-held tools and equipment.

OSHA

Tools involved in this section of the code are circular (portable) saws, abrasive wheels, explosive activated fastening tools, power lawn mowers, air machinery, etc.

FOREMAN SHOULD

Make sure that no employee should be allowed to use unsafe hand power tools.

Inspect power cords to ensure that the tool is grounded which will prevent electrical shock.

See that the correct tool is used for each job.

Make sure that all guards on portable power tools are in place.

Ensure that any employee who uses a portable power tool wear safety glasses with shields if necessary.

Welding, Cutting and Brazing. Purpose of this section is to enable management to prevent injury and/or illness, and protection of property from fire because of welding, cutting and brazing operations.

OSHA

The section is detailed and rather specific as to modes of operation regarding all types of cutting and welding procedures. If your section is involved with regular welding/cutting processes you should read this section carefully.

FOREMAN SHOULD

Be acutely aware of all the hazards involved in operations connected with welding operations.

Be alert to the accumulation of toxic gases or odors that may lead to potential hazards to the employee.

Ensure that if one of your employees is injured, he is accompanied to first aid and that the attending nurse/physician knows what he was exposed to.

Check the welding equipment to make sure it is in perfect working order at all times.

1 Make sure that the tanks are in good shape and secured properly.
2 Make sure that all electrical equipment is in good working order.

3 Make sure that all acetylene equipment is clean, free from oil and in good working order.

Electrical. This section is designated to protect employees against unwanted exposure to electrical currents. It provides guidelines for installation, maintenance and use of electrical installation and the utilization of equipment.

OSHA

The subpart is divided into 13 sections which treat: General requirements; overcurrent protection (fuses, etc.); grounding; outlets; junction boxes, etc.; flexible cords; transformers; appliances; hazardous locations.

Coverage is limited in general to building installations and services and specifically excluded is coverage for transportation (vehicles), underground mines, electric railroads, communications utilities and electric utilities. The National Electrical Code is adopted by reference.

FOREMAN SHOULD

Protect your employees against electrical hazards. Some of the things you should be constantly aware of are:

1 Is all of the equipment used in your area properly grounded.
2 Make sure there are no exposed electrical connections that people can get shocks from.
3 Never allow your employees to do electrical maintenance work that should be done by a person familiar with the hazards of working with electricity.

ELECTRICITY IS TO BE RESPECTED NOT FEARED

Recordkeeping.

OSHA

Each facility is required to keep an accurate set of records, post the yearly summary for all employees to see and to submit upon request, all such records to the Secretary of Labor.

Every serious or reportable injury must be recorded (logged) within 48 hours after it has been reported.

SUPERVISOR SHOULD

Each supervisor must fill out and submit to the personnel/safety representative, a standard foreman's report of injury form as soon as practical after any of his employees reports any injury—even if that employee does not go to the doctor.

You should investigate each incident and determine how it could have been avoided—this is very important in "curbing" major accidents before they occur.

APPENDIX FOUR

Machine Guarding Principles

The following pages are excerpted from *The Principles and Techniques of Mechanical Guarding*, published by the U.S. Department of Labor, Occupational Safety and Health Administration, OSHA 2057, revised May 1972.

THE PRINCIPLES OF MECHANICAL GUARDING

All mechanical action or motion is hazardous, but in varying degrees. Rotating members, reciprocating arms, moving belts, meshing gears, cutting teeth, and parts in impact or shear are some examples of the types of action and motion requiring protection. They are not peculiar to any one machine or industry, but are basic to the mechanical devices used for productive purposes.

Actions and Motions That Create Hazards

Since safety standards cannot be drawn which will cover every conceivable hazardous mechanical exposure, it is often necessary to use imagination and ingenuity to protect unusual situations. If the basic hazardous actions and motions are understood, it is easier to evaluate the hazard and to develop effective control measures, whatever the machine. Actions or motions involving the most hazardous exposures may be classified as:

Rotating, Reciprocating, and Transverse Motions.
In-Running Nip Points.
Cutting Actions.
Punching, Shearing, and Bending Actions.

A discussion of these actions or motions, along with typical illustrations of each, follows:

Rotating, Reciprocating, and Transverse Motions. Rotating, reciprocating, and transverse motions create hazards in two general areas—at the point-of-operation where work is being done, and at the points where power or motion is being transmitted from one part of a mechanical linkage to another. Since guarding at the point-of-operation will be discussed under the other classifications of action or motion, this section will be devoted primarily to situations where power is being transmitted or the point-of-operation is not clearly defined (extractors, mixers, etc.).

Any rotating object is dangerous. Even smooth, slowly rotating shafts can grip clothing or hair, and through mere skin contact force an arm or hand into a dangerous position. Accidents due to contact with rotating objects are not frequent, but the severity of injury is always high.

Collars, couplings, cams, clutches, flywheels, shaft ends, spindles, rotating bar stock, lead screws, and horizontal or vertical shafting are typical examples of common rotating mechanisms which are hazardous. The danger increases when bolts, oil cups, nicks, abrasions, and projecting keys or screw threads are exposed when rotating.

In many cases, the rotating mechanism is located within a stationary case or shell and consists of a revolving cylinder, a screw, agitator blades, or paddles. Washing machines, extractors, raw material mixers, and screw conveyors are typical examples of this type of hazardous rotating mechanism.

Figure A4-1   Rotating.

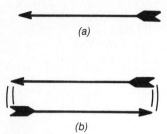

(a)

(b)

Figure A4-2   *a.* Transverse (in straight line). *b.* Reciprocating.

Reciprocating and transverse motions are hazardous because, in the back and forth or straight line action, a worker may be stuck or caught in a pinch or shear point between a fixed or other moving object.

This does not mean that certain guarding methods are not preferable to others, but the type of operation, the size or shape of stock, the method of handling, the physical layout, the type of material, and production requirements or limitations may present important considerations. A certain flexibility in operations may also determine the practicability of the method to be used.

As a general rule, power transmission apparatus can be protected by fixed enclosure guards. It is when guarding the point-of-operation, where work is being done on an object, that the most effective and practical of several means of guarding must be selected.

Classification of Guards

The methods of guarding may be grouped under four main classifications:

1   Enclosure Guards
   a   Fixed enclosures.
   b   Adjustable enclosures.

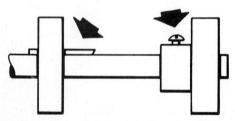

Figure A4-3   Rotating shaft and pulleys with projecting key and set screw.

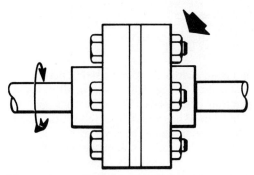

Figure A4-4 Rotating coupling with projecting bolt heads.

    2 Interlocking Guards
        a Enclosure or gate guard with electrical or mechanical interlock.
        b Barrier with electrical or mechanical interlock activating a brake.
        c Electronic or other type field or beam connected with operating and stopping mechanism.
    3 Automatic Guards
        a Moving barrier connected to operating mechanism of machine (push-away).
        b Removal device connected to operator, and operating mechanism of machine (pull-away).
        c Limitation of stroke.
        d Automatic pressure release devices.
    4 Remote Control, Placement, Feeding, Ejecting
        a Two-hand tripping devices (also multiple operation).
        b Automatic or semiautomatic feed.

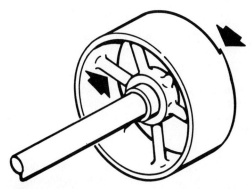

Figure A4-5 Rotating pulley with spokes and projecting burr on face of pulley.

c Special jigs or holding devices.
d Special handtools and dies.
e Special ejecting devices.

Enclosure Guards. Fixed enclosure guards should be used in preference to all other types. They prevent access to dangerous parts at all times by enclosing the hazardous operation completely. They are also used to restrain bursting machine parts from flying about. They admit the stock but will not admit hands into the danger zone because of limited feed opening size. They may be constructed so as to be adjustable to different sets of tools or dies, but once adjusted they should be fixed.

Enclosure guards may be installed at the point where cutting, bending, punching, or shearing action takes place on material being processed, and at other places where there may be a hazard to men inserting or manipulating stock. They may also be used to prevent contact with rotating, reciprocating, and transverse motion of machine members away from the point-of-operation.

Interlocking Guards. When a fixed enclosure guard is not practicable, an interlocking enclosure or barrier should be considered as the first alternative.

An interlocking enclosure guard is not fixed and may be opened or removed as the operation requires. However, due to an electrical or mechanical interlocking connection with the operating mechanism, the operation of the machine is prevented until the guard is returned to an

Methods of Guarding Actions and Motions—Enclosure Guards

Guarding Methods	*Guard Action*	*Limitations*
Fixed enclosure	Barrier or enclosure which admits the stock but which will not admit hands into danger zone because of feed opening size, remote location, or unusual shape. Also used to completely enclose power transmission apparatus. Used to contain bursting machine parts.	Limited to specific operations. May require special tools to remove jammed stock. May interfere with visibility. Require interlocks for repair or maintenance work.
Adjustable enclosure	Barrier or enclosure which is adjusted to fit around different sizes or shapes of the die. When adjusted, provides same protection as fixed enclosure.	Often requires frequent adjustment and careful maintenance.

operating position and the operator can no longer reach the point of danger.

An interlocking enclosure guard should do three things:

1 Shut off or disengage the power to prevent the starting of the machine when the guard is open.
2 Guard the danger point before the machine can be operated.
3 Keep the guard closed until the dangerous part is at rest, or stop the machine when the guard is opened.

When gate guards or hinged enclosure guards are used with interlocks, they should be so designed as to completely enclose the point-of-operation before the operating clutch can be engaged.

An interlocking barrier guard quickly stops the machine or prevents application of injurious pressure when any part of the operator's body contacts the barrier. The barrier may be a bar, a rod, a wire, or some similar device (not an enclosure), extended across the danger zone and interlocked electrically or mechanically with a braking mechanism. Electrical interlocking devices should be so designed that if they fail, they fail safe, making the guarded machine inoperative.

In-Running Nip Points. In-running nip points are a special danger existing only through action of rotating objects. Whenever machine parts rotate toward each other, or where one rotates toward a stationary object, an in-running nip point is formed. Objects or parts of the body may be drawn into this nip point and be bruised or crushed.

The in-running side of rolling mills and calendars, of rolls used for bending, printing, corrugating, embossing or feeding and conveying stock, the in-running side of a chain and sprocket, belt and pulley, a gear rack, a gear and pinion, and a belt conveyor terminal are typical examples of nip point hazards.

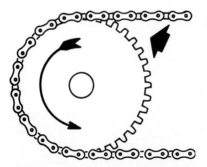

Figure A4-6 Chain and sprocket.

Figure A4-7 Belt and pulley.

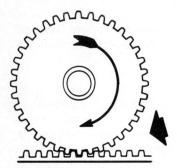

Figure A4-8 Rack and gear.

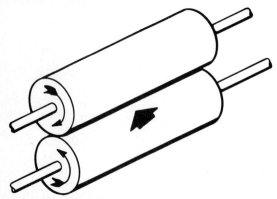

Figure A4-9 Pressure rolls.

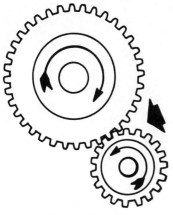

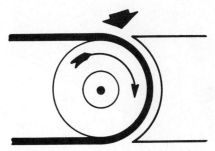

Figure A4-11 Conveyor terminal.

Figure A4-10 Gear train.

Cutting Actions. Cutting action results when rotating, reciprocating, or transverse motion is imparted to a tool so that material being removed is in the form of chips. The danger of cutting action exists at the movable cutting edge of the machine as it approaches or comes in contact with the material being cut. Such action takes place at the point-of-operation in cutting wood, metal, or other materials as differentiated from punching, shearing, or bending by press action.

Typical examples of mechanisms involving cutting action include band and circular saws, milling machines, planing or shaping machines, turning machines, boring or drilling machines, and grinding machines.

Punching, Shearing, and Bending Actions. Punching, shearing, or bending action results when power is applied to a ram (plunger) or knife

Figure A4-12 Circular saw.

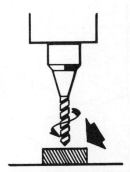

Figure A4-13 Drill.

Figure A4-14   Abrasive wheel.

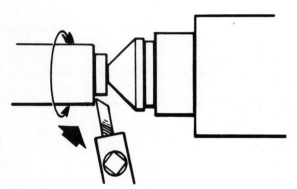

Figure A4-15   Engine lathe.

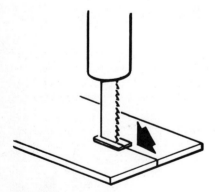

Figure A4-16   Band saw.

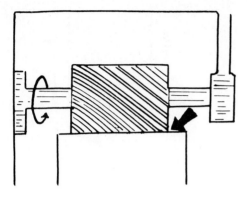

Figure A4-17 Milling machine.

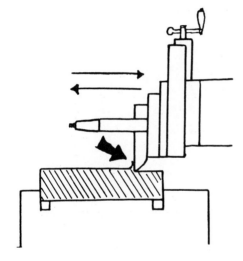

Figure A4-18 Shaper.

for the purpose of blanking, trimming, drawing, punching, shearing, or stamping metal or other materials as differentiated from removing the material in the form of chips. The danger of this type of action lies at the point-of-operation where stock is actually inserted, maintained, and withdrawn.

Typical examples of equipment involving punching, shearing, or bending action include power presses, foot and hand presses, bending presses or brakes as well as squaring, guillotine, and alligator shears.

Methods of Guarding Actions and Motions

Whenever hazardous machine actions or motions are used, a means for providing protection for the operator and fellow workers is essential. And,

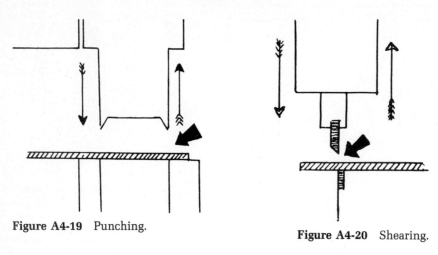

Figure A4-19 Punching.

Figure A4-20   Shearing.

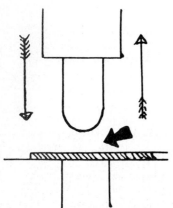

Figure A4-21   Bending.

there may be several ways to guard a situation, particularly at the point-of-operation.

Another type of interlocking barrier may be in the form of an electric-eye beam, a magnetic, radioactive or similar type circuit so designed and installed that when the operator's hand or any part of the body is in the danger zone, the machine cannot be operated, or if the hand or any part of the body is inserted while the machine is in motion, it will immediately activate a braking mechanism.

Automatic Guards.   When neither an enclosure guard nor an interlocking guard is practicable, an automatic guard may be used.

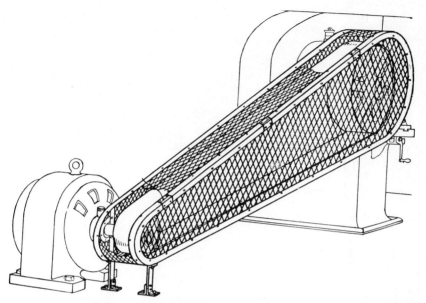

Figure A4-22   Pulleys with inclined belt.

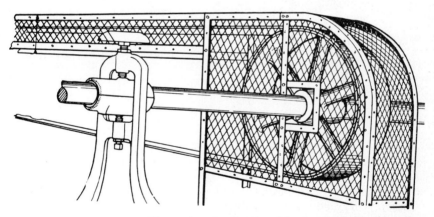

Figure A4-23   Overhead horizontal belt and pulley.

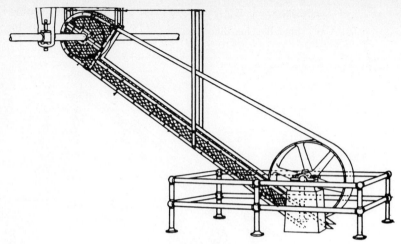

Figure A4-24 Pulleys and inclined belt.

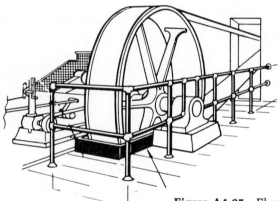

Figure A4-25 Flywheel with horizontal belt.

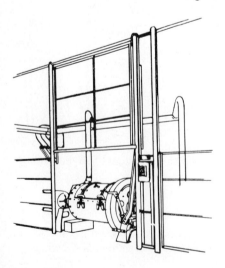

Figure A4-26 Guard for rattler or tanning drum. Angle iron gate frame is electrically interlocked so that the rattler cannot be operated unless the gate is closed.

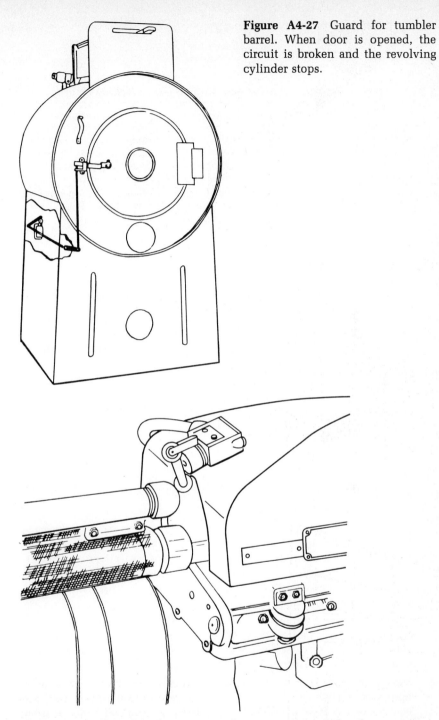

Figure A4-27 Guard for tumbler barrel. When door is opened, the circuit is broken and the revolving cylinder stops.

Figure A4-28 Guard for bar support of screw machine. When bar support or spindle cover is removed to open position, power to machine is cut off.

Methods of Guarding Actions and Motions—Interlocking Guards

Guarding Methods	Guard Action	Limitations
Enclosure with electrical or mechanical interlock.	Enclosure or barrier shuts off or disengages power and prevents starting of machine when guard is open; prevents opening of the guard while machine is under power or coasting. (Interlocks should not prevent manual operation or "inching" by remote control.)	Requires careful adjustment and maintenance. Operator may be able to make guard inoperative. Does not protect in event of mechanical repeat.
Barrier with electric contact or mechanical stop activating mechanical or electric brake.	Barrier quickly stops machine or prevents application of injurious pressure when any part of the operator's body contacts it or approaches danger zone.	Requires careful adjustment and maintenance. Possibility of minor injury before guard operates. Operator can make guard inoperative.
Electric eye	Electric-eye beam and brake quickly stops machine or prevents its starting if the hands are in the danger zone.	Expensive to install. Does not protect against mechanical repeat. Generally limited to use on slow speed machines with friction clutches or other means to stop the machine during the operating cycle.

An automatic guard acts independently of the operator, repeating its cycle as long as the machine is in motion. This type of guard removes the operator's hands, arms, or body from the danger zone as the ram, plunger, or other tool closes on the piece upon which work is being done. It is operated by the machine itself through a system of linkages connected to the operating mechanism.

Common types of automatic guards are sweep and push-away devices which are moving barriers crossing the danger zone when the machine is activated, and pull-away devices consisting of hand and arm attachments which pull the operator away from the danger zone.

Sweep and push-away devices should be designed to prevent the operator from reaching behind or across the protective device into the danger zone before the machine has completed its closing cycle. The device itself should not offer a hazard by creating a shear point between the moving guard and a stationary or moving part of the machine.

Methods of Guarding Actions and Motions—Automatic Guards

Guarding Methods	Guard Action	Limitations
Push-away device with barrier.	A movable barrier, attached to pressure part, pushes or lifts hands out of danger zone.	Hands must be in normal position for guard to be efiective. Hands may be caught by guard. Possible interference with feeding stock. Slow action may decrease production. Does not permit blanking.
Pull-away device	A cable operated attachment on ram, connected to the operator's hands or arms to pull the hands back only if they remain in danger zone; otherwise, it does not interfere with normal operation.	Requires unusually good maintenance, adjustment to each operator, and frequent inspection necessary. Limits movement of operator. May obstruct workspace around operator. Does not permit blanking from hand-fed strip stock from side of press.
Limitation of stroke	Plunger travel limited to ⅜ inch or less; fingers cannot enter between pressure points.	Small opening limits size of stock

Automatic pressure release or pivoting arm devices provide utility, yet protect in-running nip point situations.

Remote Control, Placement, Feeding, Ejecting. Although they are not guards in the technical sense, there are certain methods which can be used to accomplish the same effect, that is, of protecting the operator from the hazardous point-of-operation. They may be used to complement one of the other types of guards, or may be used in lieu of guards.

Two-handed operating devices may be used to activate the machine. These devices require simultaneous action of both hands of the operator or electrical switch buttons, air control valves, or mechanical levers. On presses with a noninterrupting stroke, two-handed operating devices should require manual operation until a point is reached in the cycle at which the hazard ceases. Hand controls may be interconnected with foot controls to permit operation of the machine. The actuating controls should be so located as to make it impossible for the operator to be able to move his hands from the controls to the danger zone before the machine has

Methods of Guarding Actions and Motions—Remote Control, Placement, Feeding, Ejecting

Guarding Methods	Guard Action	Limitations
Two-hand trip (can be adapted to multiple operation).	Simultaneous pressure of two hands on switch buttons in series actuates machine.	Operator may try to reach into danger zone after tripping machine.
	Simultaneous pressure of two hands on air control valves, mechanical levers, controls interlocked with foot controls, or the removal of solid blocks or stops permits normal operation of machine.	Does not protect against mechanical repeat unless blocks or stops are used. Some trips can be rendered unsafe by holding with the arm, blocking or tying down one control, thereby permitting one-hand operation. Cannot be used for some blanking operations.
Automatic or semiautomatic feed (with enclosure of danger points).	Stock fed by chutes, hoppers, conveyors, movable dies, dial feed rolls, etc. Enclosure will not admit any part of body.	Excessive installation cost for short run. Requires skilled maintenance. Not adaptable to variations in stock.
Special jigs or feeding devices.	Hand-operated feeding devices of metal or wood which keep the operator's hands at a safe distance from the danger point.	Machine itself not guarded; safe operation depends upon correct use of device. Requires good employee training and close supervision. Suitable for limited types of work.
Special tools or handles on dies.	Long handled tongs, vacuum lifters, or hand die holders which avoid need for operator putting his hand in the danger zone.	Operator must keep his hands out of danger zone. Requires unusually good employee training and close supervision.
Special ejecting devices	Air or mechanical ejection	Limited size of stock. Use in conjunction with other guarding methods.

completed its closing cycle. The two-handed controls should be so designed as to prevent the blocking, tying down, or holding down of one control to allow one hand free access to the point-of-operation. When more than one man is working a machine, additional controls should be installed and designed so that all men must simultaneously activate the starting mechanism from remote locations.

Automatic or semiautomatic feeding mechanisms such as roll, plunger, chute, slide and dial feeds, and revolving dies may be used in conjunction with ram enclosures. Special soft metal handtools may be used to place or remove parts in conjunction with an enclosure, interlocking or automatic guard. Special jigs, holding device and dies may be used to manipulate stock at the point-of-operation, yet keep hands safe. Mechanical or air-operated ejecting mechanisms may be used to remove parts, thus eliminating the need for the hands to be placed in the danger zone.

The theory behind these methods is that if for good reason it is impossible to completely enclose or isolate the hazard, the next best device or combination of devices should be used to keep the exposure to a minimum.

THE TECHNIQUES OF MECHANICAL GUARDING

It is recognized that a given situation—a hazard-creating motion or action—may frequently be guarded in a number of ways, several of which may be satisfactory. The selection of guarding method to be used may depend upon a number of things—space limitations, production methods, size of stock, frequency of use, and still other factors may be important in making the final decision. It is not the intent of this bulletin to suggest which method of guarding is the best for a given situation, but rather to show that there are a number of ways to guard each different condition. This will be done by illustrative typical situations which may be guarded by a variety of methods.

In the illustrations, the various motions and actions are shown with typical guards illustrating the various guarding techniques. It is not possible to apply all of the guarding techniques to all of the motions or actions, but an effort has been made to show those that are frequently found in industry.

Guarding Rotating, Reciprocating, or Transverse Motions

Rotating, reciprocating, and transverse motions are forms or directions of motion that are used for a variety of purposes. They may be used to transmit power, as for example a revolving shaft or pulley (rotating), a moving piston (reciprocating), or a moving belt (transverse). Motions are also used to perform work by giving action to a tool or die which performs a cutting or shearing action. Thus, a circular saw or milling cutter (rotating), a metal planer or shaper (reciprocating), or a band saw (transverse) are examples of using motion to perform useful work by means of a tool.

Motion is also used as a means of performing work without the cutting action of tools. A centrifugal extractor or dough mixer (rotating), a shaker screen (reciprocating), or a belt or screw conveyor (transverse) are examples of using the energy of motion to perform useful work.

The illustrations on the next nine pages are intended to show recognized techniques for guarding various forms of motion used for transmitting power and for performing work without the benefit of tools.

Where tools are used to perform useful work, a cutting, punching, bending, or shearing action is involved. Techniques of guarding these various actions will be illustrated under the general heading of the type of action, instead of under the heading of the type of motion, used to drive or power the tool.

Guarding In-Running Nip Points

In-running nip points may exist under two broad classes of conditions, namely:

1   Those purposely created to do useful work.
2   Those which are incidental to the transmission of power or the application of motion, the nip point itself doing no useful work.

Nip points that are purposely created to do useful work would include situations such as calender rolls used for finishing paper, rolls used for mixing ink or rubber, and rolling mills used for reducing the thickness of or changing the shape of metals.

Nip points created incidental to the transmission of power would include the nip point of a belt and pulley, the meshing of gears, or the take-off point of a belt conveyor. Nip points created incidental to the application of motion would include the nip or pinch point of the moving bed of a metal planer with the frame of the machine or an adjacent fixed object.

In any case, guarding is necessary and various means may be used to provide protection. The following illustrations show typical nip point situations and techniques for guarding them.

Guarding Cutting Actions

Cutting action is involved in the whole range of machine tools used for cutting metal, also for machines used for working on wood. The characteristic of a machine tool is that the metal removed is in the form of chips. This characteristic also applies to machines used for cutting wood,

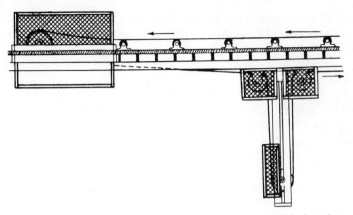

Figure A4-29   Elevated conveyor belt, pulleys, and belt tightener guard.

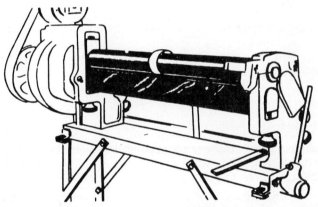

Figure A4-30   Upper roll covered with stationary cylindrical guard. Lower roll has an apron guard. Opening between guards is less than ⅜ inch.

although woodworking machines are not referred to as machine tools. Many machines used in the food, plastic, paper, or textile industries, as well as other industries, also employ cutting action. The hazard lies at the point where cutting takes place.

Various forms of motion may be involved in cutting action, but the kind of motion is not the determining factor in selecting the best guarding method. In many cases, a choice of guarding methods is available. The following illustrations show various methods of guarding typical cutting actions.

Figure A4-31 Radial saw. In addition to hood enclosing the blade, an adjustable stop should be provided to limit forward travel and head should automatically return to starting position. When used for ripping, a spreader and antikickback device should be provided.

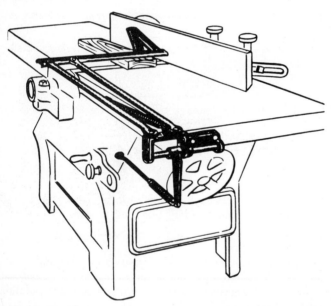

Figure A4-32 Jointer guards should automatically adjust themselves to cover all sections of the head on the working side of the fence or gauge and should remain in contact with the material at all times. The section of the cutting head back of the fence or gauge should also be guarded.

Guarding Punching, Bending, and Shearing Actions

Punching, bending, and shearing actions differ from cutting actions in that material is not removed in the form of chips. This class of action is used primarily to form or shape metals, but may be used on materials other than metal, as for example, plastics, fabrics, and papers.

Punching, bending, and shearing actions are particularly hazardous with respect to the severity of injury inflicted, many such injuries resulting in amputations or other permanent disability. As in the case of cutting actions, the hazard is at the point-of-operation.

The following are illustrations of machines which perform punching, bending, or shearing actions and techniques by which they are guarded.

General Recommendations for the Construction of Guards. Guards should be constructed in accordance with the following general recommendations. Reference should also be made to the table below.

Guards should consist of a filler of expanded metal, perforated or solid sheet metal, or wire mesh on a frame of angle iron or iron pipe securely fastened to the floor or machine frame. (Fine wire mesh is not recommended in areas where lint collects.) All metal should be free from burrs or sharp edges.

Wire mesh should be of the type in which the wires are securely fastened

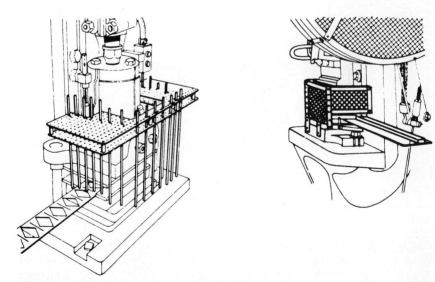

Figure A4-33 Enclosure guards for punching or forming presses for use with strip feeding.

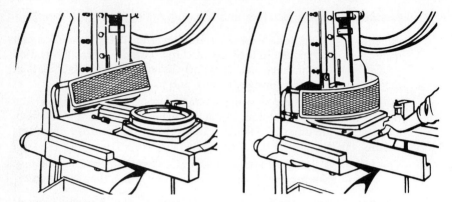

Figure A4-34 Stock is fed into die outside of danger zone. Hinged barrier guard closes automatically when sliding die is moved into position.

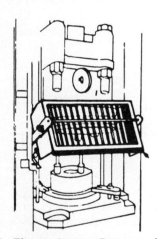

Figure A4-35 Gate guard on punch press. Guard is interlocked with tripping device of press so that plunger cannot descend until gate is in closed position.

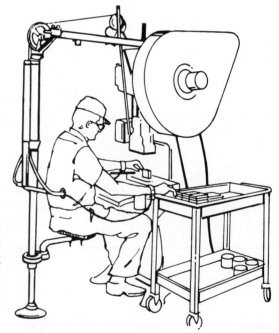

Figure A4-36 Pull-away guard. Operator's hands are pulled away from danger zone before ram descends.

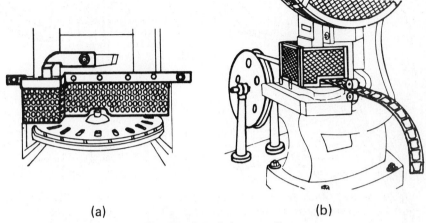

| (a) | (b) |

Figure A4-37a Rotating die. Stock is fed manually into rotating die and is carried automatically under the plunger.

Figure A4-37b Automatic feed. Stock is fed automatically to plunger, operator required only to restock feeding device.

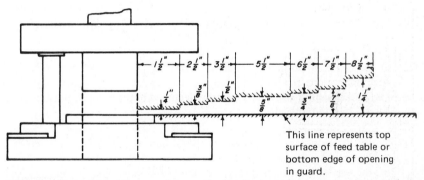

This line represents top surface of feed table or bottom edge of opening in guard.

Figure A4-38 Guarding punch press by limitation of ram stroke. If ram stroke is limited to ¼ inch, enclosure is unnecessary. When enclosure of die is necessary, size of openings should not exceed that shown for various distances.

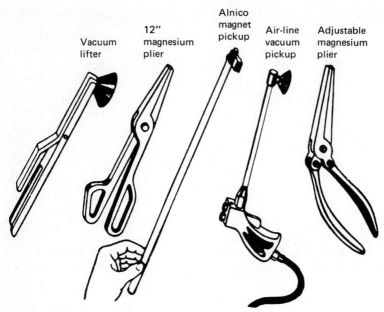

Vacuum lifter   12" magnesium plier   Alnico magnet pickup   Air-line vacuum pickup   Adjustable magnesium plier

Figure A4-39 Guarding by the use of special tools. The use of specially designed hand tools for placing material into the press is also acceptable.

at every cross point by welding, soldering, or galvanizing, excepting diamond or square wire mesh made of No. 14 gauge wire, ¾-inch mesh or heavier. The filler should be securely fastened to the guard frame by one of the following methods:

1   With rivets or bolts spaced not more than 5 inches center to center.
2   By welding to the frame every 4 inches.
3   By weaving through channel or angle frame (if No. 14 gauge ¾-inch mesh or heavier is used, by bending entirely around rod frames).
4   Where openings in pipe railings are to be filled in with expanded metal, wire mesh, or sheet metal, the filler material should be made into panels with rolled edges or bound with "V" or "U" edging of No. 24 gauge or heavier sheet metal fastened to the panels with bolts or rivets spaced not more than 5 inches center to center. The bound panels should be fastened to the railing with sheet-metal clips spaced not more than 5 inches center to center.

Where the design of guards requires filler material of greater area than 12 square feet, additional frame members should be provided to maintain panel area within this limit.

Standard Materials and Dimensions

Material	Clearance from Moving Part at All Points	Largest Mesh or Opening Allowable	Minimum Gauge (U.S. Stand.) or Thickness	Min. Height of Guard from Floor or Platform Level
Woven wire	Under 2"	3⁄8"	No. 16	7'–0"
	2"–4"	1⁄2"	No. 16	7'–0"
	Under 4"	1⁄2"	No. 16	7'–0"
	4"–15"	2"	No. 12	7'–0"
Expanded metal	Under 4"	1⁄2"	No. 18	7'–0"
	4"–15"	2"	No. 13	7'–0"
Perforated metal	Under 4"	1⁄2"	No. 20	7'–0"
	4"–15"	2"	No. 14	7'–0"
Sheet metal	Under 4"		No. 22	7'–0"
	4"–15"		No. 22	7'–0"
Wood or metal strip crossed	Under 4"	3⁄8"	Wood 3⁄4" Metal No. 16	7'–0"
	4"–15"	2"	Wood 3⁄4" Metal No. 16	7'–0"
Wood or metal strip not crossed	Under 4"	1⁄2" width	Wood 3⁄4" Metal No. 16	7'–0"
	4"–15"	1" width	Wood 3⁄4" Metal No. 16	7'–0"
Std. rail	Min. 15" Max. 20"	See Standard for railings		

All joints of framework should be made equivalent in strength to the material of the frame.

The framework of all guards should be angle iron 1 inch by 1 inch by 1⁄8 inch, metal pipe of 3⁄4-inch inside diameter, or metal construction of equivalent strength. Guards should be rigidly braced every 3 feet or fractional part of their height to some fixed part of machinery or building structure. Additional strength may be necessary where a guard is exposed to contact with moving equipment.

The framework of guards fastened to a floor or working platform without other support or bracing should consist of 1½- by 1½- by 1⁄8-inch angle iron, metal pipe of 1½-inch inside diameter or metal construction of equivalent strength.

Guards 30 inches or less in height and with a total surface area not in excess of 10 square feet may have a framework of ⅜-inch solid rod, ¾- by ¾- by ⅛-inch angle, or metal construction of equivalent strength.

The framework specifications indicated above are minimum requirements. Where guards are exposed to unusual wear, deterioration, or impact, heavier materials should be used.

Wood guards may be used in the woodworking industry and in industries where manufacturing conditions would cause the rapid deterioration of metal guards, also in construction work and in outdoor locations where metal guards and railing are undesirable because of extreme cold or heat. Wood guards should be sound, tough, and free from loose knots. They should be made of planed lumber not less than 1-inch rough-board measure, with edges and corners rounded off.

APPENDIX FIVE

Personal Protective Equipment

The following discussion of personal protective equipment is excerpted from Chap. 36, *"Personal Protective Devices,"* of *The Industrial Environment—Its Evaluation and Control*, published by the National Institute for Occupational Safety and Health in 1973. This chapter was authored by Harry F. Schulte.

GENERAL PHILOSOPHY

It is one of the fundamentals of industrial hygiene that personal protective devices are "last resort" types of controls, to be used only where engineering controls cannot be used or made adequate. It should be noted that this fundamental is stated unequivocally in the standards adopted under the Occupational Safety and Health Act.

PROTECTION AGAINST INHALATION HAZARDS

Where Used

There will always be a temptation to resort to respirators as a cheap substitute for a ventilation system. If this is done it is clear that management has not carefully considered the alternatives since reliance on, and effective use of, respirators is definitely not cheap. A careful study may be required to determine that no other effective control measures can be used. Respirators are designed to protect only against certain specific types of substances and in certain concentration ranges, depending on the type of equipment used. Many other factors should be considered carefully in making the decision that other engineering controls are not practical. Nevertheless, there are many places where respirators can and should be used with full knowledge of their limitations and requirements.

Approval Systems and Schedules

The U.S. Bureau of Mines gave its first approvals of oxygen breathing apparatus in 1918 and gradually extended this activity to include all types of respirators. In 1972 NIOSH took over the approval function from the Bureau of Mines for aboveground uses. Considerable research is now underway to expand the scope of the approval tests and to develop more satisfactory procedures to meet the changing and expanding needs.

Particle-Removing Air Purifying Respirators

Applications and functions.   These devices are designed to protect the wearer against inhalation of material dispersed in air as distinct particles — as a dust or fume in the case of solid particles or as a mist or fog in the case of liquid droplets. They consist, principally, of a facepiece with some type mechanical filter. The material to be protected against may be a nuisance dust such as sawdust, a pneumoconiosis-producing dust such as silica or coal dust, a toxic dust like lead oxide, a metal fume like cadmium, a highly toxic dust like beryllium oxide or a radioactive dust such as plutonium.

Limitations.   The user of any air purifying respirator must be certain that the atmosphere contains adequate oxygen and that the only harmful materials present are those which can be removed by the respirator to be worn. There are many ways in which air can leak around the filter, seriously reducing the protective capability of the respirator. Failure of the filter to seat properly in its holder, leaking valves and imperfect sealing of the respirator to the wearer's face are all significant. Some of these factors can be controlled by proper design but all require good and frequent inspection and maintenance.

Facepieces.   There are two basic types of facepieces used on air purifying respirators — half masks and full face masks. The half masks do not cover the eyes and hence offer no eye protection nor do they interfere as much with vision as do the full face masks. The half mask must contact a rather complex facial surface and the possibility of leaking is greater than in the case of the full face mask. A fitting program is required to provide adequately fitted respirators.

A serious problem with full face respirators is obtaining an adequate seal for the worker who must use corrective glasses. Where the temples of conventional eyeglasses emerge from the facepiece, there is a serious place of leakage. Some masks provide methods of mounting a special set of corrective lenses inside the respirator facepiece. Half masks are usually held on the face by means of a single set or a double set of elastic bands. The double sets of straps are practically a necessity in assuring a reasonable face fit and each set should fasten to separate suspension points on the respirator. Full face masks, being much heavier, are supported by a head harness which should have at least five adjustable straps.

Most respirator facepieces contain valves to direct the flow of air from outside, through the filter, into the nose and then back outside. Valveless respirators are being used, but are not approved for use with hazardous materials. The most important valve is the exhalation valve which opens during expiration allowing expired air to pass directly outside the mask. It closes during inhalation and must close positively and quickly to prevent toxic material being drawn into the facepiece.

Filters.   Filtration by fibrous filters is the method universally used in respirators for removing particles from the air. The filter material may be a loosely or tightly packed mass of fibers of cotton, wool, synthetic fibers, glass or mineral fibers, or it may be a paper made of these materials. The filter efficiency is influenced by the particle size, shape and density of the aerosol and by a number of other factors determined by the filter and the respirator.

There are approval schedules for respirators for protection against (1) pneumoconiosis-producing and nuisance dusts, (2) toxic dusts (not significantly more toxic than lead), (3) metal fumes (not significantly more toxic than lead), (4) chromic acid mist, (5) dusts significantly more toxic than lead and (6) various combinations of these. A summary of some of the requirements for approval are given in Table A5-1 from the AIHA-ACGIH Respirator Manual.

Gas and Vapor-Removing Air Purifying Respirators

Applications.   These respirators are designed to protect the wearer against the inhalation of materials in the air that are present in the form of gases or vapors. Respirators for protection against such materials are equipped with a container or canister filled with a sorbent which absorbs, adsorbs or reacts with the hazardous gas in the atmosphere. Some sorbents are highly specific for a particular compound and so the respirator containing only this sorbent gives no protection against any other material. Other sorbents take out whole classes of compounds such as acid gases or organic vapors.

Limitations.   Like the particulate removing respirator, the gas removing device is useless unless adequate oxygen is present. Unless the device is specifically equipped with a filter in addition to the sorbent, it will not offer adequate protection against any hazardous particulate substances. Leakage around canisters can also occur as around filters but this is less likely. Leakage around the facepiece is also possible but, since these respirators usually offer less resistance to breathing than filter types do, there is a slightly smaller chance of this type of leakage. Once a sorbent canister is opened and used, the sorbent may absorb moisture or other deleterious materials from the air and deteriorate even without further usage.

For respirators designed for use under highly dangerous conditions some canisters are equipped with devices which change color when sorbent depletion approaches. For respirators designed for use under less hazardous conditions the odor of the gas penetrating the canister is the only warning. For these reasons simple gas removing respirators are most frequently designed for a single usage in dealing with a specific situation.

Facepieces.   Essentially the same information relating to facepieces on particulate removing respirators applies also to gas and vapor removing respirators. Because many gases and vapors are also irritating or damaging to the eyes, full face respirators are much more frequently required than half masks. Half mask respirators usually carry relatively small sorbent canisters which can be used only for short periods in relatively low concentrations of gas. Even with the full facepiece and its head harness, the sorbent canister for heavy duty usage may be too heavy to be supported on the facepiece. It may be worn on a chest harness or strapped to the belt. The canister is then connected by a hose to the facepiece.

Since there is usually little warning of impending canister failure, the instructions of the manufacturer must be carefully followed regarding the length of time the device can be worn and the circumstances of its use. The effectiveness of a canister depends on the presence of a reactive chemical and canisters usually deteriorate with time. Here again, the manufacturer's instructions regarding shelf life should be followed and outdated canisters should be discarded.

Atmosphere Supplying Respirators

Applications.   When a lack of oxygen is known or suspected or when a very high degree of protection is required the only suitable device is the atmosphere supplying respirator. This consists of a source of air or oxygen which is fed through a hose to a mask or a helmet. Intermediate mixing and regulating equipment may be required. Supplied air devices offer essentially no resistance to breathing, and the atmosphere supplied may be cool and more acceptable than that from other types of respirators.

Facepieces.   A variety of facepieces are used with atmosphere supplying respirators. Half masks and full face masks are similar to those previously described although the valves are somewhat different. In addition to these, helmets may be used which cover the entire head and a cover may extend down to the waist as in the abrasive blasting hood. Another type of facepiece is the air supplied face shield. In this, the air is supplied by a hose to a perforated or slotted tube at the top of the face shield and the jets of air are directed downward past the eyes, nose and mouth. It is difficult to adjust this device to avoid entraining contaminated air and blowing it into the breathing zone. There are other combinations of atmosphere supplying equipment with hoods, blouses and complete clothing.

Hose Types.   There are several varieties of the hose types including the hose mask with blower, hose mask without blower and the air-line respirator. The hose mask with blower can be used in any atmosphere provided that enough respirable air is supplied to the wearer by means of the blower. The blower must be hand operated and the blower operator serves as an observer capable of rescuing the wearer in case of accident. The hose mask without blower is used under conditions not immediately dangerous to life and from which the wearer can escape without the aid of the respirator.

The air-line respirator consists of a source of compressed air to which the facepiece is connected by means of a small diameter hose. In the line is a pressure reduction valve and some type of flow regulating device. This equipment is used for protection in atmospheres that are not immediately dangerous to life or health or from which the wearer can escape without the aid of the respirator.

There are two basic types or modes of operation of air-line respirators — continuous flow and demand. In the first, the air is fed to the facepiece continuously and a positive pressure is maintained inside the mask at all times and any leakage will be outward. Considerably more air is used than is consumed in breathing. In the demand type a valve which regulates the flow of air opens only when a slight negative pressure is produced inside the mask as a result of inhalation. Thus, only air used in breathing is drawn from the source. Since a negative pressure is produced inside the facepiece it must fit tightly to the face or contaminated air will be drawn in. Usually the demand type has a by-pass valve so the wearer can switch to continuous flow if he desires. A newer variation is the pressure-demand type of flow regulation. In this, a small positive pressure is always maintained in the facepiece even on inhalation. The demand valve opens to supply air when the positive pressure decreases to a certain level as a result of inhalation. Thus leakage is always outward if a poor fit of facepiece to face is obtained.

Self-Contained Breathing Apparatus. In this apparatus the wearer carries his own supply of air or oxygen and so he is able to move about without attaching hoses which limit his travel distance and maneuverability. Since the wearer is limited to the supply of air which he can carry, several methods are used to conserve this supply. Demand and pressure-demand valves may be used in the same manner as in the air-line respirator. A bypass for continuous flow is also provided, but can be used only for very short periods or the supply will be quickly exhausted. Either air or oxygen may be used in this type of apparatus.

Another type of device uses recirculation to conserve the oxygen supply. In this unit the exhaled air is not expelled but passes through an absorber which removes carbon dioxide and then enters a breathing bag. Here it mixes with fresh oxygen from the gas cylinder and then passes back to the facepiece for rebreathing. Oxygen only enters the breathing bag when the pressure in the bag drops below a fixed value. Since the oxygen content of exhaled air is still about two-thirds of that of inhaled air, it is only necessary to make up this difference from the tank and thus the recirculating type can be used for much longer periods for the same quantity of oxygen carried. Both demand and recirculating equipment are marked for use only for a specified period of time ranging from thirty minutes to three hours.

A third type of self-contained breathing apparatus uses a chemical source of oxygen which is liberated when carbon dioxide and moisture are absorbed from the exhaled air. A breathing bag is provided to mix the incoming oxygen with the purified exhaled air. This apparatus can be used for thirty minutes only and includes a timer to warn of the approaching limit.

Once the canister is opened it cannot be resealed for further use even if it has been used for only a few minutes. Like all chemical cartridges it has a limited shelf life even if unopened.

Sources of Air or Oxygen. The hose mask simply uses air drawn from a source outside the contaminated atmosphere. The air-line respirator may use a compressed gas cylinder of air or oxygen or a compressor which picks up outside air. If the compressor is driven by a gasoline or diesel engine extreme care must be taken to see that the engine exhausts away from the pump intake and no exhaust gases enter the breathing air system. Water-lubricated compressors, or those not requiring internal lubrication, are the best sources of compressed air for these devices since heating may cause breakdown of lubricating oil forming carbon monoxide. Oil-lubricated compressors, if used, should be equipped with thermal overload switches to turn them off if overheating occurs. In any case, the compressor should be followed by a trap and filter to remove dirt, oil and water from the air line. If a regular building supply is to be used for an air-line respirator the system must be checked repeatedly to be certain that it does not contain even small concentrations of carbon monoxide.

A new alternative source of oxygen for the self-contained breathing apparatus is liquid air or liquid oxygen. Several devices using this oxygen source have been approved by the Bureau of Mines. Air from this source is cooler and may have a "fresher" odor than compressed air.

Maintenance. All types of respirators require maintenance and cleaning. The highly complex atmosphere supplying devices must be inspected for signs of wear and deterioration and to make certain that all of the parts are functioning properly. Other types of respirators are simpler but still require regular inspection. Rubber parts deteriorate with time and exposure to ozone and other gases. Hence, hoses, facepieces and valves must be inspected regularly and parts or whole respirators replaced when necessary. All respirators should be brought in for cleaning occasionally and emergency devices cleaned after every use when cylinders of gas or canisters are replaced. Emergency devices and self-contained breathing apparatus should be inspected monthly since they are usually to be used under very hazardous conditions.

When respirators are used routinely they should be brought in regularly to be dismantled, washed and dried, inspected and parts replaced as necessary, new filters or canisters put in and the whole reassembled and stored in a clean place.

A Respirator Program for Industry

The important elements of the successful plant respirator program will be summarized to provide a means for checking to be certain that all factors are covered in a particular situation.

Determination of Need. This will require knowledge of the hazards anticipated in carrying out a particular job. An estimate will be required of the possible concentrations of toxic material that could be produced; whether existing engineering controls, such as ventilation, can adequately meet the needs; the anticipated duration of the required protection and any limitations imposed by the job. The latter includes the intensity of the physical activity required and whether or not the worker must be able to move about without encumbering hoses. Obviously, the determination of the need for respiratory protection is a technical decision and can best be made by an industrial hygienist.

Selection of Equipment. Much of the same information required in establishing the need for respiratory protection is required here also. In addition, one must have a knowledge of the types of devices available for the particular circumstances encountered. Since it is impossible to expect each plant to have on hand every type of device manufactured, equipment kept on hand must be purchased in advance on a basis of surveys and studies of anticipated needs. Here, there is a requirement for close cooperation between the person in charge of the respirator program and the plant's purchasing and stores department. Preliminary education of appropriate persons in this department may be necessary since their tendency will be to stock items on a basis of price without regard to important distinctions between different pieces of equipment.

Training.   This is very important and requires continuous study and updating of knowledge by the person giving the training. The latter should be the person responsible for selection of respirators to be carried in stores, or at least he should be in very close touch with this aspect, since his direct contact with workers using protective devices makes him aware of their needs.

Superivison and Enforcement.   The support and encouragement of supervisors, such as foremen, are essential to the program. The foremen, particularly, should be asked to participate in fitting and training even though they may have little occasion to use respiratory protective devices. This provides an opportunity to demonstrate the importance of the activity and gains their support for the program. Without this support many workers will not use the equipment correctly nor care for it adequately.

Inspection.   Industrial hygienists and safety engineers in the plant must include regular inspection of the condition of respirators as one of their routine duties. It is particularly important that emergency devices such as those mounted on walls or in cabinets be checked regularly. The most important inspections, particularly of smaller devices, are those given by the worker himself. His training must include information on the importance of this and how to do it adequately.

Maintenance.   In most plants there should be a central cleaning station where respirators are brought in regularly for cleaning and replacement of worn or damaged parts. In very small plants this may be done by the worker himself with the aid of the safety engineer, industrial hygienist or plant nurse.

Storage.   New respirators should be stored in their original container in a clean, cool dry place before being issued. Cleaned respirators if not immediately reissued should be placed in a dust-tight container, such as a plastic bag. The worker also should be instructed to store his respirator properly after it is issued to him. A respirator crammed into a tool kit can be permanently distorted in shape so that it cannot fit. Dust accumulated on the interior of the mask is readily breathed in and wearing such a mask may be a cause of more exposure than failing to wear it.

Management Interest. Top management should give some evidence of support of the program. This may be done, in part, through the plant paper or magazine or through shop bulletins or other means of indicating interest or concern. It is the responsibility of the person in charge of the program to keep management informed about the program if he is to obtain their interest and support.

PROTECTION AGAINST NOISE

The previous section of this chapter has dealt at considerable length with the subject of respiratory protection. Much of this is also applicable to devices used to protect against noise and other hazards. With hearing protective devices, both ear plugs and ear muffs, there is a need for experimentation in trying different types of devices to meet specific needs. Since the noise hazard is not as acute as the respiratory hazard more experimentation is justified. Both plugs and muffs attempt to prevent the penetration of sound through the outer ear to the inner ear; however, some sound also reaches the inner ear by conduction through bone and tissue. Thus, any protective device is limited in the degree of protection which can be achieved. In sound fields in excess of 120 decibels no protective device will give adequate protection for continuous exposure.

When ear protectors are first used by a worker, he experiences a sensation that his own voice is very loud since outside noises are reduced. As a result, he tends to speak softer making it more difficult to communicate with others especially if they are wearing protectors also. Noises signaling dangers around the worker may be muffled and their warnings not heeded.

Ear Plugs or Insert Devices

Ear plugs are small conical or cylindrical devices made to fit into and seal the ear canal against the entrance of sound. The more closely the plug approximates the shape of the ear canal the more positively it will seal. Plugs are usually made of a soft pliable material like soft rubber or plastic so they can be inserted into the ear canal with positive force without being uncomfortable. Many models have soft flanges which can seal the canal even if contact with the body of the device is incomplete. Attempts have been made to improve the usefulness and acceptability of ear plugs by introducing models with "valves" or perforations which were to allow passage of sound of certain frequencies or to block loud but not soft sounds. None of these has been successful and the plain plug remains the best. Since fitting to the ear canal is important some manufacturers make plugs for individual ears by making casts or impressions of the ears and then moulding and curing a plastic material in the shape of the impression. While such devices would appear to have a great advantage in effectiveness and comfort, this is not always the case.

Malleable wax material is available which can be moulded into the ear and discarded after use. Wadded cotton has often been used, but actually is comparatively ineffective. Cotton impregnated with wax or vaseline is much more effective. A very fine glass wool material has been introduced in recent years, often called Swedish Wool after its country of origin. This material can be rolled into plugs and gives almost as good protection, if properly used, as a good commercial plug. Dispensers for this material can be installed and the material is more acceptable to many workers than regular ear plugs.

The degree of sound attenuation provided by ear plugs varies in different frequency bands or octaves. For good, well-fitted plugs this varies from 25 decibels in the low frequency or low pitched sounds to 40 decibels at frequencies over 1000 Hertz. Fitting of the plugs is very important in achieving good results. Most plugs come in several sizes and the correct size must be chosen to fit the individual's ear canal. Frequently different sizes are needed for the two ears. Fitting by the Medical Department is a good way of achieving this part of the program's purpose. The physician can examine the ear canal carefully for size and at the same time detect any ear infections or canal irregularities which may rule out the use of ear plugs. Some persons simply cannot wear ear plugs.

Ear Muffs

Ear muffs are much like communications type earphones in appearance although the ear cup is usually deeper. They are equipped with a headband which may go over the head or around the back of the neck. The latter type permits the wearing of a hat but usually offers slightly less hearing protection. They may also be mounted on a helmet or hard hat. The degree of attenuation obtained varies with the sound frequency and may range from 20 decibels at low frequency to 45 decibels at high frequency. There is a standard method of measuring attenuation by muffs, but it is complex and requires special equipment. Most users will have to rely on the attenuation data supplied by the manufacturer.

To attain good protection the muff should seal over the ear and the seal may be made of foam rubber or liquid-filled or grease-filled cushions. The latter two are somewhat more effective and comfortable. Large cup volumes and small cup openings lead to greater attenuation.

Proper fitting is important with muffs, but is not quite so individual a matter as with plugs. The worker should be offered a choice of several models to achieve the best fit and to gain his acceptance of the device. Where communication is necessary in a high noise level environment, ear muffs can be equipped with earphones and battery operated radios. The microphone can be muffled and mounted directly in front of the mouth.

Plugs or Muffs?

Good ear plugs may give slightly higher attenuation in the very low frequency range while muffs give better attenuation in the middle ranges. Above 1000 Hertz there is little from which to choose on a basis of the degree of protection afforded. Plugs are more acceptable to some workers while muffs are preferred by others. Where exposure to noise is intermittent, muffs are somewhat more easily removed and replaced when needed. Plugs are easier to carry than muffs, but for the same reason are more easily lost. They are also less expensive. Muffs are more comfortable for use in cold weather. Both plugs and muffs deteriorate with time and should be inspected frequently.

Evaluation

If ear protection is required, there should also be a medical program which includes audiometric testing of exposed workers. Results of such testing will determine whether continued exposure is advisable even if protective devices are used. Obviously, every attempt should be made to control the noise by engineering methods and eliminate the need for personal protective devices. One author puts the case as follows—"When the only possible control method is ear protection, it is important—rather it is essential—that the ear protection program be continually monitored by knowledgeable and enthusiastic people who are dedicated to the task of protecting the hearing of noise-exposed employees."

PROTECTION OF SKIN AND BODY

This section deals with the subject of what is usually called protective clothing and includes protection of the various parts of the whole body either completely or partially as may be required. The term, protective clothing, is a correct one although much protective clothing offers no more direct bodily shielding than would be offered by ordinary street dress. Since street clothing might be ruined or rendered unsuitable for street use if worn in the shop, the shop clothing provided is "protective" of one's ordinary clothing. Chemicals, dirt, heat and cold are the chief hazards against which protective clothing are used. Certain rather specialized occupations require unusual types of protection including firemen, aircraft crews, missile fuel handlers, astronauts and divers. In general, this section is not directed to these specialized vocations but much of what we have learned about protective clothing comes from experience in providing protection for workers in such unusual environments.

There is a wide variety of materials available today to meet the requirements of many types of conditions. Fabrics such as cotton, glass fibers, Orlon, Nylon, Dynel and even Teflon are available for jackets and coveralls. These can be made impervious by coating with various plastics, rubber or Neoprene. Plastics are also available in sheet form and can be made into clothing with glued or heat-sealed seams. Respiratory protective devices often are worn with protective clothing and, in many cases, become an integral part of such clothing.

While some protective clothing is supplied and cleaned by the worker it is more commonly done by the employer, especially if the wearing of such clothing is a requirement of the job. In many cases, work clothing cannot and should not be taken home since such clothing could become a hazard to persons handling it there. This is particularly true of workers handling radioactive materials, pesticides, beryllium and other highly toxic materials. If the employer is responsible for cleaning the clothing he, too, must decide whether to send it to a commercial laundry or to set up his own cleaning facility. For clothing contaminated with ordinary dirt, soil, grease and perspiration, a commercial laundry may be an acceptable method of handling the problem. In many cases, one or more plant laundries will have to be provided depending on the variety of exposures encountered.

There is also the problem of maintenance and of replacement. Garments must be inspected before or after cleaning and worn garments discarded. Those not meeting standards of cleanliness required may have to be recycled. For radioactive materials, this may mean monitoring of each garment with a special instrument. Where the exposure is to certain chemicals, it may mean periodic tests on occasional selected garments after cleaning.

Training in the use of protective clothing is important, especially with complex gear. Even with simple laboratory coats and coveralls, workers should be given some instructions in how to care for such garments and how and when to turn them in for cleaning and replacement. Here again, some one person should be in charge of selection, storage, maintenance and training in the use of such equipment. An important element in this training is that dealing with the methods of putting on and taking off this clothing. Special standardized techniques may be necessary where clothing has or could have become heavily contaminated. If care is not exercised, contaminating materials can be brought into contact with the worker's skin or transferred to his street clothing. Separate lockers for work clothing and street clothing are necessary when hazardous materials are handled.

EMERGENCIES

By its very nature, personal protective equipment is important in dealing with emergencies. Mine explosions and cave-ins, industrial fires, natural disasters and gas leaks all present occasions when men must receive protection against severe and unusual conditions while saving life and property. Engineering controls such as ventilation are completely inadequate and usually inapplicable in such situations; therefore complete reliance must be placed on protective clothing and respiratory protective devices. Some of the same devices used in the daily plant activities may be applicable but in most cases it will be necessary to use more specialized devices such as self-contained breathing apparatus or fire-resistant clothing.

Equipment for dealing with emergency situations must be stored where it is readily available and yet not be placed where it can be damaged by the accident causing the emergency. This requires careful analysis to anticipate the possible emergencies and accidents which can arise. An important characteristic of emergency protective devices is that they are *stored* which means that they are actually used very rarely. Since much equipment suffers deterioration even during storage, it is necessary to set a regular schedule of inspection and testing of each device.

Most such devices are fairly complex and require trained persons for their use. An untrained person attempting to rescue an injured man in a building filled with toxic vapors is very likely to become a casualty himself. Men must be trained in how to function in an emergency and in the use of all emergency equipment. Retraining exercises must be given at regular intervals since instructions are easily forgotten, equipment is changed, and personnel may be transferred. The essential beginning of any emergency planning is a thorough analysis of the possibilities of accidents. This analysis, in itself, may lead to the correction of unsafe conditions.

The OSHA standards are quite clear on the need for and use of most types of protective equipment. Subpart I is devoted to personal protective equipment. It states:

a *Application.* Protective equipment, including personal protective equipment for eyes, face, head and extremities, protective clothing, respiratory devices, and protective shields and barriers shall be provided, used, and maintained in a sanitary and reliable condition wherever it is necessary by reason of hazards of processes or environment, chemical hazards, radiological hazards, or mechanical irritants encountered in a manner capable of causing injury or impairment in the function of any part of the body through absorption, inhalation or physical contact.

b *Employee-owned equipment.* Where employees provide their own protective equipment, the employer shall be responsible to assure its adequacy, including proper maintenance, and sanitation of such equipment.

c *Design.* All personal protective equipment shall be of safe design and construction for the work to be performed.

RESPIRATORY PROTECTION

The requirements for respirators are lengthy and spelled out in detail. It is first necessary to determine if a hazard exists requiring the use of respirators. The compliance officer's checklist is given:

GENERAL

1 Generally, the compliance officer will look for the following:
   a That approved protection is used where there is a probability of eye, head, hearing, or respiratory injury.
   b Protective equipment is cleaned, maintained, and issued.
   c Nonapproved items, such as visitors specs, bump caps, and nuisance dust respirators, are used where approved equipment is required.

2 Safety spectacles will be examined to determine:
   a That frames support safety lenses entirely around the periphery.
   b That street-wear frames and lenses are not used on hazardous operations. (Impact-resistant street-wear lenses required by the new Food and Drug Administration Act after December 31, 1971, must withstand an impact test of *only* ⅝-inch-diameter steel ball dropped from a height of 50 inches.)
   c That full permanently attached sideshields are used where a hazard of flying particles from the side of the wearer exists. Half-sideshields, snap-on, or clip-on types are not acceptable.

3 In checking face shields, he will look for:
   a Minimum plastic window thickness of .040 inches.
   b Minimum face shield length of 6 inches.
   c In severe exposure cases, spectacles and goggles worn under face shield.

4 In checking welding helmets, he will check that the stationary plate on a lift-front helmet is impact-resistant. (Cover plates and filter plates need not be impact-resistant at this time.)

NOISE PROTECTION

Protection against occupational noise exposure is required when sound levels exceed 90 dBA. Hearing protection shall comply with ANSI Standard Z24.22–1957 Method of Measurement of Real-Ear Attenuation of Ear Protectors at Threshold.

The use of personal protective equipment or engineering controls shall be used to reduce sound levels within the . . . table.

In all cases where the sound level exceeds values shown in the table, a continuing effective hearing conservation program shall be administered.

The standards also discuss eye protection, head protection, and foot protection:

§ 1910.136 OCCUPATIONAL FOOT PROTECTION

Safety-toe footwear for employees shall meet the requirements and specifications in American National Standard for Men's Safety-Toe Footwear, Z41.1–1967.

§ 1910.135 OCCUPATIONAL HEAD PROTECTION

Helmets for the protection of heads of occupational workers from impact and penetration from falling and flying objects and from limited electric shock and burn shall meet the requirements and specifications established in American National Standard Safety Requirements for Industrial Head Protection, Z89.1–1969.

HEAD PROTECTION

The standards require that occupational head protection shall be provided and used where a hazard exists that could cause injury to workers' heads. These hazards are impact and penetration from falling and flying objects and from limited electric shock and burn.

Safety hats and caps shall meet the requirements of American National Standard Safety Requirements for Industrial Head Protection, Z89.1–1969.

Types and classes of Industrial Head Protection are listed on page 7 of ANSI Z89.1. These are:

Type 1—Helmet, full brim
Type 2—Helmet, brimless, with peak
Class A—Limited voltage protection
Class C—No voltage protection
Class D—Limited voltage protection, Firefighters' service, Type 1, only

Marking requirements are shown on page 8 of ANSI Z89.1. These consist of:

Name of manufacturer
ANSI Z89.1–1969
Class

§ 1910.133 EYE AND FACE PROTECTION

a General.

1   Protective eye and face equipment shall be required where there is a reasonable probability of injury that can be prevented by such equipment. In such cases, employers shall make conveniently available a type of protector suitable for the work to be performed, and employees shall use such protectors. No unprotected person shall knowingly be subjected to a hazardous environmental condition. Suitable eye protectors shall be provided where machines or operations present the hazard of flying objects, glare, liquids, injurious radiation, or a combination of these hazards.

2   Protectors shall meet the following minimum requirements:
   i   They shall provide adequate protection against the particular hazards for which they are designed.
   ii   They shall be reasonably comfortable when worn under the designated conditions.
   iii They shall fit snugly and shall not unduly interfere with the movements of the wearer.
   iv   They shall be durable.
   v   They shall be capable of being disinfected.
   vi   They shall be easily cleanable.
   vii Protectors should be kept clean and in good repair.

3   Persons whose vision requires the use of corrective lenses in spectacles, and who are required by this standard to wear eye protection, shall wear goggles or spectacles of one of the following types:
   i   Spectacles whose protective lenses provide optical correction.
   ii   Goggles that can be worn over corrective spectacles without disturbing the adjustment of the spectacles.
   iii Goggles that incorporate corrective lenses mounted behind the protective lenses.

4   Every protector shall be distinctly marked to facilitate identification only of the manufacturer.

5   When limitations or precautions are indicated by the manufacturer, they shall be transmitted to the user and care taken to see that such limitations and precautions are strictly observed.

6   Design, construction, testing, and use of devices for eye and face protection shall be in accordance with American National Standard for Occupational and Educational Eye and Face Protection, Z87.1–1968.

EYE PROTECTION

The standards require that employers shall supply suitable eye and face protection where machines or operations present the hazard of flying objects, glare, liquids, injurious radiation, or a combination of these hazards, and that employees shall use such protectors. These devices shall

Table A5–1
U.S. Bureau of Mines Approval Schedules and Tests
Test Conditions and Performance Requirements for Dispersoid Respirators

Dispersoids Covered by Respirator for Protection against	Test Dispersoid	Concentration of Dispersoid, mg/m³	Duration of test, hr	Maximum Allowable Leakage, mg	Maximum Final Resistance to Air Flow, mm of H₂O
Pneumoconiosis-producing and nuisance dusts	Silica dust; geometric mean not more than 0.6μ; $\sigma_g = 1.9$	50 ± 10	1.5	3.0	50
Toxic dusts (not significantly more toxic than lead)	Litharge, ~75%; free metallic lead, ~25%; geometric mean not more than 0.6 μ; $\sigma_g = 1.9$	15 ± 5	1.5	0.43 (Pb)	50
Metal fumes (not significantly more toxic than lead)	Freshly generated lead fume	15 ± 5	5.2	1.50 (Pb)	50
Chromic acid mist	Electrolytically generated chromic acid mist	15 ± 5	5.2	1.0	50
Pneumoconiosis-producing and nuisance mists	Mist formed by atomizing a silica dust-water suspension	10 ± 5	5.2	5.0	50
Various combinations of above types of dispersoids	Respirator must meet requirements for each type.[a]				

[a]For example, the protection against dusts not significantly more toxic than lead, the respirator must meet the requirements for the first two items in this table, that is, for pneumoconiosis-producing and nuisance dusts and for toxic dusts not significantly more toxic than lead.

From *Respiratory Protective Devices Manual* published by American Industrial Hygiene Association and American Conference of Governmental Industrial Hygienists, 1966.

be designed, constructed, tested, and used in accordance with the American National Standard for Occupational and Educational Eye and Face Protection Z87.1–1968.

Eye Protector Specifications

1 Frames shall:
   a Be distinctly marked so that the manufacturer can be identified.
   b Be made of materials that are able to withstand ANSI disinfection, corrosion-resistance, water-absorption, and flammability tests.
   c Contain lens supports that are able to withstand the ANSI fracture-resistance test.
   d Have metal frames that are able to withstand the ANSI joint test and both the flat and transverse tests.
   e Be made of plastic that is able to withstand the ANSI flat and edge transverse tests.
   f Have a combination metal and plastic frame that is able to withstand the ANSI joint test, and the flat and edge transverse tests.
   g Not be street-wear frames. Street-wear frames with safety lenses are prohibited.
2 Lenses shall be:
   a Not less than 3.0 mm thick.
   b Distinctively marked with the manufacturer's monogram.
   c Capable of withstanding the ANSI impact resistance test.
   d Plastic lenses shall be capable of withstanding the ANSI penetration resistance test and the ANSI flammability test as well.
3 Goggles shall:
   a Have frames which bear the manufacturer's trademark.
   b Have lenses which are distinctively marked with the manufacturer's mark.
   c Withstand the ANSI impact resistance test.
4 Face shields shall:
   a Be able to withstand the ANSI test.
5 Helmets shall be ANSI approved.

The selection chart on page 28 of ANSI Z87.1 illustrated recommended eye and face protection together with an application chart. Note that application chart lists operations, hazards, and recommended protectors. Visitor Specs are not included in the Z87.1 code and therefore are not recommended for protection of production workers.

The OSHA standards state that when limitations or precautions are indicated by the manufacturer, they shall be transmitted to the user and care taken to see that such limitations and precautions are strictly observed.

APPENDIX SIX

Bibliography

Industrial Safety

Accident Prevention Manual for Industrial Operations, 7th ed. National Safety Council, Chicago, 1974.

American Engineering Council, *Safety and Production*, Harper & Brothers, New York, 1928.

Anderson, R., *OSHA and Accident Control through Training*, Industrial Press, New York, 1975.

Berman, H. H., and H. W. McCrone, *Applied Safety Engineering*, McGraw-Hill, New York, 1943.

Bird, F., *Management Guide to Loss Control*, Institute Press, Atlanta, 1974.

Bird, F., and G. Germain, *Damage Control*, Academy Press, Macon, Ga. 1966.

Bird, F., and R. Loftus, *Loss Control Management*, Institute Press, Loganville, Ga., 1976.

Blake, R., *Industrial Safety*, 3rd ed., Prentice-Hall, Englewood Cliffs, N. J., 1963.

Brown, D., *Systems Analysis and Design for Safety*, Prentice-Hall, Englewood Cliffs, N. J., 1976.

Bush, V., *Safety in the Construction Industry: OSHA*, Reston Publishing, Reston, Va., 1975.

Calabresi, G., *The Costs of Accidents*, Yale University Press, New Haven, Conn., 1970.

DeReamer, R., *Modern Safety Practices*, Wiley, New York, 1958 (being revised).

Douglas, H., and J. Crowe, *Effective Loss Prevention*, Industrial Accident Prevention Association, Toronto, 1976.

Ferry, T., and D. Weaver, *Directions in Safety*, Charles C Thomas, Springfield, Ill., 1976.

Firenze, R., *Guide to Occupational Safety & Health Management*, Kendall Hunt Publishing Co., Dubuque, 1973.

Fletcher, J., *Total Loss Control*, National Profile Ltd., Toronto, 1974.

Fletcher, J., and H. Douglas, *Total Environmental Control*, National Profile, Ltd., Toronto, 1972.

Gardner, J., *Safety Training for the Supervisor*, Addison-Wesley, Reading, Mass., 1969.

Gilmore, C., *Accident Prevention and Loss Control*, American Management Association, New York, 1970.

Griffiths, Ernes, *Injury and Incapacity—with Special Reference to Industrial Insurance*, Wood, Baltimore, 1935.

Grimaldi, J., and R. Simonds, *Safety Management*, 3d ed., Irwin, Homewood, Ill., 1975.

Haddon, W., E. Suchman, and E. Klein, *Accident Research: Methods and Approaches*, Harper, New York, 1964.

Hammer, W., *Handbook of Systems and Product Safety*, Prentice-Hall, Englewood Cliffs, N. J., 1972.

Hammer, W., *Occupational Safety Management and Engineering*, Prentice Hall, Englewood Cliffs, N.J., 1976.

Handley, W., *Industrial Safety Handbook*, McGraw-Hill, New York, 1969.

International Labor Office, *Encyclopedia of Occupational Safety and Health*, McGraw-Hill, New York, 1970.

Johnson, W., *The Management Oversight and Risk Tree—MORT*, U.S. Atomic Energy Commission (ERDA), Washington, 1973.

Judson, Harry H., and James M. Brown, *Occupational Accident Prevention*, Wiley, New York, 1944.

Lowrance, W., *Of Acceptable Risk*, Kaufman, Los Altos, Calif., 1976.

Lykes, N., *A Psychological Approach to Accidents*, Vantage Press, New York, 1954.

Malasky, S., *Systems Safety*, Spartan Books, New York, 1974.

Manual of Accident Prevention and Construction, 6th ed., Associated General Contractors, Washington, 1971.

Margolis, B., and W. Kroes, *The Human Side of Accident Prevention*, Charles C Thomas, Springfield, Ill. 1975.

Matwes, G., *Loss Control: Safety Guidebook*, Van Nostrand, Reinhold, New York, 1973.

McCall, B., *Safety First—At Last*, Vantage Press, New York, 1975.

McGlade, F., *Adjustable Behavior and Safe Performance*, Charles C Thomas, Springfield, Ill., 1970.

Motor Fleet Safety Manual, National Safety Council, 2d ed., Chicago, 1972.

Peters, G., *Product Liability and Safety*, Coiner Publications, Washington, 1973.

Petersen, D., *Human Error Reduction and Safety Management*, McGraw-Hill, New York, 1979.

Petersen, D., *Safety Management—A Human Approach*, Aloray, Englewood, 1975.

Petersen, D., *Safety Supervision*, American Management Association, New York, 1976.

Petersen, D., *Techniques of Safety Management*, McGraw-Hill, New York, 1971 (being revised).

Pope, W., *Systems Safety Management*, Safety Management Information Systems, Alexandria, Va., 1970.

Resnick, Louis, *Eye Hazards in Industry*, Columbia, New York, 1941.

Rodgers, W., *Introduction to System Safety Engineering*, Wiley, New York, 1971.

Safety Codes and Standards, American National Standards Institute, New York.

Schenkelbach, L., *The Safety Management Primer*, Dow Jones-Irwin, Homewood, Ill., 1975.

Schulzinger, M., *Accident Syndrome*, Charles C Thomas, Springfield, Ill., 1956.

Shaw, L., and H. Sichel, *Accident Proneness*, Pergamon. Oxford, 1971.

Stack, Herbert J., Elmer B. Siebrecht, and J. Duke Elkow, *Education for Safe Living*, Prentice-Hall, Englewood Cliffs, N.J., 1949.

Supervisors Safety Manual, National Safety Council, 4th ed., Chicago, 1973.

Thygerson, A., *Safety Principles Instruction and Readings*, Prentice-Hall, Englewood Cliffs, N.J., 1972.

Tye, J., *Management Introduction to Total Loss Control*, British Safety Council, London, 1970.

Widener, J., *Selected Readings in Safety*, Academy Press, Macon, Ga., 1973.

OSHA

Binford, C., C. Fleming, and Z. Prust, *Loss Control in the OSHA Era*, McGraw-Hill, New York, 1975.

Occupation Safety and Health Administration, various publications on all aspects.

OSHA Requirements Guide, Equipment Guide Book Co., Palo Alto, Calif., 1976.

Petersen D., *The OSHA Compliance Manual*, 2d ed., McGraw-Hill, New York, 1979

Roberts, J., *OSHA Compliance Manual*, Reston, Reston, Va., 1976.

Bibliographies

Guide to Occupational Safety Literature, National Safety Council, Chicago, 1975.

A Selected Bibliography of Reference Materials in Safety Engineering, American Society of Safety Engineers, Chicago, 1967.

U. S. Department of Labor, *Occupational Safety and Health, A Bibliography*, Washington, 1974.

Supervision and Psychology

Argyris, C., *Interpersonal Competence and Organizational Effectiveness*, Irwin-Dorsey, Homewood, Ill., 1962.

Argyris, C., *Personality and Organization*, Harper, New York, 1957.

Bass, B. M., and V. A. Vaughn, *Training in Industry: The Management of Learning*, Wadsworth, Belmont, Calif., 1966.

Beaumont, Henry, *Psychology Applied to Personnel*, Longmans, New York, 1946.

Behavioral Science, National Industrial Conference Board, 1969.

Berne, E., *Games People Play*, Grove Press, New York, 1964.

Blake, R., and J. Mouton, *The Managerial Grid*, Gulf, Houston, Texas, 1964.

Carlson, S., *Executive Behavior*, Strombergs, Stockholm, 1951.

Dalton, M., *Men Who Manage*, Wiley, New York, 1959.

Drucker, P., *The Practice of Management*, Harper, 1954.

Ferster, C., and M. Perrott, *Behavior Principles*, Appleton-Century-Crofts, New York, 1968.

Gausch, J. P., "Balanced Involvement," Monograph 3, American Society of Safety Engineers, Chicago, 1973.

Gellerman, S., *Motivation and Productivity*, American Management Association, New York, 1963.

Ghiselli, E., and C. Brown, *Personnel and Industrial Psychology*, McGraw-Hill, New York, 1948.

Guion, R., *Personnel Testing*, McGraw-Hill, New York, 1965.

Harris, T., *I'm Ok—You're Ok*, Harper, New York, 1971.

Heinrich, H. W., *Basics of Supervision*, and *Supervisor's Safety Manual*, Alfred Best and Company, New York, 1944; also *Formula for Supervision*, National Foremen's Institute, New London, Conn., 1949.

Herzberg, F., *Work and the Nature of Man*, World Publishing, Cleveland, 1966.

Herzberg, F., B. Mausner, R. Peterson, and D. Capwell, *Job Attitudes: Review of Research and Opinion*, Psychological Service of Pittsburgh, Pittsburgh, 1957.

Herzberg, F., B. Mausner, and B. Snyderman, *The Motivation to Work*, Wiley, New York, 1959

Hovland, C., L. Janis, and H. Kelley, *Communication and Persuasion*, Yale University Press, New Haven, Conn., 1953.

Humble, J., *Management by Objectives in Action*, McGraw-Hill, New York, 1970.

Kalsem, Palmer J., *Practical Supervision*, McGraw-Hill, New York, 1945.

Katz, D., N. Maccoby, G. Gurin, and L. Floor, *Productivity, Supervision and Morale among Railroad Workers*, Institute for Social Research, Ann Arbor, Mich., 1951.

Katz, D., N. Maccoby, and N. Morse, *Productivity Supervision and Morale in an Office Situation*, Institute for Social Research, Ann Arbor, Mich., 1950.

Kelly, J., *Organizational Behavior*, Irwin-Dorsey, Homewood, Ill, 1969.

Larson, J., *The Human Element in Industrial Accident Prevention*, New York University, New York, 1955.

Levinson, H., *The Great Jackass Fallacy*, Harvard Business School, Cambridge, Mass., 1973.

Lewin, K., *Field Theory in Social Science*, Harper, New York, 1951.

Likert, R., *The Human Organization*, McGraw-Hill, New York, 1967.

Likert, R., *New Patterns of Management*, McGraw-Hill, New York, 1961.

Lindzay, G., and E. Aronsen, *The Handbook of Social Psychology*, Addison-Wesley, Cambridge, Mass., 1968.

Mager, R., *Developing an Attitude toward Learning*, Fearon, Belmont, Calif., 1968.

Mager, R., and P. Pipe, *Analyzing Performance Problems or You Really Oughta Wanna*, Fearon, Belmont, Calif. 1970.

Maier, N., *Psychology in Industry*, Houghton Mifflin, Boston, 1965.

McClelland, D., J. Atkinson, R. Clark, and R. Lowell, *The Achievement Motive*, Appleton-Century-Crofts, New York, 1953.

McGehee, W., and P. Thayer, *Training in Business and Industry*, Wiley, New York, 1961.

McGregor, D., *The Human Side of Enterprise*, McGraw-Hill, New York, 1960.

Odiorne, G., *Management by Objectives*, Pitman, New York, 1965.

Petersen, D., *Safety Supervision*, Amacom, New York, 1976.

Porter, L., and E. Lawler, *Managerial Attitudes and Performance*, Irwin-Dorsey, Homewood, Ill., 1968.

Safety Communications, National Safety Council, Chicago, 1975.

Sayles, L., *Managerial Behavior*, McGraw-Hill, New York, 1964.

Schulzinger, M. S., *Accident Syndrome*, Charles C Thomas, Springfield, Ill., 1956.

Skinner, B., *Science and Human Behavior*, Macmillan, New York, 1963.

Supervisors Safety Manual, National Safety Council, Chicago, 1967.

Vernon, H., *Accidents and Their Prevention*, Macmillan, New York, 1937.

Viteles, Morris S., *Industrial Psychology*, Norton, New York, 1932.

Weaver, D. A., *Strengthening Supervising Skills*, Employers Insurance of Wausau, Wausau, Wis., 1964.

Index